Lecture Notes in Mobility

Series editor

Gereon Meyer, VDI/VDE Innovation und Technik GmbH, Berlin, Germany

More information about this series at http://www.springer.com/series/11573

Beate Müller · Gereon Meyer
Editors

Towards User-Centric Transport in Europe

Challenges, Solutions and Collaborations

Springer

Editors
Beate Müller
Future Mobility and Europe
VDI/VDE Innovation und Technik GmbH
Berlin, Germany

Gereon Meyer
Future Mobility and Europe
VDI/VDE Innovation und Technik GmbH
Berlin, Germany

ISSN 2196-5544 ISSN 2196-5552 (electronic)
Lecture Notes in Mobility
ISBN 978-3-319-99755-1 ISBN 978-3-319-99756-8 (eBook)
https://doi.org/10.1007/978-3-319-99756-8

Library of Congress Control Number: 2018952899

This Springer imprint is published by the registered company Springer Nature Switzerland AG
The registered company address is: Gewerbestrasse 11, 6330 Cham, Switzerland

Preface

User-centred design has been a leading paradigm in system and product design development for quite a while, now. The main idea is to optimize a product around how people use it, rather than forcing them to change behaviour and to accommodate to the product. Originating from the idea of ergonomics, user-centred design has been particularly prominent for the development of human–computer interfaces aiming at high usability. Nowadays, almost all new everyday services and objects are being created along user experience criteria, particularly those that leave design freedom due to. e.g. digitalization; this is the case for smartphones, washing machines and car cockpits.

Recently, the user experience trend has entered the domain of mobility as the transportation systems are under fundamental transformation towards multimodality and interoperability. While public transport in the past oftentimes, has been perceived as a rigid and inflexible monolithic system dictating the passenger when and where to use it, it may turn into a more user friendly option when combined with on-demand mobility offers like car, bike or ride sharing. In a more general sense, user-centred transportation holds out the promise to provide seamless and more integrated services that better adapt to individual user needs and wishes and thereby provides more and better mobility for all, including disabled or older people. In combination with smart systems technologies, user-centred design may also be applied to gently steer usage towards more sustainable and equitable mobility options.

The EU-funded Coordination and Support Action "Action Plan for the Future of Mobility in Europe" (Mobility4EU) has put the user into the focus of its activities when developing a vision of the transport system in 2030 and mapping out the path towards it. This has been achieved on the one hand by a number of systematic multi-stakeholder consultations, and on the other hand by co-creation processes unlocking creative potential and allowing the look into the future by visual tools. The book at hand summarizes the approaches and results of the Mobility4EU project and puts them into perspective of trends in socio-economic, political and technical development. Further, experts working on the main topics discussed in the project have been invited to contribute to the book thus broadening the views.

The book shall help to initiate a further research and discussion about user-centric future mobility in Europe, to be hosted not just by the Mobility4EU project but also by the European Forum on Transport and Mobility arising from it.

As editors of this contributed volume of the Lecture Notes in Mobility book series, we are deeply grateful to the dedicated scholars and practitioners who contributed to the book. Either as project partners or external experts, they are all connected to the fantastic Mobility4EU project, which we are coordinating. The readiness of the authors to mutually and critically review the chapters is greatly acknowledged as well. There are some further people that engaged in reviews and thus greatly contributed to the improvement of this book. These are Christine Zeller (SIEMENS), Cornel Klein (SIEMENS), David Storer (CRF), George Holley-Moore (International Longevity Center—UK), Armando Carrillo Zanuy (EURNEX), Sandra Wappelhorst (ICCT) and Annette Randhahn (VDI/VDE-IT). Not least, we would like to express our gratitude to the Directorate Research and Innovation of the European Commission for funding the Mobility4EU project, and thus, the work on this book, out of the Horizon 2020 framework programme.

Berlin, Germany
August 2018

Beate Müller
Gereon Meyer

Contents

Part I
Setting the Scene—Towards a Vision for User-Centric Integrated and Sustainable Transport in 2030

Building an Action Plan for the Holistic Transformation of the European Transport System

Frauke Bierau-Delpont, Beate Müller, Linda Napoletano, Eleni Chalkia and Gereon Meyer

Abstract Global socio-economic and environmental megatrends are urging for a paradigm shift in mobility and transport. An action plan for the coherent implementation of innovative transport and mobility solutions in Europe is thus urgently needed and should be sustained by a wide range of societal stakeholders. The EU-funded Mobility4EU project developed such an action plan considering all modes of transport of passengers and freight. The action plan concentrates on user-centric issues and collaboration potential and synergies between modes. This contribution details the methodology of successful consultation with a very broad and diverse stakeholder community and summarizes major insights of the analysis and the resulting action plan towards an inclusive, seamless and sustainable transport system in Europe.

Keywords Multimodality · User-centric transport · Passenger transport Freight transport · Societal drivers

F. Bierau-Delpont (✉) · B. Müller · G. Meyer
VDI/VDE Innovation und Technik GmbH, Steinplatz 1, 10623 Berlin, Germany
e-mail: Frauke.Bierau-Delpont@vdivde-it.de

B. Müller
e-mail: Beate.Mueller@vdivde-it.de

G. Meyer
e-mail: Gereon.Meyer@vdivde-it.de

L. Napoletano
Deep Blue SRL, Piazza Buenos Aires, 20, 00198 Rome, Italy
e-mail: Linda.Napoletano@dblue.it

E. Chalkia
Centre for Research and Technology Hellas, Hellenic Institute of Transport, Egialias 52, 15121 Athens, Greece
e-mail: hchalkia@certh.gr

B. Müller and G. Meyer (eds.), *Towards User-Centric Transport in Europe*, Lecture Notes in Mobility, https://doi.org/10.1007/978-3-319-99756-8_1

1 Introduction

Already today great challenges and demands are posed on mobility and transport due to emerging megatrends as urbanization, climate protection, digitalization etc. The future transport system will be strongly affected by those trends and linked to a multitude of novel developments that can only be partly anticipated today. Novel technologies and services for passenger and freight transport provide opportunities to answer these new demands. However, in order to become disruptive and change society, markets and behaviour, these solutions have to make most efficient use of all modes and they need to be tailored to the needs of users. Thus, user-centric design concepts and interfaces between modes have to be put in focus. Collaborations and exchange are needed across various sectors and perspectives to develop efficient solutions and tap the full potential of new opportunities. Most importantly, users will have to be directly involved in a bottom-up approach. Hence, it is required to bring together a very diverse stakeholder group with a variety of individual interests to discuss a rather complex topic. A process that enables such a group to develop a common language and vision and to discuss collaboration potentials has been refined and implemented by the project Mobility4EU.

Mobility4EU is a Coordination and Support Action of the European Commission that started in January 2016 and lasts for 3 years, until 31 December 2018. The project compiled a vision for a user-centered and cross-modal European transport system in 2030 and an action plan to implement that vision. This work has been carried out engaging a broad stakeholder community including actors from research, academia, industry, operators and decision makers from all transport modes and from passenger and freight transport as well as policy makers and representatives of public authorities from community to national levels. They have been brought together within a structured process that combines scientific consultation and assessment methodologies with a more creative and interactive story mapping method.

The resulting Action Plan for Transport in Europe in 2030 details measures that address technical topics but especially refer to societal aspects and issues for multi-stakeholder interaction, as e.g. policy, user acceptance, standardization, collaboration and the integration of the user perspective into the R&D&I process. It strives to provide recommendations from a strongly user-centered and cross-modal perspective as e.g. the mainstreaming of universal design and user-centric design processes, synergies and collaboration potential between modes and the combination of transport of passengers and freight.

In the following, the methodology towards the action plan is detailed and results of the analysis are summarized and referenced. Finally, the key points and insights of the action plan are summarized.

2 Engaging a Diverse Stakeholder Group to Discuss Visions and Strategies Towards a Complex Topic

As described in the introduction, a participatory approach that enables a broad stakeholder community to engage in a consultation processes has been employed to create a common vision and also the action plan to implement that vision. This has been achieved by employing structured scientific methods and tools, namely the Multi-Actor Multi-Criteria Analysis (MAMCA) and the extended Failure Modes and Effects Analysis (FMEA), as well as an accompanying story map method that supported the process in a more creative and interactive way. The focus of this chapter lies on the story map and its contribution to successfully engage diverse stakeholders, enable them to find a common language and jointly create the vision and action plan for user-centric and cross-modal transport. Later within this article, also the extended FMEA will be shortly introduced with focus on results, while details on the MAMCA are described elsewhere (Macharis 2007; Mobility4EU: D3.3 2018) as well as in another chapter of this publication (Keseru et al. 2019).

So-called story maps (Sibbet 2012) are a rather comprehensive approach of graphic visualisation in the context of strategy development. Large murals or a series of posters are created that represent e.g. the history of a problem, challenges and opportunities, individual values and expectations. Typically, it contains a context map, a commonly drawn picture of the future vision and a roadmap describing the action plan for achieving that vision. This approach supports the alignment of goals in a participatory manner, and helps to unlock creativity in forecasting the future.

Within the project Mobility4EU, a multitude of stakeholder groups covering all transport modes from supply and demand side and especially representatives of users have been engaged into the consultation process. Step by step, the action plan has been created by interactively working on the story map within dedicated workshops which can be referred to in detail in the respective reports (Mobility4EU: D5.6 2016; Mobility4EU: Deliverable D5.7 2016; Mobility4EU: D5.11 2018; Mobility4EU: D5.13 2018; Mobility4EU: D5.14 2018).

The path along the different stages of the story map towards the action plan of the project Mobility4EU, namely the context map, the opportunity map, the vision and action plan, is shown in Fig. 1 and explained in more detail in the following. First, the status and issues of the current transportation system have been assessed from multiple standpoints. Within an interactive workshop with stakeholders from all fields of transport and mobility, expectations, economic and political factors and technologies as well as uncertainties and finally overall trends have been identified that together influence the developments towards transport in 2030. Thus, during the discussions the trends shaping the future European transport system have been put into context which was visually recorded within the context map (Mobility4EU: Deliverable D2.2 2016). Then an inventory of future transport solutions has been compiled. The solutions have been collected through the input of experts and additional desk research. Many solutions were gathered within the second interactive

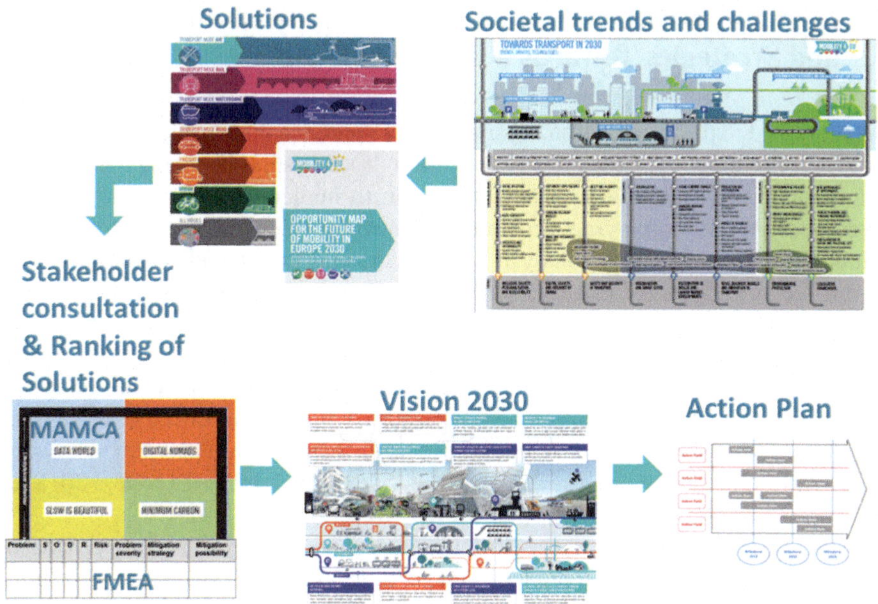

Fig. 1 Mobility4EU process towards Vision and Action Plan for transport in 2030. The story mapping methodology has been combined and complemented with structured scientific approaches, the Multi-Actor-Multi-Criteria-Analysis (MAMCA) and the Extended Failure Mode and Effects Analysis (FMEA). The parts of the story mapping are indicated from the context map (societal trends and challenges) to the opportunity map (Solutions) to the Vision and Action Plan

workshop of the project. The solutions concern all modes of transport of passenger and freight. Also walking and biking are included as active modes.

The links from the solutions to the user demands they satisfy are visualized within an opportunity map (Mobility4EU: Deliverable D2.4 2017). This opportunity map composes the second step of the story map process after the context map. The process to mediate different opinions and preferences and to prioritize between the various solutions has been supported through the more structured and scientific approach of the MAMCA. Within the MAMCA, scenarios have been built out of the solutions and preferences of various stakeholders have been gathered. For creating the vision for 2030, which again builds one part of the story map, more creative activities have been allowed to invite visionary thinking and combine possibly contradictory elements of the previous analysis. The process of action plan building was again done in various steps within interactive workshops employing visual techniques. Nevertheless, it has also been supported by a scientifically structured process for a risk analysis (extended FMEA) which is explained in more detail in a later paragraph. In a final step that action plan has been consulted and validated with a broad stakeholder community.

The approach described above greatly enabled discussions between industry and users, stakeholders of different transport modes etc. The interactive visual based methods enabled to make everybody heard and to directly record all ideas that could then be evaluated and negotiated within the group. As can be seen in Fig. 1, the outcomes have in each case been transferred into a publishable image representing the main outcomes. Thus, participants could easily follow up the process, be reminded of previous results or integrate themselves into discussions also within a later stage of the project. The next section will focus on the results and insights that have been gained throughout the process and further provide details on the complementing scientific analysis.

3 Developing a Common Vision for Transport

The work towards the action plan was based on the identification and assessment of societal challenges, requirements and needs that will influence future transport demand and supply. The interactive part of the work described above was strongly supported and complemented by a dedicated research on trends and their impact on mobility assessing their complexity and interactions (Mobility4EU: Deliverable D2.1 2016). A summary of this work can also be found within another chapter of this publication (L'Hostis et al. 2019). This research as well as the interactive discussions concerning the context map allowed the definition of user needs setting the requirements in mobility and transport in Europe in 2030. These user needs have been further validated by the collection of stakeholder objectives which is one of the first steps within the MAMCA as well as through further desk research of existing studies. In this context, the term "user" includes end users as well as transport operators and service providers or public authorities, cities governments etc.

The analyzed user needs are rather complex, but have been formulated on a higher level with as little overlap between each other as possible without losing important aspects by reducing complexity (Mobility4EU: D2.3 2018):

- End users demand efficient and intelligently organized traffic and transport flows in all modes across borders and national networks.
- While traveling, vehicles systems and services should be easy-to-use, comfortable and offer a flexible modal choice.
- Opportunities for personalization of offers and to increase productivity and leisure time should be available.
- There is an increasing demand on being informed in real-time before and during travelling.
- Users demand individually adaptable intermodal transport with less transfers and good last mile services.
- Inter-operability and reliability as well as seamless end-to-end journeys are key, but also inclusiveness, accessibility and affordability of mobility offers.
- Data security and privacy and last not least safety in all traffic modes are of highest importance to users.
- Mobility should be low-emission and low noise.

The user needs as introduced have to be satisfied by specific user-centric implementations of solutions. Hence, in the next step, a portfolio of transport and mobility solutions that have the potential to respond to these user needs has been compiled. The focus hereby lies on solutions that are either in research or concept state or have just recently been implemented but did not yet reach wide deployment. The assessment of solutions versus user needs delivers specific insights into the requirements of the future transport system. The analysis strongly point to the potential of collaborations to enable or enhance solutions or speed up their implementation. This is especially the case for those solutions that would answer to the user needs mentioned above Main ideas are reported in the following.

Many technologies or concepts can be found across several transport modes but with specified applications within the individual modes. Hence, collaboration opportunities for stakeholders of different modes and beyond the transport sector to tap potential synergies are evident. This is the case for, e.g. game changers in materials, technologies for emission reduction, electrification and efficiency in propulsion systems, solutions and concepts for low noise (waves/vibrations) in transport vehicles and aspects of modular design. Furthermore, across all modes solutions employing IoT, smart systems, big data and automation are exploited to reach higher safety and security, enable predictive maintenance or smart traffic management and forecasting, support efficiency, comfort and personalization of transport offers as well as accessibility and inclusiveness. This calls for a broader cooperation strategy also beyond the traditional transport sector. Also, since data flows across and beyond transport sectors, data security, privacy and transparency can only be implemented in cooperation between transport industry sectors and even beyond. Sharing services as well as novel business models enabling on-demand services and multimodality are needed in all modes and especially enabling seamless interfaces between modes. Naturally, the transport of freight and urban transport of passengers calls for multimodal solutions as e.g. Mobility/Logistics-as-a-service concepts enabling modal shifts. Especially in the urban context additional novel solutions as e.g. co-creation, gamification and measures facilitating active modes will be needed to transform today's transport systems. In this context again collaboration across modes, beyond the transport sector but also to local and regional planners and policy makers is needed.

As mentioned in Sect. 2, the solutions were prioritized by a broad stakeholder community within the MAMCA. This participatory process delivered as common preferences of all stakeholders: a high level of standardisation and interoperability and a regulatory framework that supports personal and corporate carbon emission reduction as well as the full digitalization of the transport system. There also emerged conflicting concepts as on the one hand the preference for personalization and on the other hand a strict focus on shared use and active mobility. Linked with this issue is the question whether capacity should be increased to answer rising demand vs. the reduction of demand to fit existing capacity.

The vision for user-centric and cross-modal transport that was jointly developed by all stakeholders describes a future of transport of passengers and freight that is decarbonized, sustainable in economic, environmental and social terms and offers tailored mobility solutions for all. Main elements are:

- Universal Design is widespread, from smart urban planning to products and services to guarantee more control and freedom especially for people with reduced mobility.
- Transport equity is enabled through innovation for efficient and economic options and supported by policy and regulatory framework.
- Co-creation and participative planning and governance with citizens are common.
- High integration between modes and thus seamless multimodality and synchromodality are enabled by interoperable interfaces, interconnected infrastructures, vehicles and services.
- Full digitalization and automation enables optimized capacity.
- Standardization of interfaces, vehicles and infrastructure and modularization especially for freight transport enables the physical internet concept and open freight networks.
- Innovation enables diverse vehicle types and mobility service options including Mobility/Logistics-as-a-service concepts and other integrated booking and ticketing services.
- Data and cybersecurity are ensured as well as visibility in the supply chain. Incentives, urban design and updated infrastructure motivate the shift to low-carbon shipping options, public transport for passengers, shared modes and walking and cycling.
- Joint approaches of passenger and freight transport enable even better capacity use.
- All this, as well as efficient management of traffic flows, also leads to less cars and more attractive public spaces.
- Safety in transport is enhanced.
- Zero- and for some applications low-emission vehicles including adequate electricity/fuel infrastructures are deployed in all modes due to push-and-pull measures.
- Vehicles and infrastructure, electricity and alternative fuels are produced sustainably.
- Higher efficiencies of energy and resources in the transport of goods and passenger system further contribute to decarbonisation and sustainability.
- Improved modal split is incentivised and vehicle ownership is discouraged, e.g. through carbon footprint accounting, measuring and verification as well as decarbonisation regulations for logistics.
- Circular economy further supports sustainability in transport of passengers and freight.

4 Assessing Implementation Risks

One step in the action plan preparation was the identification and assessment of possible implementation risks and barriers of the solutions. For this purpose, the extended Failure Modes and Effects Analysis (FMEA) was used as described in Chalkia et al. (2018). This tool was developed in the ADVISORS project (Bekiaris and Stevens

2005). It is used for design improvement of products and processes by the identification and validation of technical and non-technical risks such as behavioural, legal and organizational risks and was adapted to the needs of the Mobility4EU project, in order to identify and assess the risks of future transport trends and innovative solutions. The process was supported by a large group of experts representing all key stakeholders from the Mobility4EU context. For each of the selected solutions technical, legal, organizational and behavioural risks were identified. The risks were then validated in terms of their severity, occurrence probability, detectability, and recoverability and classified through a risk number, which indicates the significance of the risk and the possibilities to mitigate it.

In the Mobility4EU project, a total of 171 risks for implementation of solutions was evaluated using the extended FMEA methodology. For each mode, the most severe risks were selected and mitigation strategies for these risks were developed (Mobility4EU: Deliverable D4.1 2017). Some issues were raised repeatedly in various contexts in the risk assessment. One of them is data privacy and protection which is an important issue in most innovative solutions to be implemented in passenger and freight transport in order to enhance safety and security as e.g. automated and connected driving and automated maintenance raise data privacy and security concerns. Mitigation strategies mostly consider strict legislative actions. Related to the issue of data privacy and security is legal liability. This challenge also plays a major role in most solutions related to automation and in all innovative solutions where there is still a lack of clear organization framework and the participation of different stakeholder groups is required who have not yet a clearly defined role in the process; one example for this is Mobility-as-a-Service. Also for this, clear harmonized legal and regulative framework which are transparent and determine liabilities of all stakeholders can mitigate this risk. Another horizontal issue is related to the deployment of renewable energies and alternative fuels in transport. High costs of the implementation of required infrastructure and low funding accompanied by low movement towards this goal is a concern for stakeholders from all modes. Also the lack of environmental regulations that further restrict the use of conventional fuels and hence, harmful emissions, concerns the transport system across all modes. This can be mitigated for example by a balanced allocation of free carbon allowances to entities across industries and regions and harmonized emissions standards. Other mitigation strategies are the introduction of financial actions which either incentivize investments in renewable fuels or increase the costs for conventional fuels for example the internalization of externalities. Another challenge which applies to all modes and novel mobility services is the lack of business models that need to be developed for the implementation of a solution to make it feasible. New business opportunities arise from strengthened cooperation between different transport modes for facilitating seamless transport; one example is luggage check-through between modes. In some cases, though, the cooperation of governments is needed for the framing of business cases, e.g. for the on-road charging for trucks solution.

5 The Action Plan for User-Centric and Cross-Modal Transport in Europe in 2030

In order to implement the vision as described in Sect. 3, an action plan has been developed within the Mobility4EU project that strives to provide recommendations for R&D, deployment, policy and regulatory frameworks and other implementation related issues from a user-centered and cross-modal perspective (Mobility4EU: D 4.4 2018). Thus, while some recommendations clearly support technological R&D&I initiatives, most proposed action items rather concentrate on improving the collaboration of stakeholders from different modes, sectors, or organizations, give priority to mainstreaming user-centric design processes, enhancing user acceptance and focus on policy development. Starting point for the development of measures and action items is the portfolio of solutions as well as results from the risk analysis. Furthermore, two stakeholder workshops have been held to interactively collect inputs. For compiling the action plan, action items have been clustered within 6 topics, namely:

- Low-/zero emission mobility (electrification and hybridization and other alternative fuels)
- Automation and Connected Driving
- Safety and (Cyber) Security in Transport
- Mobility Planning
- Cross-modal/cross-border transport and Integration of Novel Mobility Services in Public Transport
- Putting the User in the Centre

Within these topics, action items are provided with reference to a timeframe and to the actors addressed. They are further assigned to different action fields, depending on whether they focus on technical or societal challenges or foster the interaction and collaboration of different stakeholder groups in the transport community.

Focus is on the design of a sustainable energy-efficient transport system tailored to users' needs. In order to achieve a decarbonized transport system, actions concentrate on the promotion of stakeholder collaboration from all modes and beyond the traditional transport sector to enable technology and know-how transfer regarding electrification, hybridization and automation, e.g. in task forces across European Technology Platforms. Electrification in all modes is further promoted by common planning and synergetic use of infrastructure for powering novel vehicles across modes and sectors, in particular referring to the energy and transport sector as well as heat and sewerage. Higher energy efficiency in transport can be achieved by the integration of electrified vehicles in smart cities.

User acceptance towards automation which appears as a big trend in all modes can be considerably enhanced by the improvement of safety and data security. The latter is primarily promoted through legislative actions, e.g. new General Data Protection Regulation (GDPR), and adaptations of the legal and regulatory framework in terms of enhanced transparency and clear legal liabilities of stakeholders, harmonized throughout the EU. Safety in automated and connected cooperative driving can be achieved by development of improved and secure vehicle software and electronics (e.g. sensors), technological progress in the field of artificial intelligence and cooperative ITS, and most notably, adequate testing of control software of automated vehicles under real conditions (e.g. in pilots). Automated transport services support people with reduced mobility and improve transport offers especially in rural areas.

Redesigning urban infrastructures on the one hand contributes to the improvement of safety in transport, on the other hand to emissions reduction and the improvement of air quality when new infrastructure prioritizes public transport and active modes such as walking and cycling instead of individual motorized traffic.

The implementation of push-and-pull-measures as well as the improvement of sharing services and well-working multimodal seamless transport contributes to a reduction of the total vehicle fleet. Seamless transport requires also improved stakeholder collaboration to facilitate the connection between modes and the integration of cross-modal mobility services. This is supported by the planning and deployment of cross-modal hubs in cities and around.

Much focus in the action plan is on policies and adaptations of the legal and regulatory framework. For example the wide introduction of standardized and interoperable interfaces, infrastructure and billing systems and the harmonization of laws and legal liabilities between member states are actions addressing policy makers to achieve cross-border seamless transport in Europe.

Mainstreaming accessibility and equity in transport is to be ensured by developing the appropriate legal framework, for example by making universal design a precondition for public R&D funding. The integration of the user perspective in the design process of the future transport system is facilitated by establishing co-creation processes and participatory forms of governance.

6 Conclusions

The Mobility4EU action plan seeks to answer new challenges and demands on mobility that arise from global socio-economic and environmental megatrends. Innovative solutions like novel technologies and services for passenger and freight transport can only become disruptive and change behavior, when they are sustainable and tailored to the needs of users from scratch. This is for example facilitated by the creation of a regulatory framework which enables the participation of citizens in the design process of the future transport system. It is further vital to involve the entire transport system including all modes in the transformation process and to improve the links between modes and also between sectors for tapping the full potential of new opportunities and achieving a collaborative decarbonization of the transport system. The emphasis on the user perspective and the comprehensive consideration of all traffic modes distinguishes the Mobility4EU action plan from the merely technology driven roadmaps, edited e.g. by the European Technology Platforms, and also makes is complementary to the policy driven Strategic Transport Research and Innovation Agenda (STRIA) issued by the European Commission (2017). It accordingly adds to other initiatives by providing recommendations on the mainstreaming of universal design and user-centric design processes as well as pointing out opportunities for collaboration and knowledge transfer between stakeholders and the combining of transport of passengers and freight.

Equity, data and cybersecurity as well as strict regulation to foster innovation, interoperability and zero-emission mobility are called for. The proliferation of new and especially of digital services needs planning on policy level for maximum impact benefitting all Europeans. Issues of liability and ethical questions have to be addressed and frameworks for data privacy have to be enforced. Furthermore, business models and innovation systems that are able to turn opportunities into reality are needed. Moving towards sustainability drives efficiency in passenger and freight transport. Tools and business models are needed that enable the individual stakeholders to tap this potential. Risks and uncertainties lie in the implementation of renewable and alternative fuels infrastructure in all modes. Unclear situations regarding incentives/restrictions for low emission vehicles and fuels in all modes need to be countered.

Consequent inclusion of the user in the entire innovation and development process will be imperative to achieve the goals of a sustainable and integrated transport system. At the same time, user-centric approaches have the potentially to also act as a driver for the successful development and implementation of new technologies and services. For instance, universal design putting the user in the center delivers not only inclusive transport but improves mobility offers for all. To implement user-centric approaches, methodologies, tools as well as impact assessments have to be developed. This includes models for collaboration of users and the R&D&I community, the development of digital co-creation tools to enable broad collaborations etc.

As can be seen in the individual points raised above, cross-modal approaches can enhance the impact of transport solutions within a user-centred multimodal transport system. For the development and implementation of specific solutions, cross modal approaches are even a strong enabler or they open opportunities for high additional benefits. These are mainly solutions that are enabled or greatly benefit from interoperability, standardization or technology transfer across modes which applies especially to solutions in the urban or freight context or addressing horizontal issues as e.g. safety, (cyber-)security, advanced driver assistance and automation, testing, standardization, universal design etc. As regards the integration of users in the R&D&I process the collaboration across modes and multiple disciplines requires the development of methods, tools, suitable frameworks and platforms for collaboration. In general legal issues related to open innovation and co-creation as well as regarding IP have to be solved.

Acknowledgements The authors wish to thank the consortium of the Mobility4EU project for the intense and fruitful collaborations and the European Commission for the funding of the Mobility4EU project in the framework of the Horizon2020 program (EC Contract No. 690732).

References

Bekiaris E, Stevens A (2005) Common risk assessment methodology for advanced driver assistance systems. In: Transport reviews, pp 283–292
Chalkia E, Sdoukopoulos E, Bekiaris E (2018) Risk analysis of innovative maritime transport solutions using the extended Failure Mode and Effects Analysis (FMEA) methodology. In: Proceedings MARTECH 2018
European Commission (2017) Towards clean, competitive and connected mobility: the contribution of transport. Research and innovation to the mobility package. SWD 223
Keseru I, Coosemans T, Macharis C (2019) Building scenarios for the future of transport in Europe: the Mobility4EU approach. In: Müller B, Meyer G (eds) Towards user-centric transport in Europe. Lecture notes in mobility. Springer, Switzerland, pp 15–30
L'Hostis A, Chalkia E, de la Cruz MT, Müller B, Keseru I (2019) Societal trends influencing mobility and logistics in Europe, a comprehensive analysis. In: Müller B, Meyer G (eds) Towards user-centric transport in Europe. Lecture notes in mobility. Springer, Switzerland, pp 31–49
Macharis C (2007) Multi-criteria analysis as a tool to include stakeholders in project evaluation: the MAMCA method. Transport Project Evaluation. https://doi.org/10.4337/9781847208682.00014
Mobility4EU: D5.6 (2016) Workshop on societal requirements and current challenges for transport. http://www.mobility4eu.eu/?wpdmdl=1231
Mobility4EU: D3.3 (2018) Report on MAMCA evaluation outcomes. http://www.mobility4eu.eu/?wpdmdl=2242
Mobility4EU: D5.11 (2018) Workshop on vision building. http://www.mobility4eu.eu/?wpdmdl=2073
Mobility4EU: D5.13 (2018) Workshop to initiate drafting of the action plan. http://www.mobility4eu.eu/?wpdmdl=2011
Mobility4EU: D5.14 (2018) Workshop with external experts on action plan. http://www.mobility4eu.eu/?wpdmdl=2369
Mobility4EU: D2.3 (2018) Novel and innovative mobility concepts and solutions. http://www.mobility4eu.eu/?wpdmdl=2069
Mobility4EU: D 4.4 (2018) European action plan for transport
Mobility4EU: Deliverable D5.7 (2016) Workshop on novel and innovative mobility solutions. http://www.mobility4eu.eu/wp-content/uploads/2016/12/D5.7_WS_Report_inclAnnex.pdf
Mobility4EU: Deliverable D2.2 (2016) Story map I: requirements and challenges on transport. http://www.mobility4eu.eu/wp-content/uploads/2016/12/M4EU_D2.3__final.pdf
Mobility4EU: Deliverable D2.1 (2016) Societal needs and requirements for future transportation and mobility as well as opportunities and challenges of current solutions. http://www.mobility4eu.eu/wp-content/uploads/2017/01/M4EU_WP2_D21_v2_21Dec2016_final.pdf
Mobility4EU: Deliverable D2.4 (2017) Storymap II: opportunities for transport. http://www.mobility4eu.eu/wp-content/uploads/2016/04/M4EU_D2.4_-v1_17July2017_final_DBL-1.pdf
Mobility4EU: Deliverable D4.1 (2017) Report on challenges for implementing future transport scenarios. http://www.mobility4eu.eu/?wpdmdl=2070
Sibbet D (2012) Visual leaders: new tools for visioning, management, and organization change. Wiley, Hoboken, NJ

Building Scenarios for the Future of Transport in Europe: The Mobility4EU Approach

Imre Keseru, Thierry Coosemans and Cathy Macharis

Abstract This paper outlines the scenario-building approach of the Mobility4EU project that aims to create a vision and action plan for mobility and transport in 2030. Scenario building is the first step of the Multi-Actor Multi-Criteria Analysis (MAMCA), the methodology used to conduct a broad stakeholder consultation. To emphasize the participative nature of the scenario building, the scenarios were created using the intuitive logics technique and participatory workshops. Each scenario describes future trends and technological, organisational or policy-related solutions. Based on a survey of stakeholders, "policy & legislative framework" and "lifestyle and user behaviour" emerged as pivotal uncertainties to steer the scenario building. They provided the basis for the development of four scenarios: Data World, Digital Nomads, Slow is Beautiful, and Minimum Carbon. The paper describes the trends and solutions that comprise these scenarios.

Keywords Scenario building · Mobility · Logistics · Participative · Multi-actor multi-criteria analysis

I. Keseru (✉) · C. Macharis
Mobility, Logistics and Automotive Technology Research Centre (MOBI), Department Business Technology and Operation (BUTO), Vrije Universiteit Brussel, Pleinlaan 2, 1050 Brussels, Belgium
e-mail: imre.keseru@vub.be

C. Macharis
e-mail: cathy.macharis@vub.be

T. Coosemans
Mobility, Logistics and Automotive Technology Research Centre (MOBI), Department Electric Engineering and Energy Technology (ETEC), Vrije Universiteit Brussel, Pleinlaan 2, 1050 Brussels, Belgium
e-mail: thierry.coosemans@vub.be

B. Müller and G. Meyer (eds.), *Towards User-Centric Transport in Europe*,
Lecture Notes in Mobility, https://doi.org/10.1007/978-3-319-99756-8_2

1 Introduction

Looking into the future is always an exciting task. Contemplating how we will live, work and travel 10–15 years from now can help us to prepare for several possible futures or to try to achieve our own preferred future vision. The future is indeed uncertain since there are several development paths possible. The role of strategic planning is to devise actions that are appropriate for the most probable paths so that the society is prepared for possible positive or negative events and trends.

Such a foresight is also essential in the field of mobility and transport so that European, national and local policy makers can take the necessary steps to react to ongoing or upcoming trends within and beyond the transport sector with potentially great social and economic impact. The Mobility4EU project funded by the European Commission investigates these trends, potential solutions and future developments paths aiming to create a vision for mobility and transport in 2030 and an action plan to reach that vision. While there have been many similar efforts before (Bernardino et al. 2015; Leppänen et al. 2012; Krail et al. 2014), the approach applied in this project differs from previous studies in that it explicitly involves the representatives of the users of the transport system to explore future trends, solutions and development paths in an effort to balance technocentric views and user needs.

Our participatory approach has three pillars. On the one hand, based on a study of trends and transport solutions (Mobility4EU 2016; Mobility4EU 2018), future scenarios for transport and mobility in Europe including trends and solutions have been co-created with a wide range of stakeholders from and beyond the transport sector. Furthermore, a structured evaluation process, the multi-actor multi-criteria analysis (MAMCA) (Macharis et al. 2009) is used to evaluate the scenarios and find the synergies and conflicts between stakeholder groups. Finally, the story mapping technique is used to unleash stakeholders' creativity through workshops contributing to the scenario building and the creation of the vision (for details see the following chapter in this book and Muller and Meyer 2018).

This paper focuses on the first participatory element, i.e. the co-creation of scenarios. Nevertheless, this process is closely linked to the other two participatory methods: the co-created scenarios form the basis of the stakeholder-based evaluation within the MAMCA; in addition, the story mapping process contribute to exploring dominant trends and solutions that form integral parts of the scenarios.

The goal of this paper is to outline the process of the development of the scenarios for the future of mobility in Europe and present the scenarios that were co-created with the stakeholders i.e. the consortium members, associated partners and external stakeholders of the Mobility4EU project.[1]

The paper first outlines the scenario building approach, then the four scenarios are described that were co-created with the stakeholders. In the last section, the further steps of the participatory evaluation of the scenarios are briefly outlined.

---

[1]This paper is partly based on Deliverable D3.1 of the Mobility4EU project (Keseru et al. 2016).

2 The Scenario Building Approach in Mobility4EU

2.1 What Are Scenarios?

Scenarios represent a range of possible, probable and desirable developments in the future and paths that lead to that future. Since we can never be sure how the future finally develops, scenarios are hypothetical based on assumptions. Therefore, scenarios are not capable of providing precise predictions of future development paths i.e. they do not deliver a comprehensive description of the future but rather focus on its specific elements (Kosow and Gaßner 2008).

In the Mobility4EU project, the scenarios have a *communicative function* to enhance the cooperation of different actors in the transport and related domains; a *goal setting function* to define what the European Union intends to achieve in the transport sector until 2030 and contribute to *decision-making* since a 'best' scenario is selected and transformed into a vision and action plan after evaluating several alternative scenarios with the MAMCA methodology (Bröchler et al. 1999; Greeuw 2000). In addition, scenario building helps to explore and understand the relationship between political, environmental, economic, social and technological factors which is often very complex (Wright et al. 2013).

2.2 Participative Scenario Building Approach

A multitude of methods have emerged in the past to create scenarios (Wright et al. 2013). One of the most used techniques is the intuitive logics method. Intuitive logics is based on the estimates (intuition) of experts as a reference point (Wack 1985). The process focuses on decision-making. It is called intuitive because besides relying on objective data, intuitive estimates of future trends by experts are also considered. This technique has the advantage that it considers unpredictability and covers the so-called scenario transfer, i.e. the final stage of the scenario process when the scenarios are used for strategy making (Kosow and Gaßner 2008). The intuitive logics method has been found to enhance the understanding of the relationships between major factors that define the future and it can challenge conventional thinking (Wright et al. 2013).

This technique is often criticised for being expert-led allowing little involvement of stakeholders (Wright et al. 2013). To overcome this deficiency and involve a wide range of stakeholders in the process we combined the intuitive logics method with participatory workshops. This technique allows for the involvement of stakeholders, although the process is quite time-consuming (Kosow and Gaßner 2008).

The key steps of our combined approach are (Kosow and Gaßner 2008):

1. Scenario field identification: What is the purpose of the scenarios? What is the issue to be addressed?
2. Identification of key factors or driving forces

3. Clustering of key factors
4. Analysis of key factors for unpredictability and impact
5. Scenario generation: studying the scenario logic, i.e. to create a manageable number of scenarios focusing on 'pivotal uncertainties'
6. Drafting the scenario narrative
7. Scenario workshop to co-create scenarios and provide a better understanding of the scenarios as well as increase the legitimacy of scenarios
8. Scenario writing and optimisation
9. Participatory evaluation of the scenarios with the multi-actor multi-criteria analysis
10. Scenario transfer selecting concrete strategies.

2.3 Scenario Building for the Future Vision and Action Plan for Europe

Based on the above stepwise approach, first, we defined the scenario field. The scenarios should address transport and mobility in 2030 in Europe with focus on societal trends and user needs. Then, we identified key factors and driving forces i.e. trends that will influence transport and mobility by 2030. 34 trends were identified i.e. societal challenges, requirements and needs that will influence the future transport demand and supply. This work was based on desk research and an interactive workshop with stakeholders (Berlin, 03/05/2016) (Mobility4EU 2016). The trends were clustered into 9 broader categories: distribution of wealth and labour market developments; lifestyle and user behaviour; urbanisation and smart cities; environmental protection: climate change, pollution resource and energy efficiency; digital society and internet of things; novel business models and innovation in transport; safety in transport; security in transport; legislative framework [see also following chapter in this book and Mobility4EU 2016]. Then, in October 2016, a survey was carried out among stakeholders to identify which of these trends may have the highest degree of uncertainty and impact. The survey was filled in by 33 respondents representing a wide range of stakeholder organisations. Each trend was assigned a score between 1 and 4 for both uncertainty and impact. The aggregated results are shown in Fig. 1. The dots represent the 34 trends which are plotted on a graph in which the horizontal axis represents the impact of the trend while the vertical axis shows uncertainty.

Based on their score of uncertainty and impact, the trends can be classified according to the categories in Table 1 (Kosow and Gaßner 2008).

We selected trends that have the *highest uncertainty and the highest impact* (trends that received a minimum score of 2.4 for both attributes). These trends are situated in the upper right-hand side quadrant of the graph in Fig. 1 and they are called *pivotal uncertainties*. Table 2 lists the pivotal uncertainties that we identified and their broader thematic categories.

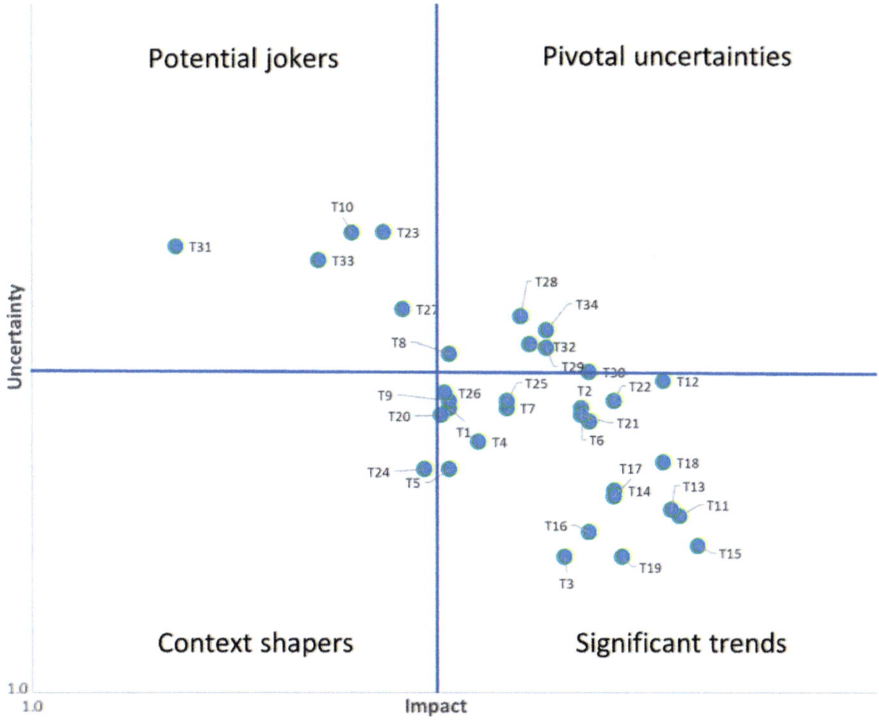

Fig. 1 Distribution of trends according to their degree of uncertainty and impact based on a survey of 33 stakeholders

Table 1 Categories of trends based on their degree of uncertainty and impact (Kosow and Gaßner 2008)

Uncertainty	Impact	Category of key factors
High	High	*Pivotal uncertainties*
High	Low	*Potential jokers*
Low	High	*Significant trends*
Low	Low	*Context shapers*

If we look at the larger thematic categories of the identified pivotal uncertainties, *policy & legislative framework* and *lifestyle & user behaviour* emerge as the key driving forces that have the highest uncertainty and greatest impact in terms of mobility demand in 2030 in Europe. These *pivotal uncertainties* define the differences between scenarios and hence provide the basis for the development of alternative scenarios. Figure 2 shows how the four possible combinations of the two extremes of the *pivotal uncertainties* define the four scenarios.

Besides major trends that influence mobility, the Mobility4EU scenarios comprise a selection of the technological, organisational and policy-related solutions that respond to these trends. 87 technological, organisational or policy-related

Table 2 List of trends with the highest score of uncertainty and impact

Trend code	Trend description	Trend category
T8	Acceleration of social life and more flexibility in spending one's time	Lifestyle and user behaviour
T28	Legislation adapts to new transport solutions and businesses	Policy and legislative framework
T29	Harmonisation of regulations at the European level to improve interoperability	Policy and legislative framework
T30	Rate of user acceptance of new technology	Lifestyle and user behaviour
T32	Increasing concern about financing transport investments	Policy and legislative framework
T34	New technologies and business models challenging legal frameworks	Policy and legislative framework

solutions were defined through a workshop with stakeholders and desk research (for details see Mobility4EU 2018).

Preliminary scenarios combining trends and solutions were created by the consortium. They were not prescriptive; rather they provided a starting ground for discussion. They were further refined at a scenario building workshop (Brussels, 05/07/2016) where the scenarios were co-created in a participative manner. The event brought together experts for passenger and freight transport across all modes. First, they validated the trends included in each scenario. Then they selected and

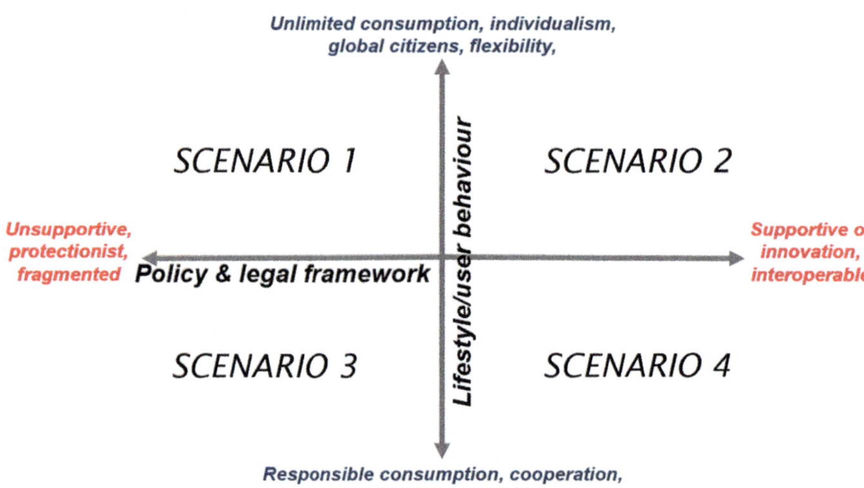

Fig. 2 Four scenarios combining the possible outcomes of the pivotal uncertainties

matched technological, organisational or policy-related solutions to the scenarios and their trends using cards depicting each solution and scenario boards representing the four preliminary scenarios (see Keseru et al. 2016 for more details). After the workshop, the input received from the stakeholders to make the scenarios more realistic and consistent was analysed and considered when drafting the next version of the scenarios.

3 Scenarios for the Future of Mobility in Europe

Based on the above methodology we identified four scenarios (Fig. 3):

1. Data world
2. Digital nomads
3. Slow is beautiful
4. Minimum carbon.

Each of the scenarios is described below with its underlying trends. Each scenario is divided into two parts: trends and solutions for freight transport and passenger mobility.

Fig. 3 The four Mobility4EU scenarios

3.1 Scenario 1: DATAWORLD

Industry and consumption increasingly rely on production outside Europe and hence demand for intercontinental freight flows is increasing. Increased trade flows from the E7 countries[2] are expected to change the scene in global supply chains and logistics. As a reaction to market demand, port operators extend seaport capacities by installing floating delivery hubs and automated container terminals, port operations and transshipments.

A growth in E-commerce stimulates intra-European freight flows as well. Supply chains become more complex, requiring tailored solutions that are industry- or even customer-specific. There is, however, little collaboration between delivery companies and shippers (e.g. retailers) to promote bundling flows and optimize deliveries.

As a response to increased demand from customers for instant deliveries and to save cost, delivery companies start to introduce personalised delivery systems using airborne drones and small autonomous freight trucks for first- and last-mile deliveries. Organised platooning of freight vehicles (road trains) will become widespread thanks to the deployment of cooperative ITS. This will increase capacity for long distance freight transport while also contributing to better safety and better fuel economy. Increased use of information and communication technologies and especially big data exploitation facilitate supply chain optimisation (i.e. cost/time reduction, load factor improvement etc.). Cybercrime becomes a great concern.

In passenger mobility, demand for information and online services to book and pay for mobility services is high. Internet connectivity and direct data collection from users is key for the management of the transport system. Technology companies provide continuous and reliable internet connection at stations and on vehicles. Their business model is to collect users' data extensively in return for free internet, travel information and entertainment.

National and local governments exercise little control over the provision of mobility services. A few large, private, multinational mobility providers emerge and compete. They own, manage and process the immense amount of mobility data collected from smart sensors in the infrastructure and vehicles and by engaging transport users through gamification using data from their connected devices. They provide real-time traffic optimisation and safety information to transport authorities. Due to the lack of expertise, governments mostly rely on these big data integrators for the management of their intelligent transport systems, road and rail infrastructure.

Travel demand continues to increase as people become increasingly mobile and flexible. The strategy of multinational mobility providers is to focus on individual needs, reduced travel time (faster travel) and specific consumer groups (e.g. young adults, families with children, medium-high income households), mainly in urban areas where demand is high. Therefore, they push governments to increase the capacity and improve maintenance of the transport network especially for roads and high-capacity public transport. Many major infrastructure investments (motorways, major roads and bridges in urban areas, high-speed railways) are implemented by private

---

[2]China, India, Brazil, Mexico, Russia, Indonesia and Turkey.

companies that levy a charge for the use of infrastructure. Due to the lack of government regulations the prices vary considerably across operators and no common payment system is introduced across major operators and across Europe.

Public and private transport services converge in Mobility as a Service (MaaS) initiated by these large mobility companies providing real-time trip planning, booking and payment services for all transport modes. Within MaaS, mobility companies promote their own transport solutions such as car-sharing (in cooperation with car manufacturers) and widely deployed ride-sourcing. Traditional public transport and taxi companies become subcontractors of the mobility companies to retain business. To serve individual needs the first commercial personal rapid transit systems are promoted in urban areas by private investors. Personalised and individualised transportation to, from and within airports are developed by mobility companies and airports. Public transport operators adopt flexible timetables and modular trains to provide flexibility to adapt to passenger needs and demand. Airlines differentiate flight cabin classes and zones according to individual needs and provide high-speed internet connection and virtual reality entertainment services. Airlines and aircraft manufacturers develop the first small on-demand aviation services to provide flexibility to customers with tight schedules.

The mobility companies focus on satisfying travel demand by high capacity and individualised transport solutions (self-driving solutions, public transport with improved comfort and commercial speed). Cycling and walking are only niche markets for them. Nevertheless, shared smart electric bicycles carrying advertisements are introduced in urban areas for which the rental fee depends on the visibility of the ads (how long the trip is).

Due to fragmented local regulations and differing priorities in transport policy in the member states, diverse proprietary solutions are developed for connected and autonomous vehicles, charging stations, data platforms and safety solutions. Due to fierce competition and the lack of supranational regulations autonomous transport systems have a low level of compatibility across regions or countries. Government regulations on vehicle emissions and infrastructure operations are less severe, the emphasis is on the voluntary sustainability initiatives of vehicle manufacturers and infrastructure managers.

Security is a major concern in all walks of life, but there are no strict regulations in place to increase security in a way that would contribute to delays and a deterioration of comfort for passengers. Pan-European cooperation for transport security is at a low level. This leads to a fragmentation of rules, regulations and procedures in cross-border traffic.

3.2 Scenario 2 Digital Nomads

Growing concerns about climate change, stricter EU-wide regulations to reduce CO_2 emissions and an increased focus on renewable energies and materials have brought innovative technologies and business models into the limelight.

A reindustrialisation takes place in Europe based on the latest technologies and innovation driven by increasing transportation costs. This mainly affects non-labour-intensive industry, which can be easily automated. Labour-intensive industries remain outside Europe. Trade barriers have been lifted.

European and national policies support and enforce cooperation between players and transport modes in the freight sector. Shippers, freight forwarders and receivers of goods are obliged to share their data on logistics operations with each other. Data sharing supports the development of the physical internet, an open global logistics system that allows for a more sustainable logistics chain where modular containers of different standardized sizes are equipped with a protocol and an interface to allow better handling and transshipment. Digital security becomes a great concern. Freight and public transport is integrated through co-modality by using available space on public transport vehicles during non-peak-hours (e.g. at night) or on duty vehicles (e.g. cleaning or maintenance vehicles). For last- and first-mile deliveries, small electric vans, electric bikes or tricycles are used because in most cities low emission zones are introduced. Full digitalization of the transport system promotes the automatization of all logistics operations at terminals and vehicles (ships, trains, organized platooning of trucks) as well as beyond terminals.

Despite increasing migration and security concerns, borders within the EU remain open and the remaining non-Schengen countries join the area of free movement. Airports introduce the 'No borders' approach i.e. the integration of passenger information and security checks. Seamless security checks and intelligent CCTV technology at airports and high-speed rail stations utilize preselection based on big-data and advanced screening equipment. Advanced digital security devices are also used to increase security in public areas through advanced face recognition and risk analysis. Since aviation relies fully on digital communication and management, protection against cyber threats is strengthened.

People are becoming increasingly flexible in their work and private lives. The boundaries between private life and work disappear as people become always online and available. Travel time is used for multitasking (e.g. working) to add useful minutes to an otherwise very crowded daily schedule.

Governments embrace the 'internet everywhere and for everyone' concept which requires transport operators and vehicle manufacturers to provide internet connection at all stations and on board all vehicles including private cars.

Governments focus on improving the efficiency and capacity of existing road and parking infrastructure and construct new high-capacity infrastructure where needed to cater for increasing demand for travel. New roads are built only where ITS and demand management cannot provide a solution to congestion problems. New infrastructure is mainly built in public-private partnership where public authorities retain control over user charges.

Governments support full digitalisation of the transport system. Strict pan-European regulations ensure interoperability of the transport infrastructure and digital interfaces as well as data privacy. Cooperative Intelligent Transport Systems (C-ITS) are developed in public-private partnership to increase road safety and road capacity. C-ITS is comprehensively implemented on motorways, the main and secondary road network and provides detailed real-time information about incidents and roadworks, speed limits and diversion routes. The vulnerable road user protection system will increase safety of pedestrians and cyclists by alerting drivers of the presence of vulnerable road users. Preventive and predictive maintenance of road and rail infrastructure is increasingly automated (remote asset inspection, autonomous operations, and integrated scheduling and system control). Local governments introduce dynamic road user and parking pricing using smart sensors.

Existing public transport and new mobility services focusing on individual needs (personal rapid transit, personalised transit to airports, ride-sharing, parking-sharing etc.) are strictly regulated and they are required to be integrated in mobility as a service systems with services that are accessible to the disabled and older people. These systems mainly promote motorized modes (road and rail rather than cycling and walking) to cater for the increasing travel demand and travel distances. Intermodal mega-hubs are built in public-private partnership to connect transport modes and host commercial activities.

Long distance travel is supported by faster and more energy efficient high-speed trains with inductive charging and a seamless integration of other modes with air transport to reduce the maximum travel time between any two points in Europe to 4 h. Flight cabins provide internet connection.

Vehicles will be produced mainly from advanced lightweight materials to increase fuel efficiency and emissions. There is a high degree of standardisation of charging, connectivity and safety systems for cars and other motorised vehicles.

There is high demand in urban areas for battery-operated personal mobility devices (e.g. Segway), local authorities adapt the walking and cycling infrastructure to accommodate these devices. The purchase of smart electric bicycles is subsidised by the government to enable longer trips even on hilly terrain.

The older population embraces digital communication technologies and benefits from easier and more accessible local and long-distance travel. Autonomous vehicles provide new opportunities for door-to-door transport for older people. Special financial aid is provided to them and the disabled who would otherwise not be able to afford such vehicles or services.

3.3 Scenario 3 Slow Is Beautiful

Governments turn inwards to guarantee national security and supply of resources to their citizens. European policy focuses on enabling local initiatives rather than supranational standardisation. Innovation is less supported due to scarce financial resources.

People more and more turn to eco-friendly local cooperative production of food and energy, urban gardens and peer-to-peer services. Citizens aim to produce what they consume within their neighbourhood. Bottom-up initiatives of local communities thrive with few legal limitations on local sharing and production initiatives. Willingness to pay for eco-friendly solutions increases.

To support self-sustaining urban communities, there is a growing market for electric cargo-bikes that are used to distribute goods within the local communities where no motorised vehicles are allowed. Logistics companies set up urban freight consolidation hubs from where goods are distributed by e-bikes and minivans. Goods deliveries also increasingly rely on sharing courier platforms connecting people who need items that are delivered with drivers and couriers 'going there anyway'. Public transport services (rail, trams, inland waterways and underground) are also used for deliveries and collection of waste.

Also, supply chain de-stressing gains momentum through different practices (i.e. synchromodality, slow steaming etc.), to reduce supply chain complexity by using the right mix of transportation modes to operate sustainably at lower cost with higher quality.

A service sector based on sharing resources such as time, space and vehicles emerges supported by local social networks. "Slow, healthy and sustainable" are the new buzzwords. People appreciate spending more time with their friends and family within their neighbourhood and rediscover their local environment. Mixed-use developments aim to decrease the distance between residential areas, jobs, education and services.

Local neighbourhood planning is initiated more and more by local citizens using social media and online co-creation platforms. More and more cities introduce car-free city neighbourhoods and various other restrictions on road traffic (e.g. limited parking). Road user charging is initiated in some urban centres using 'low-tech' solutions such as relatively low flat rates, number plate recognition or vignettes. There is, however, no EU-wide coordination of road charging schemes, payment systems and signage of restrictions. Peer-to-peer applications and online services developed by small local start-ups have replaced many of the car-sharing and ridesharing services of big corporations. Mobility applications and sharing initiatives can easily be launched with lax legal and government control concerning user rights and privacy. The popularity of cycling and other electric two-wheelers is on the rise. Cyclists are encouraged to use existing roads as shared spaces.

There is a cautious approach to the introduction of autonomous vehicles especially in urban areas. Due to societal resistance (safety concerns, lack of trust in technology, concerns over jobs in the transport sector), autonomous private vehicles are only allowed on designated motorways.

Transport providers remain primarily national and local with little cross-border activities. Financial resources to build new transport infrastructure and maintain the existing ones are scarce. Therefore, there is more focus on the optimal use of existing roads and railways by retrofitting them.

In cities, bus rapid transit corridors are built by private investors using existing road infrastructure instead of new rail or tram systems due to the lower cost.

3.4 Scenario 4 Minimum Carbon

Due to the severe pressure of climate change, governments want to fundamentally change the behaviour of their citizens and companies to steer them to reduce carbon emissions and move them away from fossil fuels.

Companies are required by government regulations to significantly reduce their environmental footprint therefore sustainable and bio-production are supported. Large international manufacturing and retail corporations adapt to the new requirements and regionalise their production. 3D printing becomes widespread boosting customised local production. Customers prefer to buy products with the smallest carbon impact. Local programmes are launched by the government to recirculate materials inside the neighbourhood and the city to reduce waste and carbon emissions associated with mass production and long-distance distribution chains.

Logistics companies are required by law to measure and report their carbon footprint through smart sensors in vehicles and infrastructure. Products transported and logistics services receive Eco-labels based on this calculation therefore logistics companies are incentivised to improve their load factors and the environmental performance of their vehicle stock. Carbon taxation is widely introduced to reflect the amount of CO_2 generated by transport activities. National and local road charging schemes are also widely introduced with electronic tolling and variable rates based on demand to reflect external costs (especially noise, air pollution, congestion). Deliveries in or to city centres are restricted to small electric vehicles and electric bicycles. Urban cross-modal logistics uses all available modes of transport to provide the lowest possible carbon footprint. Urban goods distribution regulations are harmonised in municipalities.

Strict European regulation demand good energy efficiency of freight vehicles and the use of renewable fuels, in all modes (ships, aviation, road, rail). The electrification of waterborne transport by electrifying ferries and vessels is forced by international agreements.

Burn-out from fast-paced work have turned people towards healthier and active life. Work is arranged to require less travel, in neighbourhood flexi-offices, by supporting work from home and by distributing smaller offices in city districts. Long-distance travel is expensive due to the carbon taxation introduced across Europe. People prefer to spend their free time and holidays in the proximity to their homes.

Cities embrace car-free neighbourhoods to improve liveability. Tax-incentives discourage private car ownership and support car- and ride-sharing instead. National and local road charging schemes are widely introduced with electronic tolling and variable rates with a standardised Pan-European payment and monitoring system. Use of fossil fuels is prohibited in most urban areas and heavily taxed elsewhere. The environmental footprint of aviation is strictly monitored and a carbon tax is built into ticket prices across Europe.

Smartphone or wearables apps encourage users to travel in a sustainable way and oblige them to respect the personal 'carbon budget' (the maximum amount of CO_2 they can generate per month), which is assigned to everyone by the government.

Smartphones, smart sensors and intelligent CCTV recognition technology enable the monitoring of the carbon budget for everyone.

Smart online apps empower citizens through digital technologies to participate in planning and impact monitoring for urban mobility. There is a high level of integration of mobility services steered by publicly managed transport partnerships (regional associations integrating all shared and public transport services, travel information and payments). There is a significant increase in demand for public transport due to the introduction of the personal carbon budget, the extension of capacity, however, does not keep pace with demand due to the limited availability of space and the potential negative impact of infrastructure investments. Ridesourcing and ridesharing are fully integrated with public transport to provide seamless first- and last-mile solutions.

The introduction of autonomous vehicles is limited to long distance travel. C-ITS systems are fully deployed only on motorways. Cities focus more on improving walking, cycling and public transport. Cycling is fostered by building interconnected cycle highways with added services (repairs, charging) and large capacity bicycle parking is built by local governments at public transport stations.

Travel demand is reduced through supporting densification i.e. living in densely built urban areas. Superblocks restrict road traffic to major roads around residential blocks where only local traffic is allowed with restricted speed (10 km/h). Everyone has the right to access to basic services guaranteed by the government. The walking and cycling environment is adapted to the needs of the older population, children and the disabled. Intelligent pedestrian crossings adapt to pedestrian demand. Internet of Things devices are used to enhance safety and security of pedestrians and cyclists integrated into everyday items such as wearable reflectors for children and smart monitors for older people to monitor their well-being and location. On the other hand, carbon rationing takes the activity level of citizens into account and allocates more opportunities for public transport capacity to the active population (e.g. commuters), also limiting the time periods when the non-active population can use such services (e.g. only outside peak hours).

4 Further Steps

Scenario building outlined above is the first step of the participatory evaluation of future scenarios for mobility and transport in 2030. The four scenarios then undergo a stepwise participatory evaluation. First the relevant stakeholders are identified, then their objectives and criteria. Then, each stakeholder group weights its own criteria to express their relative importance. After that, indicators and measurement methods for each criterion are identified with international experts. Indicators are used to measure the performance of a scenario i.e. how a certain future scenario would impact a criterion (e.g. air quality) compared to the business as usual. In the next step, the impact of the scenarios on the stakeholders' criteria is determined by an international expert panel. After that, based on the stakeholders' weights and the experts' evaluation, the ranking of scenarios is calculated for each stakeholder. This shows the synergies and conflicts across the stakeholder groups. Finally, stakeholders come to a consensus on the best scenario at a consensus making workshop. The consensus scenario is then used to create a vision for mobility and transport in 2030 and the action plan to fulfil that vision (Macharis et al. 2009; Keseru et al. 2018).

5 Conclusions

The scenario building and evaluation approach applied in the Mobility4EU project to develop a vision and an action plan for mobility and transport in Europe in 2030, provides a participative approach to building visions and strategic plans for the future. While it builds upon the knowledge of experts, it also opens up the scenario construction to stakeholders by linking the intuitive logic scenario technique with participatory workshops as it was demonstrated in this paper.

The scenarios described above address all transport modes for both freight and passenger transport. At the same time, scenario descriptions are meant to be short, easily comprehensible texts. Therefore, it is unavoidable that some trends or solutions that some experts or stakeholders would find important are not included. Also, the feasibility of implementing the solutions proposed for each scenario by 2030 has not been investigated in detail. In the next steps, a thorough multi-stakeholder evaluation was carried out to select the most promising scenarios which can act as visions to work towards.

Acknowledgements The authors wish to thank the European Commission for the funding of the Mobility4EU project in the framework of the Horizon2020 program (EC Contract No. 690732).

References

Bernardino J, Aggelakakis A, Reichenbach M, Vieira J, Boile M, Schippl J et al (2015) Transport demand evolution in Europe—factors of change, scenarios and challenges. Eur J Futures Res 2015(3):1–13. https://doi.org/10.1007/s40309-015-0072-y

Bröchler S, Simonis G, Sundermann K, Steinmüller K (eds) (1999) Szenarien in der Technikfolgenabschätzung. Handbuch Technikfolgenabschätzung, Edition. Sigma, Berlin

Greeuw SCH (2000) Cloudy crystal balls: an assessment of recent European and global scenario studies and models. European Environment Agency, Copenhagen. Available online: https://www.eea.europa.eu/publications/Environmental_issues_series_17/at_download/file. Accessed on 20 Sep 2017

Keseru I, Coosemans T, Macharis C, Muller (2016) Mobility4EU Deliverable D3.1 Report on MAMCA scenario descriptions. Available online: https://www.mobility4eu.eu/resource/d3-1-report-on-mamca-scenario-descriptions/#. Accessed on 20 Sep 2017

Keseru I, Coosemans T, Gagatsi E, Macharis C (2018) User-centric vision for mobility in 2030: participatory evaluation of scenarios by the multi-actor multi-criteria analysis (MAMCA). In: Procedia—Proceedings of 7th Transport Research Arena TRA 2018, 16–19 Apr 2018, Vienna, in print

Kosow H, Gaßner R (2008) Methods of future and scenario analysis : overview, assessment, and selection criteria. DIE—Deutsches Institut für Entwicklungspolitik, Bonn. Available online: http://edoc.vifapol.de/opus/volltexte/2013/4381/. Accessed on 20 Sep 2017

Krail M, Condeço-Melhorado A, Ibañez-Rivas N, Christodoulou A, Reichenbach M, Schippl J (2014) Scenario-based assessment of the competitiveness of the European transport sector, FUTRE Project Deliverable D5.3 Available online: http://publica.fraunhofer.de/eprints/urn_nbn_de_0011-n-3283527.pdf. Accessed 26 Sep 2017

Leppänen J, Neuvonen A, Ritola M, Ahola I, Hirvonen S, Hyötyläinen M et al. (2012) Future scenarios for New European social models with visualisations. Deliverable D4.1 of the SPREAD project. Available online: https://www.sustainable-lifestyles.eu/fileadmin/images/content/D4.1_FourFutureScenarios.pdf. Accessed 20 Nov 2017

Macharis C, de Witte A, Ampe J (2009) The multi-actor, multi-criteria analysis methodology (MAMCA) for the evaluation of transport projects: theory and practice. J Adv Transp 43:183–202. https://doi.org/10.1002/atr.5670430206

Mobility4EU (2016) Deliverable D2.2 Societal needs and requirements for future transportation and mobility as well as opportunities and challenges of current solutions. Available online: https://www.mobility4eu.eu/wp-content/uploads/2017/01/M4EU_WP2_D21_v2_21Dec2016_final.pdf. Accessed on 20 Sep 2017

Mobility4EU (2018) Deliverable D2.3 Novel and innovative mobility concepts and solutions. Available online: https://www.mobility4eu.eu/?wpdmdl=2069. Accessed on 20 Dec 2017

Muller B, Meyer G (2018) Mobility4EU—action plan for the future of mobility in Europe. In: Procedia—Proceedings of 7th Transport Research Arena TRA 2018, 16–19 Apr 2018, Vienna, in print

Wack P (1985) Scenarios: shooting the rapids. Harvard Bus Rev (November)

Wright G, Bradfield R, Cairns G (2013) Does the intuitive logics method—and its recent enhancements—produce "effective" scenarios? Technol Forecast Soc Chang 80:631–42. https://doi.org/10.1016/j.techfore.2012.09.003

Societal Trends Influencing Mobility and Logistics in Europe: A Comprehensive Analysis

Alain L'Hostis, Eleni Chalkia, M. Teresa de la Cruz, Beate Müller and Imre Keseru

Abstract The objective of this paper is to establish a comprehensive view of societal trends that have an impact on mobility and logistics in the future. Based on a review of scientific literature, the output of European research projects and reports from consultancies, the result of this investigation provides a broad and comprehensive set of factors that influence, and will influence in the future, mobility and logistics. The set is composed of 29 trends organised under 9 larger categories covering economic issues, societal issues, urbanisation, the environment, the digital society, new business models, safety, security and the legislative framework. The present analysis has greatly benefitted from the concept of liquid modernity developed by Bauman that enabled a broad and complete view on the dynamics of society and mobility. It allows describing linkages between social and economic trends, and between society and technology, especially information and communication technology. Thus, the added

A. L'Hostis (✉)
Université Paris-Est, LVMT (UMR_T 9403), Ecole des Ponts ParisTech, IFSTTAR, UPEMLV, 14-20 Boulevard Newton, Cité Descartes, Champs sur Marne, 77455 Marne La Vallée Cedex 2, France
e-mail: alain.lhostis@ifsttar.fr

E. Chalkia
Center for Research and Technology Hellas (CERTH), 6th km Charilaou-Thermi Rd, P.O. Box 60361, GR 570 01 Thermi Thessaloniki, Greece
e-mail: hchalkia@certh.gr

M. T. de la Cruz
Zaragoza Logistics Center, C/Bari 55, Edificio Náyade 5 (PLAZA), 50197 Zaragoza, Spain
e-mail: mdelacruz@zlc.edu.es

B. Müller
VDI/VDE Innovation und Technik GmbH, Steinplatz 1, 10623 Berlin, Germany
e-mail: Beate.Mueller@vdivde-it.de

I. Keseru
Department Business Technology and Operations (BUTO), Mobility, Logistics and Automotive Technology Research Centre (MOBI), Vrije Universiteit Brussel, Pleinlaan 2, BE-1050 Brussels, Belgium
e-mail: Imre.Keseru@vub.be

B. Müller and G. Meyer (eds.), *Towards User-Centric Transport in Europe*, Lecture Notes in Mobility, https://doi.org/10.1007/978-3-319-99756-8_3

value of this contribution is its systematic approach allowing to describe this complex topic in an exhaustive manner while focussing on links and dependencies.

Keywords Mobility · Logistics · Societal trends · Liquid modernity

1 Introduction

Understanding current and future mobility and logistics is a key element in order to shape transport policies and orientate future research. Therefore, establishing a comprehensive view of societal trends that have an impact on mobility and logistics, represents a significant step. In the context of the European research project Mobility4EU,[1] such an analysis was carried out (L'Hostis et al. 2016). The aim here is to present societal factors, in the broad sense, that influence mobility and logistics. Studying mobility of people and logistics together remains a challenging task. The "last mile" of logistics is very often a matter of individual mobility, which shows how interrelated the two topics can be. Mobility of persons is the most prominent topic in the study of societal trends, but logistics issues can nevertheless be found in several places in this contribution.

There is a broad consensus within scientific literature and in policies that mobility is of crucial importance for society. According to a study by the International Transport Forum in 2011 (Wilson 2011; 'ITF Transport Outlook 2017' 2017), by 2050 passenger mobility will increase by a staggering 200–300% and freight activity by as much as 150–250%. Mobility is increasingly becoming so important that several authors proposed to replace the study of society by the study of mobility (Urry 2007). We chose not to separate society from mobility and transport, but rather to take an integrated approach.

A societal trend is here described as an emerging pattern, movement, and evolution in society that leads to change, and potentially has implications for mobility and transport (e.g. ageing, social networks). Societal trends impact transport infrastructure and demand, thus they can be considered key elements for transport related change. The presented analysis had to find a compromise between the need to identify and hence separate, and the need to highlight interactions between individual trends. In this perspective trends have been organised in groups of individual trends.

In this work, it has proven difficult to dissociate clearly societal trends from political trends. For example, it is difficult to identify if environmental issues have been led by changes within patterns of behaviour in citizens or if it has been led by legislation. Likewise, it is difficult to dissociate societal trends from technological trends as expressed in the concept of *digital society*. Consequently, our analysis of societal

[1] The Mobility4EU project aims at producing a roadmap for mobility and logistics in 2030 starting from societal needs. The overall objective consists in linking present and future societal trends and needs to existing and emerging transport and mobility solutions. http://www.mobility4eu.eu.

trends that impact transport uses five domains of investigation: this analysis covers sociocultural, political, technological, environmental, legal and economic trends.

Another difficulty of the present approach has been to produce a coherent set of trends, shaped around a general view of present and future social behaviour. In order to avoid bias, we chose not to rely only on the sociologists of mobility such as Kaufmann (2002), Urry (2007) or Kellerman (2012), who tend to consider mobility (and immobility) as central societal values. Instead, we refer to a more general view on society, as contained in the idea of *liquid modernity* introduced by the sociologist Bauman (2000). This proposal formulates many ideas described in the *post-modern* concept. This new phase of modernity can be seen as being characterised by five trends:

- a continuing movement of individualism;
- the development of fluidity, which can be seen positively with the ideas of change and innovation, but which also has a darker side with respect to the ideas of rupture and precariousness;
- the principle of an acceleration of the pace of existence, especially as it is felt through the experience of individuals;
- the emergence of social networks that gain more significance, as compared to the more stable and strong (Granovetter 1983) social ties of the family, or of the workplace;
- the introduction of the technologies of information and communication in almost all aspects of social life.

The emergence of a consensus among social scientists can be observed (Clegg and Baumeler 2010) on the idea of a shift of society towards *liquid modernity* introduced by Bauman (2000). In consequence, this view provides a sound basis for the analysis of societal trends having an impact on mobility and logistics.

One of the merits of this approach is to establish links between domains. Liquid modernity links social dynamics with technology, with social and economic dynamics. In addition, if Bauman (2000) intends to describe society and not mobility itself, building an analysis of mobility out of these elements is very straightforward. Liquid modernity provides a strong support to the list of trends formulated here especially through the linkages it creates between the individual sociocultural trends and between sociocultural and technology trends.

Liquid modernity ideas provide direct insights for understanding persons' mobility, and also gives indications of how freight transport is and should be organised in the future. Needs are more individualised, personalised and hence logistic flow tends to be individual based. Fluidity and the feeling of acceleration converts into the need for immediacy as is reflected in the development of so-called "instant delivery" (Dablanc et al. 2017). These aspects are devised in the relevant trends.

Several researchers have focused on trends in favour of sustainable transport (Rudinger et al. 2004; Boschmann and Kwan 2008), while others have studied trends from travel surveys in a single country (Frändberg and Vilhelmson 2011). As compared to other similar exercises in European research, the approach reported here is rooted in the study of societal trends as opposed to a more classic transport demand analysis, found in for example TransForum (Anderton et al. 2015), TRANSvisions (Petersen et al. 2009), FUTRE (Bernardino et al. 2013), RACE2050 (Sena e Silva et al. 2013), EUTransportGHG (Sessa and Enei 2009), ORIGAMI (Lemmerer and Pfaffenbichler 2012) and VOYAGER (Brög et al. 2005). The latter approaches tend to separate trends along classic transport modes and transport markets, without extending into the study of societal trends. In the study presented here, firstly societal dynamics have been analysed in a broad sense, and secondly trends that have an interaction with mobility and logistics have been identified. Thus, societal trends instead of the identification of transport solutions or markets are the starting point thereby complementing previous research. Further, more thematic approaches to studying societal trends in relation to mobility, for instance CITYLAB on freight (Dablanc et al. 2016) or TransForum on several targeted transport sectors (Anderton et al. 2015), the presented analysis intends to cover all transport modes, all geographical scales and freight as well as passenger transport.

2 Method

The objective of this chapter is to establish a comprehensive view of societal trends that impact mobility and logistics. In this aim, a literature review has been conducted, encompassing published scientific sources, research reports and statistics sources.

In order to set up a comprehensive view, a list of trends has been created. A set of rules has been defined for the identification of these trends in relation to mobility and logistics. Firstly, a trend is coherent, not equivocal. A trend can carry a paradox, like the trend about the acceleration of liquid modernity, and a trend can also move in a single direction.

Secondly, a trend is not redundant. However, bearing in mind the complex nature of societal issues, the task of building a set with little overlap has proved challenging.

Thirdly, the set of trends should not omit factors or tendencies in society in the broad sense that may exert an influence on mobility and logistics.

And finally, a trend must be described and supported by evidence found in the literature. Statistics, surveys, and figures are provided for each trend. These analytical elements determine the direction of the influence exerted on mobility and logistics. In cases where an idea was formulated without evidence, it was not considered as a trend.

3 Sociocultural, Political, Technological, Environmental, Legal and Economic Trends Interacting with Mobility and Logistics

The analysis of this system of trends is developed in the present with a temporal horizon of 2030. Consequently, the description of trends aims at capturing the present situation and the dynamics of these trends in the future.

The result of this process is the following list. This list can be seen as the shortest possible set of trends that allows describing present and emerging societal factors, in a broad sense, that have an impact on mobility, both for freight and passengers' mobility.

The list is composed of 29 trends organised under 9 larger categories. These larger categories cover issues relating to economy and issues relating to demography and lifestyle as well as urbanisation, the environment, digital society, new business models, safety, security and finally, the legislative framework. Table 1 lists the categories and the trends, and indicates the type of each trend: sociocultural, political, economic, technological and legal.

Considering that trends can be of several types, it can be observed that 16 trends are sociocultural, 8 trends are economic, 6 trends are political, 4 are technological, 4 are legal, and only one is environmental. If, as expected, more than half of the trends refer directly to social issues, the other aspects play a significant role in the description of trends.

The next section describes the trend categories in detail.[2]

3.1 *Trends Relating to Economic Development*

Economic climate and economic conditions play a major role in shaping the demand for mobility. In the economic domain two trends have been identified. Firstly, it is foreseen that the adaptation of Europe's economy in the global context will have a relative decline of GDP (Bassanini and Reviglio 2011). European GDP and population should grow but much slower than the rest of the world on average. The direct consequence of global growth will be an increase in flows, particularly freight, but also the possibility of a re-industrialisation of Europe, which could lead to significantly modified freight flows ('ITF Transport Outlook 2017' 2017). At the individual level, economic growth usually converts into more mobility, as illustrated by the growth of tourism.

The second trend within this category refers to the restructuring of working arrangements. Telework and part-time work are the two major foreseeable tendencies already at work that should grow in the future (Isusi and Corral 2004; Jagger et al. 2014). They have direct and indirect effects on mobility. Although there is

[2]For a detailed description of trends see L'Hostis et al. (2016).

Table 1 The 29 societal trends having an impact on mobility and logistics and their categories

Categories	Trends	Type of trend
3.1 Trends relating to economic development	Share of the European economy in world GDP declines	Economic
	Restructuring working arrangements	Economic, Sociocultural
3.2 Trends relating to demography and lifestyle	Increasing life expectancy of the population	Sociocultural
	Migration trend generating long distance flows	Sociocultural
	Trend towards inclusion of vulnerable to exclusion groups	Political
	Less car usage by younger generations	Sociocultural
	Move towards more active and healthy lifestyles	Sociocultural
	Acceleration and flexibility of liquid modern society	Sociocultural
	Personalisation of liquid modern society	Sociocultural
	European integration facilitating flows	Political
3.3 Urbanisation	Rising and expanding urbanisation	Economic, Sociocultural
	The emergence of Smart cities	Technological
3.4 Environmental protection	Stricter regulations for environmental protection	Sociocultural, Political
	Limited resources require more resource efficiency and circular economy in transport	Economic, Sociocultural
	Move away from fossil fuels towards energy efficiency and renewable energies	Political, Economic, Technological
	Impact of climate change on transport	Environmental
3.5 Digital society and the Internet of things	Rise of the Internet of Things and big data	Technological
	More automation	Technological
	Expectation of customers and digitisation of mobility	Sociocultural
	New uses of travel-time	Sociocultural

(continued)

Table 1 (continued)

Categories	Trends	Type of trend
3.6 Novel business models in transport	New models challenging the individual vehicle ownership model	Economic, Sociocultural
	New players and new business models	Economic
	Emerging co-development and co-creation of new systems by users and economic actors	Economic
3.7 Safety in transport	The persisting issue of transport safety	Sociocultural, Political
	The emerging safety issue in complex networks with new vehicles	Sociocultural, Legal
3.8 Security in transport	Growing concern over security threats	Sociocultural
3.9 Legislative framework	Diversifying approaches of governance	Political, Legal
	Legislative models adapts to new transport solutions and businesses	Legal
	Trend toward harmonisation in legislative frameworks	Legal

a decreasing quantity of home-to-work flows, those trips could take longer due to the urban sprawl made easier by telework and less peak hour traffic. There may be more trips for other purposes (Jackson and Victor 2011; Bernardino et al. 2013). A likely consequence of the last effect has been identified as a demand for more flexible tickets for public transport.

3.2 *Trends Relating to Demography and Lifestyle*

The second category of trends refers to sociocultural dynamics that interact with mobility and freight demand such as; "Increasing life expectancy of the population" or "Migration trend generating long distance flows", but some of them are also supported by "Move towards more active and healthy lifestyles" or driven by "Trend towards inclusion of vulnerable to exclusion groups" and "European integration facilitating flows".

"Increasing life expectancy of the population" is an essential dynamic of European societies (Eurostat 2015; World Health Organization and The World Bank 2011). The interaction with mobility and logistics are complex though: the population of car drivers is likely to grow, and less active mobility is expected from the "oldest old", but maybe more active mobility will come from those who will stay or move back to denser urban areas; a need for proximity in goods and service deliveries for the urban elders contrasts with specific and costly mobility demand in ageing rural areas (Anderton et al. 2015; Velaga et al. 2012).

"Migration trend generating long distance flows" (European Commission 2011) introduces or develops specific patterns of mobility demand: the foreseen increasing migration in Europe will generate longer distance flows of persons and goods with the countries of origin of migrants ('Asylum Statistics—Statistics Explained' 2016; Vasileva 2009).

There is a consensus on the fact that policies support a "Trend towards inclusion of vulnerable to exclusion groups" (Martens 2012). This trend has a direct impact on transport policy in terms of accessibility for all. In addition to known vulnerable groups, the digitisation of mobility carries the risk of creating new exclusion, for instance among those who do not own a smartphone (Pauzié 2013; Velaga et al. 2012).

Forming part of the explanation of the *peak car* phenomenon, the tendency of "Less car usage by younger generations "has been observed recently since the mid-2000 (Davis et al. 2012; Goodwin 2012; Newman et al. 2013; Metz 2013). Connecting to the social network, whatever the means, physically based or telecommunication based, seems to have replaced the car ownership dream observed in previous generations (Corwin et al. 2015; McKinsey et al. 2012).

Another consequence of the societal awareness for environmental issues is the "Move towards more active and healthy lifestyles". This trend is fuelled by individual awareness and by policies aimed at influencing individual behaviour. Health is likely to become a major concern in the future with direct implications for the policies aimed at orientating mobility behaviour (Krzyzanowski et al. 2005).

There is an emerging consensus among social scientists around the idea of *liquid modernity* introduced by Bauman (2000). In order to characterise the interactions of liquid modernity with mobility and logistics two trends have been established: "Acceleration and flexibility of liquid modern society" and "Personalisation of liquid modern society". The first trend refers to the ideas of acceleration and flexibility and provides an explanation to the increase of leisure time and its associated mobility patterns (Harvey 1990; Levine 1998; Rosa 2003). It also entails that transport users need less time for planning their trips, and have access to immediate and seamless information. The second trend of liquid modernity highlights the personalisation aspect. The individualisation process, illustrated among other indicators by the decreasing size of households (Euromonitor 2013; Capros et al. 2013), favours individual transport modes of cars but also bike and walking, and also favours models of the type "one-stop-shop" for mobility services.

Essentially driven by policies and political choices "European integration facilitating flows" is still, despite the recent reverse movement by the United Kingdom, an ongoing process. Its impacts on mobility are straightforward through the increase in tourism and freight flows in Europe (Kester 2014).

3.3 *Urbanisation*

Urbanisation is the major trend of human settlement. Despite an already high level of urbanisation in Europe, it is foreseen that urbanisation is set to increase from 73% in 2014 to 84% in 2050 (United Nations 2014).

Cities and city-regions, which are densifying and spatially extending, are increasingly the dominant forms of settlement (Fujita et al. 2001). These trends lead to more intense and longer urban flows, both for passengers and goods (Sena e Silva et al. 2013).

The emerging model of the *smart city* aims at articulating human and social development with information and communication technologies in cities (European Investment Bank Institute EIB 2013). Equipping cities with ICT infrastructure is being led by the introduction of new technologies of mobility (mainly electric vehicles, car sharing, car-pooling) and should lead to new social interactions and to new uses of city spaces, and hence should have significant impact on mobility behaviour and freight demand.

3.4 *Environmental Protection*

In the domain of environmental protection, four trends can be identified. The first three are led or encouraged by policies, while the last one refers to the management of the consequences of climate change.

Rising awareness for environmental issues leads to the adoption of "Stricter regulations for environmental protection". Transport has a large impact on the environment, and is confronted with the strategic policy goals of decarbonisation (Pachauri and Meyer 2014; Anderton et al. 2015).

The economy is adapting in the context of increasingly limited resources available ("Limited resources require more resource efficiency and circular economy in transport"). A "sustainable consumption" culture emerges among citizens and firms tend to conform to social and environmental rules and approaches: corporate responsibility, circular economy, life cycle assessment (Petersen et al. 2009). All these elements will require the reconsideration of the organisation of logistics, as for instance in the case of local food consumption that need short supply chains (Blanke and Burdick 2005; Coley et al. 2009; Meisterling et al. 2009; Kulak et al. 2015; Dablanc et al. 2016).

In the domain of energy, policy goals support a movement to "Move away from fossil fuels towards energy efficiency and renewable energies" (European Commission 2011; Harrison 2013). The current dependence of transport on fossil fuels is expected to be replaced by more electricity and biofuels (Pachauri and Meyer 2014).

Finally, regarding environmental trends, the "Impact of climate change on transport" is direct and significant. Extreme weather events cause damages to transport systems of road, rail and aviation (Doll et al. 2014). Global warming could have one positive effect though, to open the North-West passage for freight between Europe and Asia (Anderton et al. 2015).

3.5 Digital Society and the Internet of Things

In this section on digital society, two types of trends are covered. Regarding the digital world, technology, as an enabler, exerts a real influence and drives individual and social uses. Two technological trends of "Internet of things and big data", and "automation" form the supply side. But at the same time, technology is sometimes used for a slightly different purpose than what was foreseen by the designers. In this sense, individuals and social groups can be seen as actors of digital society, and able to fuel trends that are not driven by technology. This is the demand side of digital society.

The technological trends of "Rise of the Internet of Things and big data" is impacting many aspects of the production of goods and services, and particularly in the transport domains. Vehicles, transport infrastructures, ICT devices, parcels will all be able to communicate in real-time. Dealing with the masses of data produced require new methods, the so called big data approaches, but promise to improve many transport issues like transport operations planning, traffic management, or safety (Löffler and Tschiesner 2016; Jeske et al. 2013; Zakir et al. 2015).

The trend of "More automation" is driven by the development of artificial intelligence, sensors and information and communication technologies (Frisoni et al. 2016). This technological development has ambitious road safety promises, but also raises difficulties expressed in another trend ("The emerging safety issue in complex networks with new vehicles"). Automation is also developing in the air and rail transport domains (Verstraeten and Kirwan 2014).

Regarding the demand side of the digital society, the first trend refers to the "Expectation of customers and digitisation of mobility". Travellers in the digital world expect to be able to connect their mobile devices, and expect to receive accurate and real-time information about their trips (Pauzié 2013). All these expectations are challenging for transport providers (Goodall et al. 2015). Quite ambivalently, travellers also want more data privacy (Pauzié 2013).

Mostly driven by the development of the digital society, "New uses of travel-time" can be observed (Jain and Lyons 2008; Lyons et al. 2013). Usually seen as a burden, travel time can become a positive moment for users. This trend is able to influence the transport mode choice in favour of public transport (Russell 2012), until automation is introduced.

3.6 Novel Business Models in Transport

The transport sector witnesses the emergence of new players and new business models interacting with—if not fuelled by—new behaviour. New business models are closely related to the previously mentioned trend of "Rise of the Internet of Things and big data". The main issue is the currently dominating individual vehicle ownership model, described in the first trend "New models challenging the individual vehicle ownership model" (Shaheen et al. 2012; Hardesty 2014; Cirstea 2015). The second trend covers the other cases where new players and new business models emerge, in batteries, in data, in freight "New players and new business models" (Forbes 2015; Leminen et al. 2015; Rantasila et al. 2014; Casey and Valovirta 2016). The last trend highlights the emergence of the co-development model and its implications for mobility: "Emerging co-development and co-creation of new systems by users and economic actors" (Chang and Yen 2012; Finnish Prime Minister's Office 2015; Kostiainen et al. 2016).

3.7 Safety in Transport

Despite significant improvement of the levels of safety, especially in the road transport domain, and encouraging perspectives linked to automation, transport safety will most probably remain a pressing issue in the future ('European Commission Press Release—2015 Road Safety Statistics: What Is behind the Figures?' 2015).

Considering the long-term promise of the decrease in road casualty through the introduction of automated cars, a new safety issue emerges with the coexistence of automatic and non-automatic vehicles, creating complex networks and environments (Lazakis 2014). Safety will become a far more complex issue than today with new insurance and liability issues (Smith and Svensson 2015).

3.8 Security in Transport

Terrorism is a growing concern in our societies and for governments (Zellner 2014). Attacks often target transportation infrastructure, and hence the interaction between this trend and mobility and freight is straightforward (Jenkins 2007). More security is expected which raises the security and accessibility tension: the provision of more security in transport by introducing controls and barriers reduces accessibility.

3.9 Legislative Framework

The legislative dimension converts societal demand, through the production of laws and rules by public authorities and jurisprudence. This legislative dimension is an expression of the broader policy process and institutional environment that directly affects the transport sector. Nevertheless, beyond the mere role of translation of societal demand, the legislative dimension can be considered as a dynamic on its own, and hence can be considered as a societal trend in the broad sense. Three legislative trends exert influence in the domains of mobility and logistics.

We observe in the legislative domain a trend of "Diversifying approaches of governance"(European Environment Agency 2015). More actors are invited to contribute to the governance of transport and mobility. In particular, with the association of citizens in decision processes, more transparency is required in governance models (Albrechts 2010). The innovation at play in the domain of legislation and governance, leads to a diversification of governance models.

Secondly, with the trend "Legislative models adapting to new transport solutions and businesses "an interaction occurs between new business models and the legislative framework. The legislative framework has to adapt to new solutions (Azevedo and Maciejewski 2015), but newcomers must also make sure their business can sustain in a given and changing legislative framework.

The "Trend toward harmonisation in legislative frameworks" of the legislative framework in Europe has direct implications for transport, in the aims of interoperability of transport systems ('Road Transport: Harmonisation of Legislation | EU Fact Sheets | European Parliament' 2016). This trend refers also to the fact that legislative adaptations to new models and solutions in a given European country will inspire other countries' reactions.

4 Interrelations Between Trends

In this section, the main interrelations between the trends discussed above are presented. Even though this work mostly identified and isolate individual factors interacting with mobility and logistics in an unequivocal sense, the intrinsically interdependent nature of the material cannot be ignored.

The main interactions between trends form three groups. A small group links sociocultural dynamics to policies aimed at correcting or accompanying them. A large group includes all the links between sociocultural trends and digital technologies. Finally, a few remaining interrelations are identified beyond these two categories.

The first group of interrelations links policies, expressing the will of policy makers, and sociocultural trends as actual transformations in the social field. Several emerging trends in transport demand are pushed by policies, but can also be seen as reflecting societal demand. This is the case in the environmental protection domain where a sustainable consumption culture interacts with extensive sets of policies dedicated to environment protection. Hence, the present analysis mixes socioculturall trends, like ageing, and political responses to identified issues like policies aiming at inclusion of vulnerable groups.

The second group of interrelations links society and digital technologies.

The *liquid modern* society described by Bauman (2000), is intimately shaped by digital technologies of information and communication. There is hence clearly a strong link between the two trends of liquid modernity and the four trends of the digital society. The current and future digital society would not exist without the technology developments in communication and information, but, on the other hand, the use by individuals and social groups is not always envisaged by the creators of services, yet contribute significantly to shaping the digital society. In this sense, digital society is shaped by technology and by societal factors. As is well known in the domain of transport infrastructure, a new supply generates a demand that was not expressed before: this is the so called induced demand. Digital society is rooted in the societal trends of liquid modernity, and also influences the transport demand. The expectations of the customers excert a specific dynamic in the digital world. And the digital society does not limit to transport user requirement, but has deep interactions with mobility in general, including avoiding mobility.

The observed reduction in car use by younger generations has been seen by several analysts as linked to a change in values. The possession of a car tends to be replaced by the idea of connecting to the social network; this points to the use of ICT and hence to several trends identified in the digital society section.

A trend towards the inclusion of vulnerable groups can be identified. But in the emerging digital society new forms of exclusion arise; these forms are particularly of concern in the transport and mobility sectors. Here, a clear issue lies at the intersection of public policies aiming at inclusion and the development of new transport services making use of ICT.

We have identified a trend of new uses of travel time, mainly by means of ICT that currently favours public transport. But, in the future car, automation is likely to erase this comparative advantage. A complex interaction of trends exists here with evolving developments over time.

Beyond the two categories of interrelations, three other linkages can be identified.

Complex interactions link automation and transport safety. Automation comes with the promise of significant improvement of the levels of safety in transport, but it also introduces new kinds of safety problems; an illustration is the self-driven Tesla car casualty accident in 2016 and the many similar events since then. This represents a new type of accident raising significant liability and insurance issues most likely to impact the legislative and regulatory frameworks.

The responses to the security threat perception, as noted, carries the risk of introducing more controls that may be detrimental to the ease of access and use of collective transport systems. The tension between higher security versus accessibility is clearly contrary to the fluidity, acceleration and flexibility features of the *liquid modern* society.

Liquid modernity is both a consequence, or a symptom, and a source for the observed restructuring working arrangements. Indeed, the observation of acceleration is paradoxically based on the growing mobility for non-work purposes, and hence, directly related to the idea of growing part-time work described in the restructuring arrangement trend. This forms an example of the links between economy and society.

This review of linkages was necessary to reveal the complexity of the system of trends. It also contributes to validating the list of trends, by distinguishing many factors for the understanding of the present and future dynamics between society and transport.

5 Conclusion

This paper focuses on the identification and description of societal trends shaping the demand of mobility and logistics. The work has led to the identification of a list of 29 societal trends covering sociocultural, political, economic, technological and legal trends. The result of this investigation forms a broad and comprehensive set of factors that influence, and will influence future mobility and logistics. This approach has been organised so that, even if trends are linked one to another, as can be seen in the last section, no duplication can be found in the set of trends. This set borrows characteristics of a system in the sense that individual sub-parts, once assembled, form a coherent set.

This work has been strongly inspired by the ideas of the liquid modernity developed by Bauman. Liquid modernity highlights key societal trends of individualism, of the fluidity of society, of the feeling of acceleration of the pace of life, of the growing importance of social networks, and of the role of information and communication technologies. This broad and complete view on societal trends has proved to be very supportive of the present analysis. It allows to describe linkages between social and economic trends, and between society and technology, especially information and communication technology. Basing the analysis on liquid modernity provides coherence and exhaustiveness in covering the topic.

The added value of this contribution is its systematic approach and that evidence is provided for each identified trend. The proposal forms a broad and comprehensive view of societal trends that play, and will play, a role in shaping the demand for mobility and logistics in Europe at the horizon 2030. The trends identified in this paper form the bases of the further work of scenario building and the creation of an action plan in the Mobility4EU project which is the subject of the previous papers within this book publication (Keseru et al. 2018; Bierau-Delpont 2018).

Acknowledgements The authors thank, Corinne Blanquart, Freek Bos, Annette Brückner, Thierry Coosemans, Laetitia Dablanc, Erzsébet Földesi, Alessia Golfetti, Stefania Grosso, George Holley-Moore, Juho Kostiainen, Jochen Langheim, Gereon Meyer, Linda Napoletano, Cristina Pou, Annette Randhahn, Joachim Skoogberg, Yves Stans, Anu Tuominen, Marcia Urban, Susana Val, Ineke van der Werf for their inputs in this work.
The authors thank Amna Diaz from ILC for language check.
The authors wish to thank the European Commission for the funding of the Mobility4EU project in the framework of the Horizon2020 program (EC Contract No. 690732).

References

Albrechts L (2010) More of the same is not enough! How could strategic spatial planning be instrumental in dealing with the challenges ahead? Environ Plan 37(6):1115–1127
Anderton K, Brand R, Leiren MD, Gudmundsson H, Reichenbach M, Schippl J (2015) TRANSFORuM urban mobility roadmap. European Comission. http://www.transforum-project.eu/fileadmin/user_upload/08_resources/08-01_library/TRANSFORuM_Roadmap_Urban.pdf
'Asylum Statistics—Statistics Explained'. 2016. Accessed 29 July 2016. http://ec.europa.eu/eurostat/statistics-explained/index.php/Asylum_statistics
Azevedo F, Maciejewski M (2015) Social, economic and legal consequences of Uber and similar transportation network companies (TNCs)—think tank. Briefing. European Parliament. http://www.europarl.europa.eu/thinktank/en/document.html?reference=IPOL_BRI(2015)563398
Bassanini Franco, Reviglio Edoardo (2011) Financial stability, fiscal consolidation and long-TERM investment after the crisis. OECD J Financ Market Trends 2011(1):31–75
Bauman Z (2000) Liquid modernity, vol 9. Polity Press, Cambridge. http://neilsquire.pbworks.com/w/file/fetch/35116162/Bauman-Liquid%EE%80%80Modernity%EE%80%81.pdf
Bernardino J, Vieira J, Garcia H (2013) FUTRE Deliverable 3.1: factors of evolution of demand and pathways. European Comission. http://www.futre.eu/Publications/Deliverables.aspx
Bierau-Delpont F (2018) Building an action plan for the holistic transformation of the European transport system. In: Towards user-centric transport in Europe. Challenges, solutions and collaborations. Springer
Blanke Michael, Burdick Bernhard (2005) Food (miles) for thought-energy balance for locally-grown versus imported apple fruit (3 Pp). Environ Sci Pollut Res 12(3):125–127
Boschmann E Eric, Kwan Mei-Po (2008) Toward socially sustainable urban transportation: progress and potentials. Int J Sustain Transp 2(3):138–157. https://doi.org/10.1080/15568310701517265
Brög W, Barta F, Erl E (2005) Societal megatrends: like it or not, the framework is set. In: 56th UITP world congress. http://socialdata.de/info/Societal%20Megatrends%20-%20UITP%20Plenary%20Presentation.pdf
Capros P, De Vita S, Tasios N, Papadopoulos D, Siskos P, Apostolaki E, Zampara M et al (2013) EU energy, transport and GHG emissions: trends to 2050, reference scenario 2013. https://trid.trb.org/view.aspx?id=1285618
Casey T, Valovirta V (2016) Towards an open ecosystem model for smart mobility services. VTT, Espoo. http://www.vtt.fi/inf/pdf/technology/2016/T255.pdf
Chang Y-C, Yen HR (2012) Introduction to the special cluster on managing technology–service fusion innovation. Technovation 32(7):415–418
Cirstea A (2015) The implications of mobile commerce applications: the case study of Uber in Romania. Int J Sci Knowl 6(2):1–5
Clegg S, Baumeler C (2010) Essai: from iron cages to liquid modernity in organization analysis. Organ Stud 31(12):1713–1733. https://doi.org/10.1177/0170840610387240
Coley D, Howard M, Winter M (2009) Local food, food miles and carbon emissions: a comparison of farm shop and mass distribution approaches. Food Policy 34(2):150–155
Corwin S, Vitale J, Kelly E, Cathles E (2015) The future of mobility, how Transportation technology and social trends are creating a new business ecosystem. Deloitte http://www2.deloitte.com/ru/en/pages/manufacturing/articles/future-of-mobility.html
Dablanc L, Blanquart C, Combes F, Heitz A, Klausberg J, Koning M, Liu Z, Seidel S (2016) CITYLAB observatory of strategic developments impacting urban logistics (2016 Version). Deliverable 2-1 CITYLAB European Project. http://www.citylab-project.eu/deliverables/D2_1.pdf: European Commission H2020 Programme. http://www.citylab-project.eu/deliverables/D2_1.pdf
Dablanc L, Morganti E, Arvidsson N, Woxenius J, Browne M, Saidi N (2017) The rise of on-demand "instant deliveries" in European cities. Supply Chain Forum: Int J, 1–15
Davis B, Dutzik T, Baxandall P (2012). Transportation and the new generation: why young people are driving less and what it means for transportation policy. http://trid.trb.org/view.aspx?id=1141470

Doll C, Klug S, Enei R (2014) Large and small numbers: options for quantifying the costs of extremes on transport now and in 40 years. Nat Hazards 72(1):211–239
Euromonitor (2013) Downsizing globally: the impact of changing household structure on global consumer markets. http://www.euromonitor.com/downsizing-globally-the-impact-of-changing-household-structure-on-global-consumer-markets/report
European Commission (2011) Impact assessment accompanying document to the white paper: roadmap to a single European transport area—towards a competitive and resource efficient transport system. http://eur-lex.europa.eu/legal-content/EN/ALL/?uri=CELEX:52011DC0144
'European Commission Press Release— 2015 Road Safety Statistics: What Is behind the Figures?' (2016) Accessed 27 June 2016. http://europa.eu/rapid/press-release_MEMO-16-864_en.htm
European Environment Agency (2015) Assessment of global megatrends—extended background analysis—European Environment Agency. Publication 11/2015. EEA Technical Report. European Environment Agency, Luxembourg. http://www.eea.europa.eu/publications/global-megatrends-assessment-extended-background-analysis
European Investment Bank Institute EIB (2013) Smart cities. Concept and challenges, assessing smart city initiatives for the mediterranean Region (ASCIMER). 2013. http://www.eiburs-ascimer.transyt-projects.com/
Eurostat (2015) Key figures on Europe, 2015 Edition. Statistical Books, Eurostat
Finnish Prime Minister's Office (2015) Finland, a land of solutions strategic programme of Prime Minister Juha Sipilä's Government. Prime Minister's Office Finland
Forbes (2015) Tesla's business model highlights what the shift to electric means for the auto industry. 2015. http://www.forbes.com/sites/greatspeculations/2015/09/01/teslas-business-model-highlights-what-the-shift-to-electric-means-for-the-auto-industry/#4b8daf945029
Frändberg L, Vilhelmson B (2011) More or less travel: personal mobility trends in the Swedish population focusing gender and cohort. J Transp Geogr, Special section on Luxembourg Travel Futures 19(6):1235–1244. https://doi.org/10.1016/j.jtrangeo.2011.06.004
Frisoni R, Dall'Oglio A, Nelson C, Long J, Vollath C, Ranghetti D, McMinimy S, Gleave SD (2016) 'Research for TRAN committee—self-piloted cars: The future of road transport?' European Parliament, Directorate-General for internal policies, policy department B: Structural and Cohesion Policies, Transport and Tourism
Fujita M, Krugman PR, Venables A (2001) The spatial economy: cities, regions, and international trade. MIT press. https://books.google.fr/books?hl=fr&lr=&id=07Mzawou-8EC&oi=fnd&pg=PR11&dq=+europe+cities+regions+power&ots=LI6X4g_N9J&sig=Ip-T9_iv4kcKQecmWrV03AqRbvU
Goodall W, Fishman T, Dixon S, Perricos C (2015) Transport in the digital age, disruptive trends for smart mobility. Deloitte
Goodwin P (2012) Three views on peak car. World Transport, Policy and Practice 17. http://eprints.uwe.ac.uk/16119/21/wtpp17.4.pdf#page=9
Granovetter M (1983) The strength of weak ties: a network theory revisited. Sociol Theory 1:201–233
Hardesty, L (2014) Ride-sharing could cut cabs' road time by 30 Percent. MIT News, 1 September 2014. http://news.mit.edu/2014/rideshare-data-cut-taxi-time-0901
Harrison P (2013) Fuelling Europe's future. http://www.camecon.com/EnergyEnvironment/EnergyEnvironmentEurope/FuellingEuropesFuture.aspx
Harvey D (1990) The condition of postmodernity: an enquiry into the conditions of cultural change. http://www.citeulike.org/group/14819/article/9699561
Isusi I, Corral A (2004) Part-time work in Europe. EurWORK. http://www.eurofound.europa.eu/observatories/eurwork/comparative-information/part-time-work-in-europe
'ITF Transport Outlook 2017' (2017) International transport forum/OECD. http://www.oecd.org/about/publishing/itf-transport-outlook-2017-9789282108000-en.htm
Jackson T, Victor P (2011) Productivity and work in the "green economy": some theoretical reflections and empirical tests. Environ Innov Societal Transitions 1(1):101–108

Jagger C, Wohland P, Fouweather T, Kirkwood T (2014) Raising the retirement age: implications for UK and Europe
Jain J, Lyons G (2008) The gift of travel time. J Transp Geogr 16(2):81–89. https://doi.org/10.1016/j.jtrangeo.2007.05.001
Jenkins BM (2007) The terrorist threat to surface transportation. National Transportation Security Center. Mineta Transportation Institute. http://scotsem.transportation.org/Documents/Jenkins–TheTerroristThreattoSurfaceTransportation.pdf
Jeske M, Grüner M, Weiss F (2013) Big data in logistics: a DHL perspective on how to move beyond the hype. DHL Customer Solutions & Innovation 12
Kaufmann V (2002) Re-thinking mobility. http://hal.archives-ouvertes.fr/halshs-00439011/
Kellerman A (2012) Daily spatial mobilities: physical and virtual. Ashgate Publishing Limited
Keseru I, Coosemans T, Macharis C (2018) Building scenarios for the future of transport in Europe: The Mobility4EU approach. In: Towards user-centric transport in Europe. Challenges, solutions and collaborations. Springer
Kester JGC (2014) 2013 International tourism results and prospects for 2014. UNWTO News. http://cf.cdn.unwto.org/sites/all/files/pdf/unwto_fitur_2014_hq_jk_2pp.pdf
Kostiainen J, Aapaoja A, Hautala R (2016) Public transport ITS test environment for a smart city. Presented at the 11th ITS European Congress, Glasgow, June. http://glasgow2016.itsineurope.com/
Krzyzanowski M, Kuna-Dibbert B, Schneider J (2005) Health effects of transport-related air pollution. WHO Regional Office Europe. https://books.google.fr/books?hl=fr&lr=&id=b2G3k51rd0oC&oi=fnd&pg=PR1&dq=Health+effects+of+transport-related+air+pollution&ots=O65x2DGs9z&sig=5IKuTQngcA2LTFbzrGmfwhUs4ZY
Kulak M, Nemecek T, Frossard E, Chable V, Gaillard G (2015) Life cycle assessment of bread from several alternative food networks in Europe. J Clean Prod 90:104–113
L'Hostis A, Müller B, Meyer G, Brückner A, Foldesi E, Dablanc L, Blanquart C et al (2016) 'MOBILITY4EU—D2.1—societal needs and requirements for future transportation and mobility as well as opportunities and challenges of current solutions. Research Report. IFSTTAR—Institut Français des Sciences et Technologies des Transports, de l'Aménagement et des Réseaux. https://hal.archives-ouvertes.fr/hal-01486783
Lazakis I (2014) EXCROSS final report on synergies and opportunities. European Commission. http://www.excross.eu/deliverables.htm
Leminen S, Rajahonka M, Westerlund M, Siuruainen R (2015) Ecosystem business models for the internet of things. In: Internet of things Finland. Digile, 10–13. https://www.google.fr/url?sa=t&rct=j&q=&esrc=s&source=web&cd=1&ved=0ahUKEwjt9bSQpazMAhULXhoKHaS-BHwQFggdMAA&url=http%3A%2F%2Fwww.internetofthings.fi%2Fextras%2FIoTMagazine2015.pdf&usg=AFQjCNGXos4lcQk1GCkVVtpZFkm6Q8bMgw&sig2=Gmp3uRCtXifJaxsUtE6CdQ
Lemmerer H, Pfaffenbichler P (2012) 'ORIGAMI Deliverable 3.1 current travel behaviour, future trends and their likely impact. European Commission, Edinburgh. www.origami-project.eu
Levine RV (1998) A geography of time: the temporal misadventures of a social psychologist, Revised edn. Basic Books
Löffler M, Tschiesner R (2016) The internet of things and the future of manufacturing | McKinsey & Company. Accessed 8 July 2016. http://www.mckinsey.com/business-functions/business-technology/our-insights/the-internet-of-things-and-the-future-of-manufacturing
Lyons Glenn, Jain Juliet, Susilo Yusak, Atkins Stephen (2013) Comparing rail passengers' travel time use in Great Britain between 2004 and 2010. Mobilities 8(4):560–579
Martens K (2012) Justice in transport as justice in accessibility: applying Walzer's 'spheres of justice' to the transport sector. Transportation 39(6):1035–1053
McKinsey AC, Moh D, Weig F, Zerlin B, Hein A-P (2012) Mobility of the future, opportunities for automotive OEMs

Meisterling K, Samaras C, Schweizer V (2009) Decisions to reduce greenhouse gases from agriculture and product transport: LCA case study of organic and conventional wheat. J Clean Prod 17(2):222–230

Metz D (2013) Peak car and beyond: the fourth era of travel. Transp Rev 33(3):255–270

Newman P, Kenworthy J, Glazebrook G (2013) Peak car use and the rise of global rail: why this is happening and what it means for large and small cities. J Transp Technol 03(04):272–287. https://doi.org/10.4236/jtts.2013.34029

Pachauri RK, Meyer LA (2014) Fifth assessment report—synthesis report contribution of working groups I, II and III to the fifth assessment report of the Intergovernmental Panel on Climate Change. https://www.ipcc.ch/report/ar5/syr/

Pauzié A (2013) 'DECOMOBIL Deliverable 3.4 nomadic transport services for multimodal mobility: issues and perspectives. European Commission. http://decomobil.humanist-vce.eu/Downloads.html

Petersen MS, Sessa C, Enei R, Ulied A, Larrea E, Obisco O, Timms P, Hansen C (2009) 'TRANSvisions final report on transport scenarios with a 20 and 40 year horizon. European Commission. http://81.47.175.201/flagship/attachments/2009_02_transvisions_report.pdf

Rantasila K, Mantsinen H, Casey T, Hautala R, Lankinen M (2014) Development of ITS multiservice from idea to deployment. Presented at the 10th ITS European Congress, Helsinki, June 16

'Road Transport: Harmonisation of Legislation | EU Fact Sheets | European Parliament' (2016) Accessed 27 June 2016. http://www.europarl.europa.eu/atyourservice/en/displayFtu.html?ftuId=FTU_5.6.4.html

Rosa H (2003) Social acceleration: ethical and political consequences of a desynchronized high-speed society. Constellations 10(1):3–33

Rudinger G, D K, Poppelreuter S (2004) Societal trends, mobility behaviour and sustainable transport in Europe and North America: The European Union Network STELLA. Eur J Ageing 1(1):95–101

Russell ML (2012) Travel time use on public transport: what passengers do and how it affects their wellbeing. Thesis, University of Otago. http://otago.ourarchive.ac.nz/handle/10523/2367

Sena e Silva M, Oliveira M, Ribeiro NS, Moraglio M, Ludvigsen J, Christ A, Seppänen T-M et al (2013) 'RACE2050 D5.1—Current transport demand and global transport outlook. European Commission. http://www.race2050.org/index.php?id=4#news

Sessa C, Enei R (2009) EUTransportGHG report EU transport demand: trends and drivers. European Commission. http://www.eutransportghg2050.eu/cms/assets/EU-Transport-GHG-2050-Task3-Paper-EU-Transport-Trends-and-Drivers-22-12-09-FINAL.pdf

Shaheen SA, Mallery MA, Kingsley KJ (2012) Personal vehicle sharing services in North America. Flex Transp Serv 3(August):71–81. https://doi.org/10.1016/j.rtbm.2012.04.005

Smith BW, Svensson J (2015) Automated and autonomous driving: regulation under uncertainty. https://trid.trb.org/view.aspx?id=1358502

United Nations (2014) World urbanization prospects: the 2014 revision. UN

Urry J (2007) Mobilities. Polity, London

Vasileva K (2009) Citizens of European countries account for the majority of the foreign population in EU-27 in 2008—Issue Number 94/2009—Product—Eurostat'. Eurostat. http://ec.europa.eu/eurostat/en/web/products-statistics-in-focus/-/KS-SF-09-094

Velaga N, Beecroft M, Nelson J (2012) Transport poverty meets the digital divide: accessibility and connectivity in rural communities. J Transp Geogr 21:102–112

Verstraeten J, Kirwan B (2014) OPTICS 2nd expert workshop: from hazard management to operational resilience. European Commission

Wilson S (2011) Transport outlook 2011: meeting the needs of 9 billion people. Text. http://www.itf-oecd.org/transport-outlook-2011-meeting-needs-9-billion-people

World Health Organization, and The World Bank (2011) World report on disability 2011. http://apps.who.int/iris/handle/10665/44575

Zakir J, Seymour T, Berg K (2015) Big data analytics. Issues Inf Syst 16(2). http://www.iacis.org/iis/2015/2_iis_2015_81-90.pdf

Zellner W (2014) Threat perceptions in the OSCE area. Institut für Friedensforschung und Sicherheitspolitik an der Universität Hamburg. https://ifsh.de/file-CORE/documents/core_news/CORE_News_Spring_2014.pdf

Pathways Towards Decarbonising the Transportation Sector

Oliver Lah and Barbara Lah

Abstract Transport plays a key role in delivering on the Paris Agreement, the Sustainable Development Goals and the New Urban Agenda. While providing essential services to society and economy, transport is also an important part of the economy and it is at the core of a number of major sustainability challenges, in particular climate change, air quality, safety, energy security and efficiency in the use of resources. This chapter identifies the linkages between decarbonisation pathways, policy design, coalition building and institutional frameworks. The analysis shows that there are critical interlinkages between these aspects. Decarbonisation of the transport sector is not possible through isolated measures. A broad range of local and national actions are needed to bring the sectors on to low-carbon development path. Furthermore, a holistic policy approach is needed to deliver on wider sustainable development objectives. Addressing a broader range of policy objectives can help forming coalitions and consensus among key political and societal actors. Finally, Consensus oriented institutions are needed to maintain a stable policy environment that enables the long-term transitions towards a low-carbon development path. The chapter identifies the potential for land transport climate change mitigation actions at the local and national level, opportunities for synergies of sustainable development and climate change objectives and governance and institutional issues affecting the implementation of measures.

Keywords Transport policy · Climate change · Decarbonization pathways
Governance

O. Lah (✉)
Wuppertal Institute for Climate, Environment and Energy, Neue Promenade 6,
10178 Berlin, Germany
e-mail: oliver.lah@wupperinst.org

B. Lah
Climate Action Implementation Facility (CAIF) gGmbH, Schwedter Str 225,
10435 Berlin, Germany
e-mail: barbara.lah@caif.eu

B. Müller and G. Meyer (eds.), *Towards User-Centric Transport in Europe*,
Lecture Notes in Mobility, https://doi.org/10.1007/978-3-319-99756-8_4

1 Chapter: Pathways Towards Decarbonising the Transportation Sector

Internationally agreed under the United Nations Framework Conventions on Climate Change (UNFCCC) is the necessity to stabilise global warming below 2 °C and the transport sector plays a key role in achieving this target. The Paris Agreement, the Sustainable Development Goals and the New Urban Agenda can only be achieved by recognising the importance of transportation as key enabler of economic activity, participation and social cohesion, while being at the core of major environmental and sustainability challenges, such as air quality, safety, energy security and resource use efficiency.

According to the IPCC, greenhouse gas (GHG) emissions from the transport sector are set to double by 2050 if no changes will be made (IPCC 2014). To achieve the 2 °C target (UNFCCC) developed countries need to rapidly decarbonise the transport sector until 2050 by at least 80% and developing and emerging countries need to curb growth by 2050 by plus 70% compared to 2016 levels (ITF 2017).

None of the current measures taken have the potential to curb emissions to a sustainable level even under very optimistic scenarios (Fulton et al. 2013) because the likely growth in the transportation sector will outpace efficiency gains. Without rapid action in is very unlikely that the transport sector the global warming stabilisation can move a pathway in line with the target stipulated in the Paris Agreement.

The submitted Nationally Determined Contributions (NDCs) by most countries do not provide adequate policy actions necessary to curb the current trend, as seen in passenger transport and freight transport, both of which are vital to tackle to decarbonise the sector, but still no policy action has been developed to mitigate their emissions sufficiently (Cooper 2016; Antimiani et al. 2016; Zhang and Pan 2016; Cassen and Gracceva 2016). Freight transport is expected to increase due to growing demand to 60% of total transport emissions by 2050 (Cooper 2016).

Emissions of freight transport in China and India are expected to grow three times faster than in Europe. However, this does not reflect the whole product lifecycle in which it is necessary to think about freight decarbonization not as a single event yet an integral part of the decarbonization process of the global economy. Growing demand in transportation in developing and emerging countries will continue to rise as vital component of economic development (Berry et al. 2016; Gschwender et al. 2016; Spyra and Salmhofer 2016).

While the challenges the transport sector faces are great, there are a large number of measures available to curb the current trends and to reduce CO_2 emissions, improve air quality and safety, reduce energy costs and make transport more accessible. However, vital for a successful transition towards sustainable mobility is the adoption of a holistic approach at the political level and to:

1. Create synergies between sustainable development policy objectives and avoid trade-offs
2. Consider institutional issues and create coalitions among key actors in the policy process.

2 Policy Integration, Coalitions and Continuity as Drivers of Long-Term Change

The transport sector has a very high potential to reduce transport sector GHG emissions cost effectively. However political and institutional change, progress and change of view in this area needs to start. It is essential to identify the institutional barriers, which hinder the process of low-carbon transport measures not only in industrialised countries, but also in emerging and developing countries, where economic growth and unsustainable urban planning can create emission increase from the transport sector resulting in an irrevocable pathway away from the 1.5° scenario.

Decarbonizing the transport sector is highly visible and policy actions can have wide-ranging impacts on daily mobility behaviour, but also industries, access to jobs and markets and on other sectors, in particular energy. This creates a challenging policy environment, for transport sector climate change mitigation measures as actions can provide opportunities for co-benefits, buts also risks for trade-offs. In this regard, two steps are vital to decarbonise the transport sector:

1. An integrated policy approach that combines various measures to provide a basis for political coalitions, and
2. Political continuity and coalitions that enable take-up of policies and ensure stability.

Policy integration of all available local and national policy interventions is vital to achieve low-carbon stabilisation pathways as single measures do not provide sufficient CO_2 mitigation potential and have the risk of creating trade-offs for other policy objectives. An integrated approach can also create the basis for coalition building if policy objectives of key stakeholders and veto players are taken into account. Hence, an integrated policy approach that aims to generate synergies (rather than trade-offs) between policy objectives can help to maximise socio-economic benefits and can help to form coalitions that endure and are resilient in the face of political volatility (Fig. 1).

While individual measures can potentially be implemented faster in a political environment that is based on minimal majorities, a broader, multi-actor coalition is needed to implement integrated decarbonization strategies, which are necessary to achieve a low-carbon pathway for the sector.

If current trends persist, the growth of transport sector carbon emissions outpaces all efficiency gains and is likely to continue to increase GHG emissions in the future. Even with substantial technological improvements, electric mobility and modal shifts CO_2 levels from the transport sector in 2050 will still be at 2015 levels of around 7.5 Giga tonnes of CO_2, while in a business-as-usual scenario is bound to double by 2050 (Fulton et al. 2013; Harvey 2013; ITF 2017).

Any change of transport policies is intertwined with nearly all other sectors of the economy and society and can lead to unintended consequences, positive and negative, which makes it necessary to generate links and synergies between measures, combining transport policies in a wider policy package. To enable a sustainable

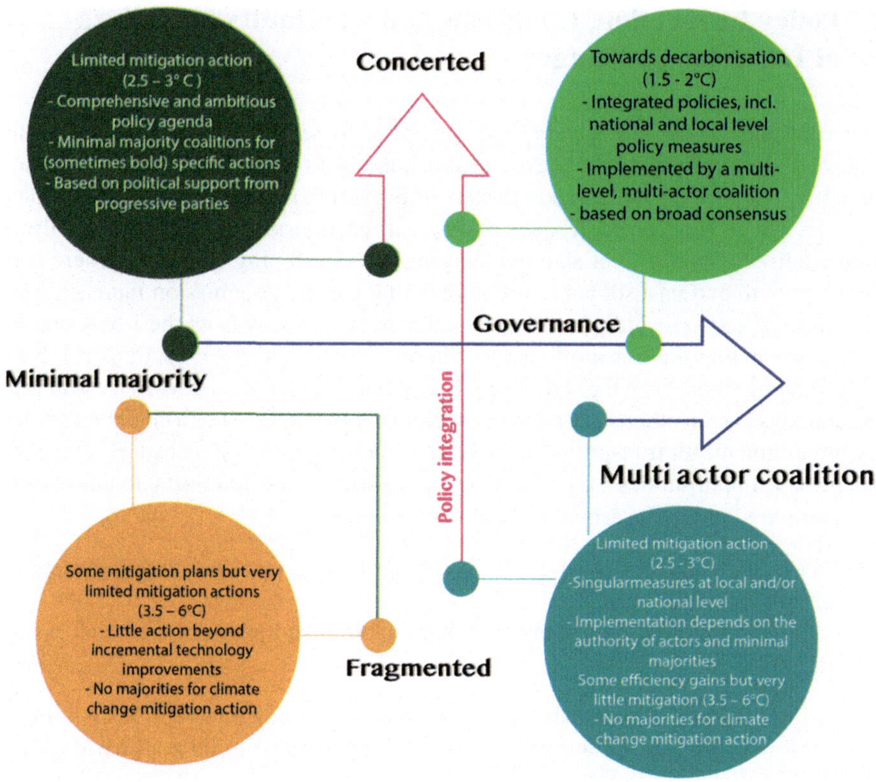

Fig. 1 Policy and governance framework for a low-carbon transport transition pathway with a concerted governance, fragmented governance, multi actor coalition and minimal majority

development pathway in the transport sector two factors for policy integration and governance factors need to be identified.

Industrialised countries need almost entirely decarbonize the transport sector in the coming decades to be on a sustainable emission pathway (IEA 2014/2016; ITF 2017). Initial costs will be high for mitigation actions in the transport sector, yet evidence shows a direct and indirect benefit outweighing the initial costs, saving between USD 50 trillion and 100 trillion in fuel savings, reduced vehicle purchases, needed infrastructure and fuel costs (IEA 2014/2016), while also providing improvements of safety, air quality, reduced travel time etc.

From a climate change perspective vehicle technology and fuel switch options provide the biggest mitigation potential. However, to reach 80–100% emission reductions, but also enable wider sustainable development a broader multimodal approach is necessary, which also manages motorized travel demand, with a strong focus on transport mode switch and shift, e.g. non-motorized transportation (IEA 2014/2016; Fulton et al. 2013).

3 Using Co-benefits to Find Consensus

While the transportation sector is evidently an area with many players and therefore a challenging area for consensus building, recent papers investigated the role of an integrated policy approach utilize co-benefits of policy objectives such as health, safety, efficiency, cost savings, access, environment and other areas to focus on the socio-economic benefits that provide the basis for a coalition to support policy implementation (Vale 2016; Cai et al. 2015; van Vuuren et al. 2015; Bollen 2015; Dhar and Shukla 2015; Lah 2015; Schwanitz et al. 2015; Dhar et al. 2017; Lah 2017). In this sense, energy and climate change policies can thrive by building on a broader consensus, which helps creating a more coherent political framework, but also requires more time for the coalition building and a more stable environment to deliver on the long-term policy objectives. Transport policies need to be linked and packaged to create synergies with other sustainable development goals, such as energy security, road safety, public health (Kanda et al. 2016; Wen et al. 2016).

A survey carried out by Lah (2017) on barriers for low-carbon transport policy implementation shows that technology, public support and lack of funding are the not perceived by policy advisors as main barriers for political decision making (Lah 2017b). Main barrier for transportation policy uptake is considered to be the lack of knowledge about co-benefits within political decision makers and institutional barriers (Fig. 2).

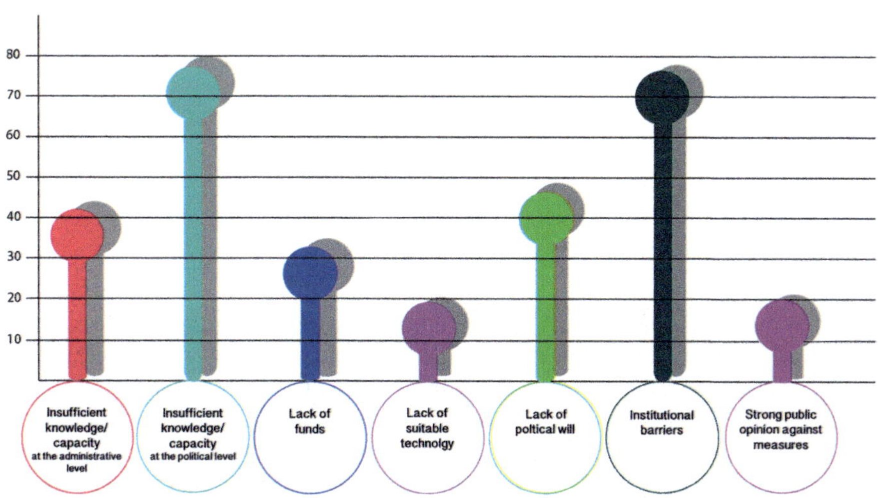

Fig. 2 Main barriers for sustainable mobility policy implementation within political decision makers and institutional barriers (based on Lah 2017)

4 Overcoming Institutional Barriers Through the Consensus Building and Integration

Isolated policy and technology shift measures cannot reduce GHG sufficiently to reach a 2 or 1.5 °C pathway, which is why it is essential to use an integrated policy approach to combine emission reductions of various measures, create consensus, establish a basis for wider economic, social and environmental co-benefits (Lah 2014; Creutzig 2016). Policy in practical terms means creating incentives for individuals shift towards more sustainable transport modes, e.g. through the provision of high quality public transport and compact city design, combined with measures such as road user charging, parking pricing, fuel and vehicle tax that create disincentives for individual car use, but also provide revenue for low-carbon infrastructure investments (Cuenot et al. 2012). A comprehensive low-carbon transport policy package includes improvements of public and non-motorized modes (walking, cycling and public transit services), fuel pricing, vehicle fuel efficiency regulation and taxation, and integrated transport and land use planning, which creates more compact, mixed and better connected communities with less need to travel (Figueroa Meza et al. 2014; Sims et al. 2014). With such a broad and strategic approach, integrated measures have the potential to provide a basis for wider public acceptance, which is a vital factor for political support and long-term sustainability.

Integrated policy approaches not only create synergies but also minimise rebound effects. For example fuel efficiency standards for light duty vehicles (LDV) will improve the fuel economy of vehicles, but can lead to an increase in travel demand, which may off-set the efficiency gains (Sorrell 2010; Lah 2014). Hence, managing travel demand is vital, which can be done through fuel taxation and road user charging to minimise rebound effects (IPCC 2014).

5 Pathways, Policy and Governance Approaches for Low-Carbon Transport

Energy and climate change policies for the transport sector require a stable political operating environment to enable long-term investment decisions by industry and consumers (Lakshmanan 2011; Fais et al. 2016; Spataru et al. 2015). Policies to reduce energy consumption in the transport sector require a strong political commitment to appear on the policy agenda and to remain in place as they rely on investments that are only cost-effective over the medium to long-term (ITF 2017), as policy interventions in the transport sector can be highly visible and politically sensitive.

Political processes are complex, and it is challenging to draw direct conclusions from institutional settings to climate policy performance. Particularly obvious becomes the relationship between institutional structures and climate policy performance when assessing the stability and continuity (or the lack) of specific policies in different countries. To establish a relationship between institutional structures

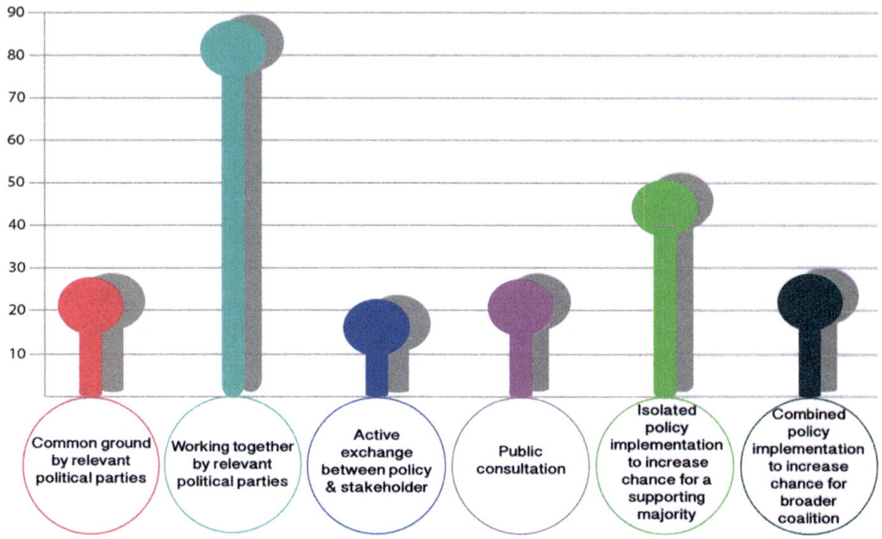

Fig. 3 Factors of governance and policy integration affecting the implementation of sustainable mobility measures (based on Lah 2017b)

and capacities, and climate policy impacts, i.e. emission reductions, it is first vital to establish the link between a certain set of policies and their ability to deliver substantial emission reduction impacts.

By reviewing the main policy measures in G20 countries, the improvement of (LDV) fleet highlights that a combination of vehicle fuel efficiency standards along with vehicle and fuel taxation can improve the efficiency of the vehicle fleet and decrease emissions (Yang et al. 2017). When correlating these measures with institutional features of consensus democracies a first indication of a positive relationship between the presence of institutions that enable a stable and consensus oriented policy environment and the presence of key national level policy measures can be derived (Lah 2017a). The reliance on particular political parties may deliver swift and more ambitious immediate policy action, which may, however, be overturned after elections (Lah 2017b). The risk of political volatility is also reflected in Fig. 3, where only 20% of the respondents consider it feasible that political parties can find common ground on sustainable transport issues and that policies are often implemented in an isolated way. However, relevant authorities at the local and national level are considered more prone to cooperate with counterparts on sustainable transport policy.

6 Steps Towards the Decarbonization of the Transport Sector

The objective of this chapter was to identify some of the main factors that affect policy implementation in the area of sustainable transport and develop a framework that can guide the policy development and coalition building. The following sections summarise some of the key steps towards a long-term transition towards sustainable and low-carbon mobility.

Policy continuity and consensus—are key aspects for policy agenda setting, which are products of political and intuitional relationships (Fankhauser et al. 2015; Marquardt 2017) at different levels of government (local, state, federal, national) and knowledge about scientific research on climate change, which can vary widely (Never and Betz 2014). Finding common ground on the understanding of climate change impact is the first necessary step for consensus on mitigation and adaptation strategies. While political environments are different from location to location, progress in reducing GHG emissions are limited from political volatility, as transport policy measures are long-term and often costly. Continues support from diverse political and public stakeholders is necessary for the long-term success of any policy, especially in the transport sector. Public perception and the influence of epistemic communities also plays an important role for the political agenda setting and consensus on major policy issues such as climate change and energy efficiency (Hagen et al. 2016; Cook and Rinfret 2015).

Policy integration and coalition building—The policy environment, or context in which decisions are made, is as important as the combination of policy decisions and infrastructure investments that make up a low-carbon transport strategy (Justen et al. 2014). This policy environment includes socio-economic and political aspects of the institutional structures of countries. These structures help build coalitions, but can also increase the risk that a policy package fails because one measure faces strong opposition (Sørensen et al. 2014). A core element of success is the involvement at an early stage of potential veto players and the incorporation of their policy objectives in the agenda setting (Tsebelis and Garrett 1996).

Institutional context—The political and institutional context in which policies are pursued is a factor to be considered for the success or failure of implementation (Jänicke 1992). Institutional aspects, the respective roles of departments and ministries in the process are likely to have an effect on the implementation of (primarily) climate related transport measures (Fredriksson et al. 2016). The legal power, budget and political influence of these agencies are equally important (Jänicke 2002). Hence it is vital to make a thorough case for suitable mobility measures that goes beyond one policy objective and also include social and economic considerations.

Acknowledgements Research that led to the publication of this paper has been supported by the European Union's Horizon 2020 Framework Programme, Grant Agreement No. 723970 (FUTURE RADAR).

References

Antimiani A, Costantini V, Kuik O, Paglialunga E (2016) Mitigation of adverse effects on competitiveness and leakage of unilateral EU climate policy: an assessment of policy instruments. Ecol Econ 128(August):246–259. https://doi.org/10.1016/j.ecolecon.2016.05.003

Berry A, Jouffe Y, Coulombel N, Guivarch C (2016) Investigating fuel poverty in the transport sector: toward a composite indicator of vulnerability. In: Energy demand for mobility and domestic life: new insights from energy justice, vol 18, pp 7–20, Aug. https://doi.org/10.1016/j.erss.2016.02.001

Bollen J (2015) The value of air pollution co-benefits of climate policies: analysis with a global sector-trade CGE model called worldscan. Technol Forecast Soc Chang 90, Part A(January): 178–191. https://doi.org/10.1016/j.techfore.2014.10.008

Cai Y, Newth D, Finnigan J, Gunasekera D (2015) A hybrid energy-economy model for global integrated assessment of climate change, carbon mitigation and energy transformation. Appl Energy 148(June):381–395. https://doi.org/10.1016/j.apenergy.2015.03.106

Cassen C, Gracceva F (2016) Chapter 7—energy security in low-carbon pathways A2—Lombardi, Patrizia. In: Gruenig M (ed) Low-carbon energy security from a european perspective, Academic Press, Oxford, pp 181–205. http://www.sciencedirect.com/science/article/pii/B9780128029701000073

Cook J, Rinfret S (2015) are they really so different? climate change rule development in the USA and UK. J Public Aff 15(1):79–90. https://doi.org/10.1002/pa.1512

Cooper M (2016) Renewable and distributed resources in a post-paris low carbon future: the key role and political economy of sustainable electricity. Energy Res Soc Sci 19(September):66–93. https://doi.org/10.1016/j.erss.2016.05.008

Creutzig F (2016) Evolving narratives of low-carbon futures in transportation. Transp Rev 36(3):341–360. https://doi.org/10.1080/01441647.2015.1079277

Cuenot F, Fulton L, Staub J (2012) The prospect for modal shifts in passenger transport worldwide and impacts on energy use and CO_2. Energy Policy 41(February):98–106. https://doi.org/10.1016/j.enpol.2010.07.017

Dhar S, Shukla PR (2015) low carbon scenarios for transport in India: co-benefits analysis. Energy Policy 81(June):186–198. https://doi.org/10.1016/j.enpol.2014.11.026

Dhar S, Pathak M, Shukla PR (2017) Electric vehicles and India's low carbon passenger transport: a long-term co-benefits assessment. In: Bridging the gaps for accelerating low carbon actions in Asia, vol 146, pp 139–148, Mar. https://doi.org/10.1016/j.jclepro.2016.05.111

Fais B, Sabio N, Strachan N (2016) The critical role of the industrial sector in reaching long-term emission reduction, energy efficiency and renewable targets. Appl Energy 162(January):699–712. https://doi.org/10.1016/j.apenergy.2015.10.112

Fankhauser S, Gennaioli C, Collins M (2015) The political economy of passing climate change legislation: evidence from a survey. Glob Environ Change 35(November):52–61. https://doi.org/10.1016/j.gloenvcha.2015.08.008

Figueroa Meza MJ, Lah O, Fulton LM, McKinnon AC, Tiwari G (2014) Energy for transport. Ann Rev Env Resour 39(1):null

Fredriksson PG, Sauquet A, Wollscheid JR (2016) Democracy, political institutions, and environmental policy☆. In: Reference module in earth systems and environmental sciences. Elsevier. https://doi.org/10.1016/B978-0-12-409548-9.09714-1

Fulton L, Lah O, Cuenot F (2013) Transport pathways for light duty vehicles: towards a 2° scenario. Sustainability 5(5):1863–1874. https://doi.org/10.3390/su5051863

Gschwender A, Jara-Díaz S, Bravo C (2016) Feeder-trunk or direct lines? economies of density, transfer costs and transit structure in an urban context. Transp Res Part A: Policy Pract 88(June):209–222. https://doi.org/10.1016/j.tra.2016.03.001

Hagen B, Middel A, Pijawka D (2016) European climate change perceptions: public support for mitigation and adaptation policies. Environ Policy Governance 26(3):170–183. https://doi.org/10.1002/eet.1701

Harvey LDD (2013) Global climate-oriented transportation scenarios. Energy Policy 54(March):87–103. https://doi.org/10.1016/j.enpol.2012.10.053

International Energy Agenca (IEA) (2014/2016) Energy technology perspectives. Paris, IEA

Jänicke M (1992) Conditions for environmental policy success: an international comparison. Environmentalist 12(1):47–58. https://doi.org/10.1007/BF01267594

Jänicke M (2002) The political system's capacity for environmental policy: the framework for comparison. In: Weidner H, Jänicke M (eds) Capacity building in national environmental policy, Springer Berlin Heidelberg, pp 1–18. http://dx.doi.org/10.1007/978-3-662-04794-1_1

Justen A, Schippl J, Lenz B, Fleischer T (2014) Assessment of policies and detection of unintended effects: guiding principles for the consideration of methods and tools in policy-packaging. Policy Packag 60:19–30. https://doi.org/10.1016/j.tra.2013.10.015

Kanda W, Sakao T, Hjelm O (2016) Components of business concepts for the diffusion of large scaled environmental technology systems. In: New approaches for transitions to low fossil carbon societies: promoting opportunities for effective development, diffusion and implementation of technologies, policies and strategies, vol 128, pp 156–67, Aug. https://doi.org/10.1016/j.jclepro.2015.10.040

Lah O (2014) The barriers to vehicle fuel efficiency and policies to overcome them. Eur Transp Res Rev. http://link.springer.com/journal/12544

Lah O (2015) Sustainable development benefits of low-carbon transport measures. Deutsche Gesellschaft für Internationale Zusammenarbeit (GIZ) GmbH, Eschborn, Germany. http://transport-namas.org/wp-content/uploads/2015/12/giz_TRANSfer_2015_Sustainable-developement-benefits-of-low-carbon-transport-measures_web.pdf

Lah O (2017a) Factors of change: the influence of policy environment factors on climate change mitigation strategies in the transport sector." Transp Res Procedia (WCTR Special Issue):0–17

Lah O (2017b) Factors of change: the influence of policy environment factors on climate change mitigation strategies in the transport sector. WIRE Energy Environ (Special Issue):0–17

Lakshmanan TR (2011) The broader economic consequences of transport infrastructure investments. J Transp Geogr 19(1):1–12. https://doi.org/10.1016/j.jtrangeo.2010.01.001

Marquardt J (2017) Conceptualizing power in multi-level climate governance. J Clean Prod. https://doi.org/10.1016/j.jclepro.2017.03.176

Never B, Betz J (2014) Comparing the climate policy performance of emerging economies. World Dev 59(July):1–15. https://doi.org/10.1016/j.worlddev.2014.01.016

Schwanitz VJ, Longden T, Knopf B, Capros P (2015) The implications of initiating immediate climate change mitigation—a potential for co-benefits? Technol Forecast Soc Chang 90, Part A(January):166–177. https://doi.org/10.1016/j.techfore.2014.01.003

Sims R, Schaeffer R, Creutzig F, Nunez X, D'Agosto M, Dimitriu D, Meza M, Fulton L, Kobayashi S, Lah O (2014) Transport In: Mitigation. Contribution of working group III to the fifth assessment report of the intergovernmental panel on climate change

Sorrell S (2010) Energy, economic growth and environmental sustainability: five propositions. Sustain 2(6):1784–1809

Sørensen HH, Isaksson K, Macmillen J, Åkerman J, Kressler F (2014) Strategies to manage barriers in policy formation and implementation of road pricing packages. Policy Packag 60:40–52

Spataru C, Drummond P, Zafeiratou E, Barrett M (2015) Long-term scenarios for reaching climate targets and energy security in UK. Sustain Cities Soc 17(September):95–109. https://doi.org/10.1016/j.scs.2015.03.010

Spyra H, Salmhofer H-J (2016) The politics of decarbonisation—a case study. Transport Research Arena TRA 2016 14:4050–4059. https://doi.org/10.1016/j.trpro.2016.05.502

Tsebelis G, Garrett G (1996) Agenda setting power, power indices, and decision making in the european union. Int Rev Law Econ 16(3):345–361

Vale PM (2016) The changing climate of climate change economics. Ecol Econ 121(January):12–19. https://doi.org/10.1016/j.ecolecon.2015.10.018

van Vuuren DP, Kok M, Lucas PL, Prins AG, Alkemade R, van den Berg M, Bouwman L et al (2015) Pathways to achieve a set of ambitious global sustainability objectives by 2050: explorations using

the IMAGE integrated assessment model. Technol Forecast Soc Chang 98(September):303–323. https://doi.org/10.1016/j.techfore.2015.03.005
Wen J, Hao Yu, Feng G-F, Chang C-P (2016) Does government ideology influence environmental performance? Evidence based on a new dataset. Openness Institutions Long-Run Socio-Economic Dev 40(2):232–246. https://doi.org/10.1016/j.ecosys.2016.04.001
Yang Z, Mock P, German J, Bandivadekar A, Lah O (2017) On a pathway to de-carbonization–a comparison of new passenger car CO_2 emission standards and taxation measures in the G20 countries. Transp Res Part D: Transp Environ (Special Issue Climate Change and Transport)
Zhang W, Pan X (2016) Study on the demand of climate finance for developing countries based on submitted INDC. Adv Clim Change Res. https://doi.org/10.1016/j.accre.2016.05.002

Part II
Making Transport Accessible for All

Mainstreaming the Needs of People with Disabilities in Transport Research

Erzsébet Földesi and Erzsébet Fördős-Hódy

Abstract Lack of accessible transport vehicles and services prevents people with disabilities from actively and fully participating in the society, thus depriving them of the freedom of movement. Mainstreaming disability aspects following the universal design concept guarantees that the deliverables of a transport-related research project do not result in new barriers for people with disabilities and they can enjoying the benefits of the innovation and development on equal basis with other passengers. Using the method of mainstreaming disability does not exclude the necessity of conducting special disability-related transport research. This twin-track approach can significantly increase the accessibility of transport for all.

Keywords Accessibility · Mainstreaming disability · Disability-specific transport research · Universal design

1 Introduction

Around 80 million people live with disabilities in the European Union's member states. This number is forecast to increase to 120 million by 2020 due to the ageing population.

In 2006 the United Nations adopted the Convention on the Rights of Persons with Disabilities (CRPD) addressing disability as a human rights issue. CRPD has been ratified by all EU member states and the European Union itself. The new approach enshrined in the CRPD confirms that disabled people do have human rights, as well as civil, political, economic, social and cultural rights on an equal basis with non-disabled people. To be able to exercise these rights, disabled people shall be provided

E. Földesi (✉) · E. Fördős-Hódy
Budapest Association of Persons with Physical Disability, Universal Design Information and Research Center, Hegedüs Gyula u. 43, Budapest 1136, Hungary
e-mail: erzsebet.foldesi@etikk.hu

E. Fördős-Hódy
e-mail: etikk@etikk.hu

B. Müller and G. Meyer (eds.), *Towards User-Centric Transport in Europe*, Lecture Notes in Mobility, https://doi.org/10.1007/978-3-319-99756-8_5

with physical and information and communication environment responding to their needs.

The group of disabled people is very diverse and represents a wide range of needs. Persons with disabilities include those who have long-term physical, mental, intellectual or sensory impairments which in interaction with various barriers may hinder their full and effective participation in society on an equal basis with others (United Nations 2006a).

According to the CRPD disability is an evolving concept. The CRPD tells that disability is not caused by the hearing, visual, physical, intellectual, mental impairment of the person, but it is the result of the interaction between a person with his/her impairments and the attitudinal and environmental barriers in the society. This interaction may hinder full participation of persons with disabilities in the society on an equal basis with others (United Nations 2006b).

State Parties to the CRPD (e.g. EU and its member states) are obliged to take appropriate measures to ensure that people with disabilities can access, on an equal basis with others, the physical environment, transportation, information and communications, including technologies and systems, and to other facilities and services open or provided to the public, both in urban and in rural areas. This means for transportation, measures shall be taken to ensure that transport infrastructure can be reached and vehicles, information and services are available and usable for disabled people on equal basis with others. These measures shall include the identification and elimination of obstacles and barriers to accessibility. Research should aim to ensure participation and identify barriers that are developed by society and lead to disability and exclusion.

The European Union's Directive on public procurement as of 2014, provides that technical specifications of call for tenders made by the contracting authorities shall take into account accessibility and universal design criteria. Member States of the European Union had the obligation to transpose the directive into their national law by April, 2016. The directive provides that for all procurements subject to the directive the technical specifications shall consider accessibility criteria for persons with disabilities or design for all users European Commission (2014).

2 Mainstreaming Disability in Transport Research

There is a lack of accessible transport vehicles and services which prevents people with disabilities from actively and fully participating in the society, thus depriving them of the freedom of movement. Inaccessible transportation makes it impossible or extremely difficult for persons with disabilities to take part in mainstream education, to find work or to visit healthcare services, etc.

World Health Organization (2011) addresses the barriers in public transportation for persons with disabilities and says that there are initiatives worldwide to improve accessibility to public transportation infrastructure and services. Inaccessible timetable information, lack of ramps for vehicles, large gaps between platforms and vehicles, inaccessible stations and stops are the most typical transportation according to the report. The report highlights that without accessibility throughout the travel chain e.g. inaccessible routes create longer travel times for disabled people than for other passengers.

The European Commission Guidance Note on Disability and Development states that 'removing obstacles for participating in social and economic life strengthens people with disabilities and enhances poverty reduction of the whole community' (European Commission 2004).

CRPD emphasizes the importance of mainstreaming disability issues as an integral part of relevant strategies of sustainable development (United Nations 2006c). There is a strong need for a systematic mechanism to be put in place to mainstream accessibility in all research in the transport area. Effective mainstreaming of accessibility in all research is essential. Mainstreaming is the only guarantee that the deliverables of the research project do not result in new barriers for people with disabilities in enjoying the benefits of the innovation and development in the transport area.

2.1 Why Is Involving Disability Aspects in Mainstream Transport Research so Important?

After the ratification of the CRPD by the EU and its member states, accessibility and specifically accessibility to transportation is very often in the political agenda. The CRPD brought a paradigm shift from a medical to a social approach. This new approach shall be applied also to transport research. To put this approach into practice it means that transport research shall focus on removing barriers and prevention of new ones instead of focusing on impairments and redress them. The real implication of social model of disability is that, if society by building barriers disables, then the response should be to dismantle barriers and include disabled people into the society by creating a transport environment that includes everybody regardless of disability.

People with disabilities do not require special transportation since it does segregate them from the non-disabled passengers. With accessible transportation they are able to use the transport services on equal basis with others. It is more sustainable for the whole society than to maintain expensive special services designed only for disabled people.

And finally, though accessibility to transport vehicles and infrastructure as well as access to travel information is indispensable for people with disabilities, there are elderly passengers, people with baby prams, small children or travelers with heavy luggage who are also benefitting from accessible transportation. Accessibility benefits the whole population, as everyone will get a higher comfort in accessible transport service. Step free entrances to the transport infrastructure, wide doors, low floor buses, elevators, non-slippery surfaces, short and easy to understand information on pictograms, information both in audible and visible form, etc. are elements of accessibility criteria in transport. Low floor vehicles provide quicker boarding for non-disabled passengers as well due to lack of steps. Use of elevators is more convenient than escalators or stairs for passengers with heavy luggage. In a noisy airport visible information can be utilized when information from loudspeakers are not hearable. On non slippery surfaces the possible accidents can be avoided also by non-disabled passengers.

2.2 How to Mainstream Disability in Transport Research?

In regards to transport, mainstreaming disability in the research and development area will ensure that implications for people with disabilities are considered as follows:

a. Research aimed at improving transport opportunities for the society at large shall take into account the needs of people with disabilities as well. This research is aimed to improve quality of transport for each member of the society, including people with disabilities. With taking into account their needs in the research, disabled people are put in the position to be able to exercise the same rights to mobility as any other member of the society. It is to be ensured that disabled people should benefit equally from the research, thus preventing that inequality is perpetuated or increased.
b. In conducting research and development on transport, at all key stages of research it should be avoided that the future product, environment or service as the result of the research ignores the needs of people with disabilities thus creating discrimination in their rights to mobility.
c. It also needs to be investigated how the transport related research could support increasing inclusion of people with disabilities.

2.3 Universal Design

Inclusion of accessibility following the "Universal Design" approach in transport research projects for transport infrastructure, vehicles and services, as well as alternative means of communication such as easy-to-read, Braille, speech-to-text support and sign language are essential elements to enhance rights of people with disabilities to mobility. Application of universal design in transport related research and transport services will be discussed in another article of this publication.

2.4 Mainstreaming with the Involvement of Persons with Disabilities

In order to conduct a research which is useful for all, it is vital to include the user-perspective of persons with disabilities from the beginning onward in all research projects affecting their life. The contribution of disabled peoples' organizations can significantly enhance the quality and provide added value to research.

This obligation comes also from the undertaking of States Parties to the CRPD. CRPD provides that states shall closely consult and actively involve people with disabilities, incl. children with disabilities through their representative organisations into the development and implementation of legislation on issues that affect their lives (United Nations 2006d).

Since findings of transport research could serve as a basis for future transport legislation and policies thus affecting the life of disabled people as well, disabled people shall be part of the research teams in the field of transport.

In 2008–2009, the European Research Agendas for Disability Equality (EuRADE) project's (http://www.eurade.eu/) aim was to review priorities defined at the policy level or by organisations of disabled people (DPOs) on disability equality. A broad survey was developed to identify priorities that DPOs saw as necessary. DPOs identified research on the accessibility to disabled people of existing and future transport systems among the priorities.

The EuRADE project sought also to increase and enhance the full participation of disabled people's organisations as equal and active partners in future research initiatives.

In the EuRADE survey, National and European organizations of disabled people were also asked to rate the importance for DPOs of collaborating with academic researchers. DPOs thought that there was positive potential in academic research collaboration. However, DPOs had negative comments as well, saying that research agendas are lacking responsiveness to the real needs of disabled people, but more often reflecting the priorities of academic researchers and the bodies that fund them.

The comments also reflected that disabled people wish to be engaged as equal partners (rather than the 'subjects' of research) and ongoing collaboration with disabled people's organisations can help universities to be more responsive (EuRADE 2008).

People with disabilities understand their needs the best. They are aware of innovative ways to solve everyday situations with their limited capacity. Their practical knowledge can be utilized by researcher often possessing either none or only theoretical knowledge about the skill and needs of people with disabilities. DPOs are aware of the social and human rights model of disability, therefore, their involvement in the research could increase the commitment of the academic researchers to a new approach of research grounded in human rights with regard to persons with disabilities. With their involvement, people with disabilities could enhance creation of a new research culture and contribute disability inclusive research findings and results.

2.5 Twin Track Approach

There is a need for research on the benefits of accessibility for all passengers, though it is challenging to quantify. How many more passengers could a railway company have, if they made their trains and stations accessible and what would their economic benefit be. Clearly shown economic benefit of accessibility for the long run could help in changing the mindset of investors and would result in requirements for accessibility for all in their contracts.

Besides mainstreaming disability in transport research for the society at large, research aimed specifically at the needs of persons with disabilities is also essential.

Use of mobility scooters for disabled people is more and more often prohibited on public transport. More research is needed to look for safe solutions to take the mobility scooters on buses and trains without restricting the rights of persons with disabilities to personal mobility.

Another area of possible research is how to design more accessible transport vehicles, especially airplanes. Often the focus is on saving space and there are not sufficient technological solutions, e.g. to allow all wheelchairs to fit through the doors of aircrafts. More intensive research is needed on how to enable passengers to stay seated in their own wheelchair during the flight instead of transferring into a seat.

Conflict between safety requirements versus the right to transport shall be also addressed. In air travel, though an EU regulation on the rights of disabled persons and persons with reduced mobility when travelling by air prohibits operators from refusing reservation or boarding to persons because of their disability, denied boarding on the ground of disability can be an exemption if airline staff considers it "unsafe" to take them on the flight or if the size of the aircraft or its doors make their embarkation or carriage physically impossible Regulation (EC) No 1107/2006. Research is needed on how safety can be guaranteed while safeguarding the rights of all passengers (European Commission 2006).

The list is exemplary and does not cover all areas where disability specific transport projects should be launched.

A twin-track approach shall be used to increase the usability of transport vehicles: mainstreaming disability aspects following the universal design concept in mainstream transport research and conducting special disability-related transport research.

2.6 Good Practice—Involving Organization of Persons with Disabilities

Involvement of persons with disabilities both in the process of mainstreaming disability in transport development as well as in finding specific solutions for disabled people's needs is very important. In Hungary an organization of disabled persons with the initial funding of the Ministry of Human Resources made an initiative which can be considered a good practice.

The Universal Design Information and Research Center (ETIKK) was established in 2013 by the Budapest Association of Persons with Physical Disability (MBE). ETIKK promotes universal design, its human rights aspects and economic and social advantages. Universal design is an efficient tool for social inclusion Földesi (2017). With the application of this method, mainstream products, built environment, transport and other services, applications will be accessible for all regardless of age, ability or disability, gender, stature, etc. reducing the need for special solutions.

Persons with different disabilities work together with rehabilitation engineers in the Center.

In the field of transport, ETIKK works closely with the Budapest Transport Centre (BTC) that is committed to apply universal design in its future developments. In collaboration with the BTC the users' group of ETIKK tested urban buses in Budapest with different wheelchairs and mobility scooters and made suggestions for internal adaptation of the buses to get more space e.g. for entering the buses or parking in the designated place. Position of newly installed ticketing machine onboard was also discussed and tested with and by the test group of disabled. These tests gave good feedback for BTC on the current accessibility status of its bus fleet and recommendations for aspects for future procurements.

Safety is of priority for every passenger regardless their ability. Therefore safety belts for mechanical, electrical wheelchairs and mobility scooters were installed onboard at their designated places. The aim was to develop a proper seatbelt design and use on trams ensuring barrier free access to wheelchair user passengers. As a result safety belts were manufactured and installed within easy reach and accessible position thus enabling the widest possible scope of disabled passengers to use the belts.

ETIKK took also part in testing the first electronic access gate to be used in Budapest metro stations. Width of the gates for wheelchair users, baby prams, as well as contrast colors, tactility and easy reach of the ticket machine were tested by passengers with mobility and visual disabilities. As a result: the electronic access gates can be universally usable.

ETIKK has been also consulted on accessibility criteria for several other new transport infrastructures or renovation plans of infrastructures or vehicles.

Recently ETIKK made a contract with BTC for guideline for accessible transport infrastructure and vehicles following the universal design method. The aim of the guidelines is to establish uniform requirements for surface reconstruction, architectural or internal design and development of public transport vehicles for the investments of the transport company in the future.

In 2015 the Budapest Association of Persons with Physical Disability became the consortium member of the Mobility4EU project. The project (2016–2018) aims to deliver a vision for the European Transport system covering all transport modes in 2030. The task of ETIKK within the project is to mainstream the aspects of disabled people's with the principles of universal design. Car sharing and driverless cars as potential elements of future transport can be of great benefits for disabled people in case their needs are taken into account. ETIKK held workshops with disabled people on the driverless cars. The accessibility requirements of the workshop participants for driverless cars were forwarded to the participants of the 'User-centered aspects of automotive transport' session of the 21st International Forum on Advanced Microsystems for Automotive Applications (AMAA 2017). Beyond the necessity of involving disabled from the onset of the development, the attention of developers to untapped potential users/customers was also called for.

The examples above show that there are already initiatives to mainstream disabled users' aspects by involving persons with disabilities themselves in new transport related investments or reconstruction. Transport vehicles where disability was mainstreamed are accessible for disabled passengers. However, even if vehicles or infrastructure are designed for everybody, disabled people might still need special solutions to use e.g. safety belts, ticketing machine. It needs to be also emphasized that with a universally designed transport vehicle or infrastructure the need for special solutions is decreased. Therefore twin track approach leads to the usability of transport for all.

3 Conclusions

Transport shall provide equal access to education, work, social interactions, shopping, healthcare, etc. for all, including persons with disability. This can be reached only if we use twin track approach to consider the needs of disabled people.

1. ***Mainstreaming disability following the universal design*** should be part of all transport researches in all phases of the work thus resulting in vehicles, infrastructures, applications, information systems that are not limited for usage by 'only' non-disabled passengers. Without mainstreaming disabled people's aspects in transport related research, their mobility needs have to be fulfilled by special solutions. If e.g. public transport vehicles are not accessible, special transportation means shall be developed for disabled passengers which are not only segregating but also more expensive due to the fact that special transport solutions will serve much less passengers since disabled people form minority group of the population. With mainstreaming disability in transport research creating barriers for disabled people can be avoided.
2. In cases where equal solutions can not be reached through mainstreaming, ***disability-specific research/development*** is the only effective way to meet disabled people's needs. Examples for this have been given in relation to air transport and public transport vehicles.

Since findings and deliverables of transport research could serve basis for future policies, legally binding legislation, as well as future development or manufacturing activities, it is essential that transport research shall be user-led, including people with disabilities as users of transport services on an equal basis with others.

There is a lack of broad institutional support mainstreaming disability in research in general. Mainstreaming disability in transport research shall be supported by binding legislation with strict provisions stipulating that transport research shall be performed with human rights approach brought by the UN Convention on the rights of persons with disabilities. Selection and award criteria for transport research project funded by the Member States of the European Union or by the European Union itself should also include references to accessibility and universal design to avoid creation of new barriers for persons with disabilities or to dismantle the existing ones. Active involvement of people with disabilities in the transport research is for the benefit of the whole society, since result would be resulted in solutions that will be usable by everyone. Participation of people with disabilities and their representative organisations in transport research as equal partners can bring a new research culture and can guarantee that the deliverables and result of the research will be inclusive to each member of the society, including persons with disabilities as well.

Acknowledgements The authors wish to thank the consortium of the Mobility4EU project for the intense and fruitful collaborations and the European Commission for the funding of the Mobility4EU project in the framework of the Horizon2020 program (EC Contract No. 690732).

References

EuRADE (2008) New priorities for disability research in Europe, Report of the European Disability Forum Consultation Survey 'European Research Agendas for Disability Equality' prepared by Priestly M, Waddington L for the European Disability Forum, December, 2008

European Commission (2004) EC guidance note on disability and development for the European Union delegations and services, July, 2004

European Commission (2006) Regulation (EC) No 1107/2006 of the European Parliament and of the Council of 5 July 2006 concerning the rights of disabled persons and persons with reduced mobility when travelling by air, Article 4

European Commission (2014) Directive 2014/24/EU of the European Parliament and the Council as of 26 February 2014 on public procurement and repealing Directive 2004/18/EC, Article 42

Földesi E (2017) Working with the method of universal design—the effective tool of social inclusion. https://medium.com/@mobility4eu/working-with-the-method-of-universal-design-9ee546b83da8. Jan 2017

United Nations Convention on the Rights of Persons with disabilities, Article 1, (2006a)

United Nations Convention on the Rights of Persons with disabilities, Article 1, (2006b), Preamble (e)

United Nations Convention on the Rights of Persons with disabilities, Article 1, (2006c), Preamble (g)

United Nations Convention on the Rights of Persons with disabilities, Article 1, (2006d), Article 4, p 3

World Health Organization and World Bank (2011) World Report on Disability, pp 178–181

Universal Design as a Way of Thinking About Mobility

Jørgen Aarhaug

Abstract The concept of universal design in reference to a strategy to counter social exclusion was first coined by the architect Ronald Mace. He defined Universal design (UD) as "the design of products and environments to be usable by all people, to the greatest extent possible, without the need for adaptation or specialized design". This paper will look into the use of UD as a policy objective for transport policy, using Norwegian experience as an example. UD was adopted as one of the four major policy objectives in Norwegian transport policy in 2009. However, from 2018 onwards UD is no longer a main policy objective. This experience with UD as a policy objective is used as an empirical backdrop for a more principal discussion on the usefulness of UD in transport and mobility. I conclude by pointing at UD as a useful vision, but difficult policy objective.

Keywords Universal design · Disability · Social inclusion · Policy objectives Accessibility

1 Introduction

With changing structures of work and family life, mobility has become an increasingly important precondition of the fully functioning citizen. The last decades have seen an interest in how mobility restrictions can be a cause of social exclusion (Cass et al. 2005; Preston and Rajé 2007; Preston 2009; Priya and Uteng 2009) in which individuals cannot fully participate in the normal activities of society even though they would like to (Burchardt et al. 1999). The concept of universal design (UD), when used in the context of transport, is a way of thinking about these issues mostly as an alternative and complement to 'accessibility'. Here, the difference can be interpreted as accessibility with a focus on solutions created for individuals with impairments, while universal design is a focus on providing a solution by which impairments

J. Aarhaug (✉)
Institute of Transport Economics, Gaustadalleen 21, 0349 Oslo, Norway
e-mail: jaa@toi.no

B. Müller and G. Meyer (eds.), *Towards User-Centric Transport in Europe*,
Lecture Notes in Mobility, https://doi.org/10.1007/978-3-319-99756-8_6

become irrelevant; in other words, that the main solution is usable by as many people as possible. An example; tactile tiles and braille writing, are not considered universal design. As these elements are useful in the case of people with reduced vision but not for many others. A walkway with clear natural guidelines, where orientation is easy and where signs can be readily understood by the visually impaired and by the rest of the population, is universal design.

I start by looking briefly at the concept of universal design. I examine the term disability—as this can be seen as both a medical issue and a societal issue—and provide a link between UD and public transport. Further I discuss universal design as a concept and policy objective and discuss this drawing on Norwegian experience. I then point to upcoming issues in mobility and how these relate to universal design at a general level. My main findings are summed up in a final section.

2 The Concept of Universal Design

The concept of universal design (UD) in reference to a strategy towards promoting social inclusion was first coined by the architect Ronald Mace, who defined it as "the design of products and environments to be usable by all people, to the greatest extent possible, without the need for adaptation or specialized design" (1997). The term is used primarily in the United States, Scandinavia and Japan, whereas the expression 'design for all' is used with a similar meaning elsewhere (Audirac 2008). The term is used in the UN convention on the rights of persons with disabilities, with the definition: "the design of products, environments, *programs and services* to be usable by all people, to the greatest extent possible, without the need for adaptation or specialized design. "Universal design" shall not exclude assistive devices for particular groups of persons with disabilities where this is needed" (United Nations 2006).[1] The concept has been superimposed on transport from the built environment, where concepts similar to universal design can draw on history back to the 1970s. According to Story et al. (1998), early efforts to render environments accessible were frequently dependent on segregated measures that were "more expensive and usually ugly" compared to universal design, which includes accessibility for all in early design phases.

The objective of universal design is an environment where people with disabilities can function as natural members of society, and a guiding notion is that accessibility solutions benefit everyone, not just people with disabilities. Rebstock (2017) states that an accessible environment is essential for 10% of the population, necessary for between 20 and 40%, and comfortable for 100% of the population. It is therefore often included in the broader social inclusion paradigm or, as Audirac (2008:4) states: "UD is a philosophy of design that not only subscribes to the ideals of accessible and barrier-free design and assistive technology, it also professes to be a broader paradigm of design that celebrates diversity and is inclusive of all users regardless

[1] Authors emphasize.

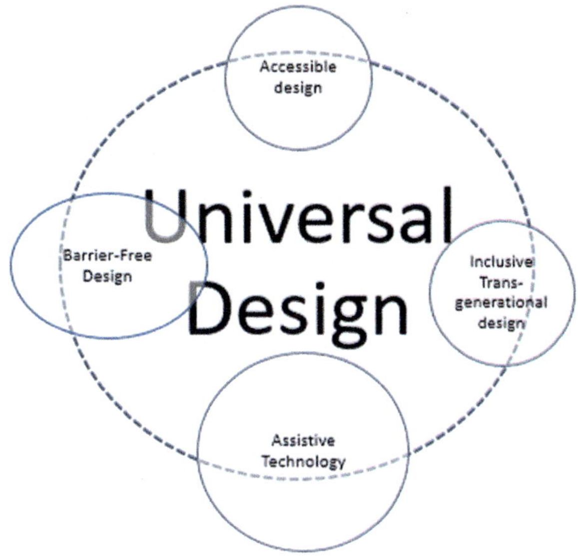

Fig. 1 Universal design in relation to other design philosophies (Audirac 2008)

of age or ability". The relationship between UD and other design philosophies is illustrated in Fig. 1.

Accessible design promotes equal opportunity of access to mobility and services for people with disabilities. Inclusive Transgenerational design is products and services designed for the widest possible audience, irrespective of age and ability. This includes improving the quality of life of people of all ages and abilities. Assistive technology is rehabilitative engineering that enables people with disabilities to carry out more tasks with their physical, sensory and/or cognitive abilities enhanced. Barrier-free design is retrofitting buildings, facilities or services to accommodate physically impaired people (Audirac 2008). All of these concepts are, at least in part, included in the larger universal design or "design for all" concept.

2.1 The Concept of Disability

Although universal design is targeted at 'everyone', it is widely promoted and used by organizations fighting for the interests of people with mobility impairments or disabilities. This links universal design with another challenging concept, namely disability. The classic conception of disability is often referred to as the "medical model" (Shakespeare 2006), according to which disability is caused by impairment and is a characteristic of the individual. It has to be cured or ameliorated (Hanson 2004). By contrast, the social model of disability suggests disability is a social construction produced in the interplay between the individual and society (Shakespeare 2006).

If one chooses to use the term "disability" within the social model, it may not necessarily be a permanent feature of the individual; passengers travelling with a pram or heavy baggage might be seen as transitorily disabled. A society where attitudes, standards and technologies are adapted only to the needs of the young and healthy thus produces a large number of disabled people, whereas one where solutions are adapted to the abilities and requirements of a larger group will produce fewer. The proponents of this model concede that there is a medical reality underlying disability, but emphasize that society contributes to marginalizing the disabled through its implicit endorsement of a certain norm. This approach draws attention to how physical design may create barriers to participation. From this perspective, poorly designed public transport may produce disability through preventing certain groups from using the public transport system, and thus from full participation in society.

In a quantitative study, Aarhaug and Gregersen (2016), found that people with disabilities not only travel less, but are also more affected by adverse weather conditions when travelling than the rest of the population. This illustrates that the disabled are more affected by hostile environments than the rest of the population. In extension, it points towards universal design as being important in facilitating their inclusion in society.

2.2 *Applying Universal Design to Transport*

In applying the general concepts of universal design to the mobility and transport sector, public transport becomes a key issue, particularly access to it. Private-car-based transport is not universal and therefore is problematic from a universal design point of view. The use of a private car is exclusive. It is not shared. Large sections of the population in a car-based society are excluded from mobility as they cannot access private car transport. From a user perspective, the private car may be overly expensive, requiring a license or skills that are not suitable for the potential user's mobility needs.

Within public transport, UD can be many different things. Some examples include having real time information on an easy to read format, as opposed to having only printed schedules (Fig. 2), having a snow and ice removed from bus-stops frequently as opposed to a low level of winter maintenance (Fig. 3) and having level access to vehicles as opposed to having stairs (Fig. 4).

Figures 2, 3 and 4 illustrate some of the public transport infrastructure measures included under the universal design umbrella. Other examples and illustrations are available in NPRA (2014).[2]

---

[2]https://www.vegvesen.no/_attachment/118984/binary/963983.

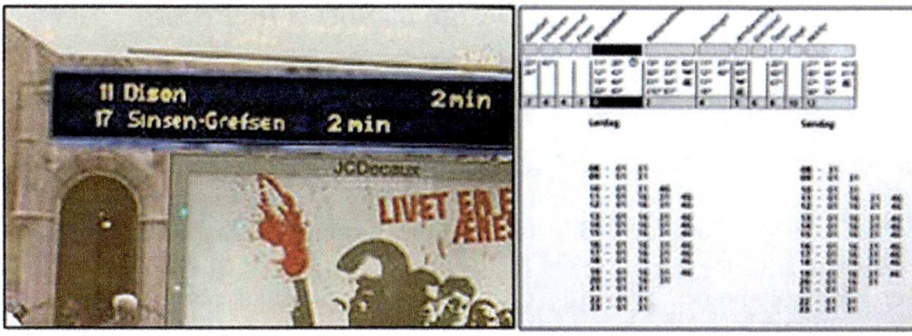

Fig. 2 Real-time information display and printed public transport schedule. Illustration Thomas Tveter

Fig. 3 Snow removal on a bus stop. Illustration Thomas Tveter

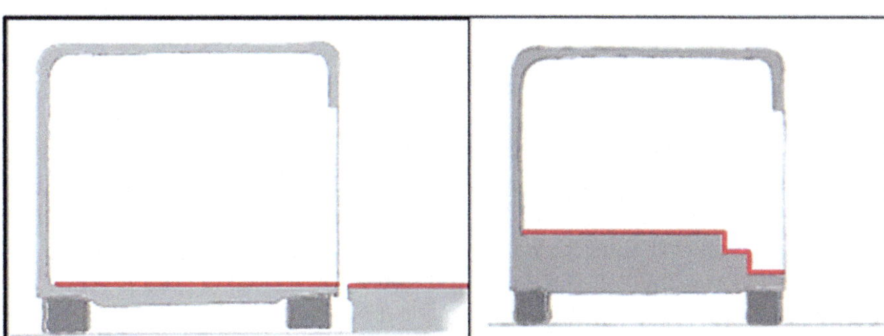

Fig. 4 Level access to public transport vehicles. Illustration Thomas Tveter

3 Transport and Universal Design as Social Inclusion

According to Nordbakke and Schwanen (2015) access to transport services is a pre-condition to being a functioning member of society. This forms part of the conceptual link between universal design and transport. Providing mobility for disabled persons is not a marginal problem. Also, it is a challenge expected to grow as time goes by. Although there are many difficulties in measuring the prevalence of disability in society, estimates tend to be in the region of one in five adults having a form of disability or another. Thompson (2017) summarizes and discusses different estimates of disability prevalence finding global estimates ranging between 15 and 20%. Higher among older people. Nordbakke and Skollerud (2016) uses survey data that identify 19% of the Norwegian population above 18 years of age as having a form of disability. Disability is correlated with age, and as the population in Europe and the Western world in general is ageing the challenges will increase. This is the case even if the older population is healthier than ever, as ageing often implies reduced functional capabilities. According to Eurostat (2013), those aged 65+ will account for 29.5% of the EU's population by 2060 compared to 17.5% in 2011, whilst the share of those aged 80+ will almost triple between 2011 and 2060.

Disability is not a static situation. Increasing numbers of people with disabilities can be linked to the mobility options that are available, not only their health status. Disability is not the only issue, the geographic and economic limitations of public transport in low density areas are expensive and so 'forced car ownership' becomes an issue, too. Delbusc and Currie (2011) point to the high levels of car reliance in city suburbs as resulting in social exclusion. Even though their empirical study was conducted in suburban Melbourne in Australia, which is extremely car-reliant, their findings are transferable to many European settings. Car-reliance has a bearing on both work and leisure activities, particularly the latter. People who, for whatever reason, do not have full access to a car and live in car-reliant areas do not participate in as many activities as they would like to. This highlights one of the issues of a private-car-based transport system. What happens with people who for one reason or another—means, age or disability—do not have their own car? Delbosc and Currie's (2011) answer is that they are less able to participate meaningfully in society.

From a universal design perspective, the car as a dominant mode of transport cannot be universal design as it is not available to everyone. Cars are for the most part privately owned and operated and as a consequence public transport is often an object of study when linking the idea of universal design with transport. Provision of public transport is motivated by a set of different criteria. In the US, in particular, it is seen as a way of providing mobility for those who cannot afford to pay for these services themselves. In Europe, public transport is often more focused on providing transport volumes, or as an environmentally friendly and area-efficient form of transport, i.e. public transport in Europe often has a wider user group than is the case in the US. In all settings, public transport will provide better accessibility for the population if the majority find it attractive to use. The concept of universal design in transport is therefore linked to access to public transport.

Universal design as a concept is linked to rendering impairments irrelevant. Not all disabilities are relevant when it comes to public transport. Still, an increasing number of older people indicate that accessibility to public transport is an important issue. With populations ageing, there is reason to believe that an increasing number of people will have difficulty using public transport in the future. In many Western countries, the share of the population aged above 65 is now approaching 15% (Crews and Zavotka 2006) and in Norway this has already been passed, with the figure expected to rise substantially (Folkehelseinstitutttet 2013).

Universal design or accessibility for all is a feature of public transport legislation in many countries, with government and transport providers often obliged to make public transport available to most groups of passengers. According to Øksenholt and Aarhaug (2018) the non-discrimination aspects of universal design is mostly addressed within EU legislation. This does not in itself guarantee accessibility, however, since implementation is often patchy (Arsenjeva 2017).

4 Experience with Universal Design as a Policy Objective

Research on the concepts of universal design and its implications for mobility and transport is limited. Universal design can be seen as being different things from different points of view. To look closer at the merits of universal design it can be useful to divide the analyses in three levels; strategic, what do we want to achieve?; tactical, what product can help us achieve the aims?; and operational, how do we produce that product?, following Anthony (1988) adopted for the transport sector by Van de Velde (2004).

At a strategic level, the vision of a universally designed society is a target to work towards, and creating a society that is designed according to universal design principles is an ambitious objective.

On a tactical level, universal design can be used in setting priorities between different groups. In terms of public transport this would be to give priority to making more parts of the city accessible, and in planning new developments in a way that makes them accessible to all.

On an operational level, universal design helps form guidelines for design of features in the transport network; that can be low-floor vehicles, high kerbs and a sufficient number of seats and resting places. Universal design of these different levels with different objectives is difficult to evaluate.

In the Norwegian case, universal design was one of the four major objectives of the national transport plan for the period 2010–2019. This was set as a strategic objective. However, in the current national transport plan (2014–2023), experience has shown that the results have fallen short of the objectives. Out of a planned improvement of 100 transport nodes only about 25 were improved and out of 6500 bus stops along national roads, only 480 were improved. The reason given is that universal design is a relatively new field, that the ambitions were too optimistic, and that UD-standard has proved to be more difficult to implement than expected.

Also, measures to implement universal design have turned out to be more expensive than expected (Meld.st.26 2013:32–33). Not stated in the document is the critical fact that although universal design is a national ambition, most of the features that need to be improved, such as bus-stops, are within the regional government's area of responsibility. Consequently, even if the national government had managed to operationalize universal design within all its areas of responsibility, it would still fall short of the policy objectives. This illustrates one of the challenges with universal design as a strategic policy objective. It is ambitious; it includes many different areas of responsibility and it is difficult to reach within a set timeframe.

At the tactical level, Herriot (2011) observed that implementation of universal design is sketchy, while pointing out that one reason could be user consultation frequently taking place at too late a stage in the design process. Even if efforts are made to make systems accessible, this does not in itself guarantee that the measures work as intended. Tennøy et al. (2015) finds problems related to; mandatory consultations with disabled, where the disabled are not, qualified to give advice on mobility impairments in general, only impairments due to their own disability. A wheel chair user is not automatically qualified to give advice related to the issues faced by sight impaired and vice versa. Handbooks and manuals that are not practical enough and real-life considerations make implementation of universal design difficult. Skartland and Skollerud (2017) using a case study looking at user involvement related to implementation of UD in transport infrastructure investments find that although the mandatory user involvement was perceived as a positive factor by both the user groups, project owners and entrepreneurs, there are still issues related to lack of competence among user representatives. In particular, the issues related to user representatives not understanding fully the implications of different impairments. In other words, having a disability does not make you an expert of all types of disabilities. Mandatory user involvement helps in making terminals and other infrastructure more universally designed, but it does not solve all issues.

Still, universal design has been more successful at a tactical level, compared with at strategic level. Studies by Fearnley et al. (2010) and by Odeck et al. (2010) indicate that accessibility measures used within the universal design framework have positive side effects. They facilitate travelling in the case of passengers with prams or heavy baggage, thus contribute significantly to a positive valuation of universal design elements. In fact, Fearnley et al. (2011) and Odeck et al. (2010) found that the measures studied for creating universal design had a higher benefit-cost ratio compared to most other public transport investment opportunities. That said, although the valuation of these measures is high, it has not translated into a significant demand effect (Fearnley et al. 2015). In extension, universal design measures have proved to be cost-efficient investments in public transport. They increase welfare, but they do not result in more passengers—at least not by themselves.

At the operational level, universal design is useful in that the concept provides a framework from which guidelines can be formulated and, by extension, a design of the physical infrastructure. It is also useful in that it is user-centric, focusing on the travel chain, door-to-door, rather than on the responsibilities of each party.

On an operational level, despite the fact that large sums of money are spent on making transport systems more accessible and well-designed, very little research has been done on how universal design and accessibility measures work for people with disabilities, or indeed for passengers in general (Øksenholt and Aarhaug 2018). Of several studies on this subject, one by Øksenholt and Aarhaug (2018) points to challenging issues with the operationalization of universal design. They look at how people with different disabilities respond to the challenges faced by public transport users, and conclude that public transport, even though it has reached a high level of universal design in the case area (Oslo), is not practical for many of their respondents. Respondents who are not able to use public transport have impairments that either make things difficult or are difficult for the untrained observer to observe, and therefore do not provide the level of service intended. Also, several of the respondents had "better options", e.g. a personalized private car or supported access to taxis over the discomforts of public transport.

In summary, the Norwegian experience points towards mixed results in using universal design as a policy objective. It has advantages on all levels, but also challenges. Universal design is very useful in creating a direction at strategic level, but this direction has proven difficult to set as a reachable quantifiable policy objective. At a tactical level, the benefits are well documented, however implementation is sometimes not as good as it could have been. On an operational level, questions can be raised to what degrees the ten percent of the population, for whom these measures are essential, are receiving the benefiting from universal design. The studies rather suggest that implementation in public transport in Norway, so far fall short here. Looking at the Norwegian experience universal design is successful in making transport better and more accessible, but not as successful as it was hoped to be.

5 Universal Design in a Changing Mobility Setting

At present, and at global level, there are several societal trends influencing the transport sector. In particular, digitalization and digitally facilitated mobility are influencing universal design—directly, in that the use of smart phone technology is changing the user interface between the transport provider and the user, and indirectly with digitalization changing the availabilities of transport options.

In the development between 2012 and 2017, the major change has come in the form of new ride-sourcing services. These have changed the modal split in many major cities by generating new trips and taking market shares from traditional taxis and public transport (Rayle et al. 2016; Schaller 2017). From a universal design perspective, this has been positive in many ways in that people get access to door-to-door transport at a lower cost than previously. It also means that people who were unable to use either public transport or their own vehicle now have access to transport services in a more inclusive society. However, there are also downsides. Some people are not able to use these services and are left out, particularly if they have an impairment that makes use of a private car, even as a passenger, problematic (such as being wheelchair bound). A question remaining with crowd-based services is who it is setting the standard. The provision of the service is positive, but in a universal design perspective a precondition is that someone sets the standards and requirements to optimize use by as wide a range of individuals as possible.

A second development is Mobility-as-a-Service, or MaaS for short, which is a system assembling the products of different transport providers within one common package. The relationship between the user of transport and that person's transport provider is simplified, as everything goes through an intermediary, namely the MaaS provider. In principle, this should allow more accessibility and more universal design and it should be easier to tailor and be included in a universal design perspective, as the service providers are companies rather than private individuals.

A third development is self-driving vehicles, and although these are not yet fully in operation, potentially they will make the public transport system much more universally designed. They will offer a hybrid between a door-to-door taxi service and a scheduled public transport system at a lower cost than today's demand responsive services. However, the question remains: Who will set the standards and requirements, for the vehicles and operators? And who will design the way the system will be implemented?

6 Conclusions

The origins of universal design as a concept can be traced to the built environment and architecture. Can UD be applied to the transport sector? Universal design adds perspective, at least if we consider access to transportation a fundamental building block in today's society. Universal design can be seen in the context of enabling people with a disability to function in society. As a concept, it fits more with a disability model, where disability is a construction of society rather than a characteristic of the individual. Universal design is a holistic approach to problems associated with people whose abilities to function in society are different.

Looking at the empirical experiences from Norway, the results are mixed. The vision is clear, but difficult to reach. Many of the measures necessary in making a society universally designed are themselves good investments from a cost benefit point of view. Still, the experience so far is that efforts made to promote universal design have not radically changed the day-to-day lives of people with disabilities, even if their disabilities are minor or merely temporal. There is still a long way to go before the Norwegian society is up to universal design standard.

At a global level, present trends in new modes of transport can at least be part of the solution, but even here there are problems. To fully utilize the possibilities that these services create, new technology often requires a certain technical knowhow. In other words, we cannot expect new technology to render society universally accessible without regulation and dedicated policy.

Acknowledgements Sections of this paper draws on the work of Aarhaug and Elvebakk (2015).

References

Aarhaug J, Elvebakk B (2015) The impact of Universally accessible public transport—a before and after study. Transp Policy 44(November):143–150. https://doi.org/10.1016/j.tranpol.2015.08.003

Aarhaug J, Gregersen FA (2016) Vinter, vær og funksjonsnedsettelser—en dybdeanalyse av RVU. TOI report 1543/2016

Anthony RN (1988) The management control function. Harvard Business School Press, Boston

Arsenjeva J (2017) Annotated review of European Union law and policy with reference todisability. Academic network of European Disability experts (ANED)

Audirac I (2008) Accessing transit as universal design. J Plann Lit 23(1):4–16

Burchardt T, Le Grand J, Piachaud D (1999) Social exclusion in Britain 1991–1995. Soc Policy Adm 33(3):227–244

Cass N, Shove E, Urry J (2005) Social exclusion, mobility and access. Sociol Rev 539–555

Crews DE, Zavotka S (2006) Aging, disability and frailty: implications for universal design. J Phys Anthropol 25:113–118

Delbusc A, Currie G (2011) The spatial context of transport disadvantage, social exclusion and well-being. J Transp Geogr 19(2011):1130–1137

Eurostat (2013) Population structure and ageing, available at http://epp.eurostat.ec.europa.eu/statistics_explained/index.php/Population_structure_and_ageing. Last Accessed on Nov 2013

Fearnley N, Leiren MD, Skollerud KH, Aarhaug J (2010) Nytte av tiltak for universell utforming i kollektivtransporten. In: Selected Proceedings from the annual transport conference at Aalborg University. ISSN 1903-1092

Fearnley N, Flügel S, Ramjerdi F (2011) Passengers' valuations of universal design measures in public transport. In: Research in transportation business & management, vol 2. Elsevier, pp 83–91. http://dx.doi.org/10.1016/j.rtbm.2011.07.004

Fearnley N, Aarhaug J, Flügel S, Eliasson J, Madslien A (2015) Measuring the patronage impact of soft quality factors in urban public transport. Paper presented to ITEA Annual conference and summer school (Kuhmo Nectar), Oslo, Norway, June 2015

Folkehelseinstitutttet (2013) Eldre—andelen over 65 år i befolkningen. http://www.fhi.no/eway/default.aspx?pid=233&trg=MainLeft_6039&MainArea_5661=6039:0:15,4576:1:0:0:::0:0&MainLeft_6039=6041:70828::1:6043:8:::0:0

Hanson J (2004) The Inclusive City: delivering a more accessible human environment through inclusive design. Institute of Transport Studies, Monash University, Social Research in Transport (SORT) Clearinghouse, Jan 2004

Herriot R (2011) Complexity and consultation—inclusive design in public transport projects. Include2011, London

Mace R (1997) What is universal design. The center for universal design at North Carolina State University. Retrieved 19 Nov 1997

Meld.St.26 (2013) Nasjonal transportplan 2014–23. http://www.ntp.dep.no/Nasjonale+transportplaner/2014-2023

Nordbakke S, Schwanen T (2015) Transport, unmet activity needs and wellbeing in later life: exploring the links. Transportation 42(6):1129–1151

Nordbakke S, Skollerud K (2016) Transport, udekket aktivitetsbehov og velferd blant personer med nedsatt bevegelsesevne. TOI-report 1465/2016

NPRA (2014) Universell utforming av veger og gater, håndbok V129. SVV. https://www.vegvesen.no/_attachment/118984/binary/963983

Odeck J, Hagen T, Fearnley N (2010) Economic appraisal of universal design in transport: experiences from Norway. In: Research in transportation economics, vol 30. Elsevier, pp 304–311. http://dx.doi.org/10.1016/j.retrec.2010.07.038

Øksenholt KV, Aarhaug J (2018) Public transport and people with impairments—exploring non-use of public transport through the case of Oslo, Norway. Disabil Society (in press). https://doi.org/10.1080/09687599.2018.1481015

Preston J (2009) Epilogue: transport policy and social exclusion—some reflections. Transp Policy 16:140–142

Preston J, Rajé F (2007) Accessibility, mobility, and transport-related social exclusion. J Transp Geogr 15:151–160

Priya T, Uteng A (2009) Dynamics of transport and social exclusion: effects of expensive driver's license. Transp policy 16(3):130–139

Rayle L, Dai D, Chan N, Cervero R, Shaheen S (2016) Just a better taxi? A survey-based comparison of taxis, transit, and ride-sourcing services in San Francisco. Transp Policy 45:168–178

Rebstock M (2017) Economic benefits of improved accessibility to the transport systems and the role of transport in fostering tourism for all. Disscussion Paper 2017-04 ITF/OECD

Schaller B (2017) Unsustainable? The growth of app-based ride services and traffic, travel, and the future of New York City. Schaller Consulting

Shakespeare T (2006) The social model of disability. In: The disability studies reader, vol 2. pp 197–204

Skartland E-G, Skollerud K (2017). *Universell utforming og brukermedvirkning i transportsektoren—en casestudie*. TØI rapport 1570/2017

Story M, Mueller J, Mace R (1998) The universal design file: designing for people of all ages and abilities. Des Res Methods J 1.1

Tennøy A, Øksenholt KV, Fearnley N, Matthews B (2015) Standards for usable and safe environments for sight impaired. In: Municipal Engineer. pp 24–31. http://dx.doi.org/10.1680/muen.13.00043

Thompson S (2017) Disability prevalence and trends. K4D Helpdesk Report. Institute of Development Studies, Brighton

United Nations (2006) convention on the rights of persons with disabilities. https://www.un.org/development/desa/disabilities/convention-on-the-rights-of-persons-with-disabilities.html accessed 15 Nov 2017

Van de Velde D (2004) Reference framework for analyzing targeted competitive tendering in public transport. TØI-report 730/2004

Older People's Mobility, New Transport Technologies and User-Centred Innovation

Charles Musselwhite

Abstract People are fitter and more mobile than ever before, but transport can still be an issue in later life due to physiological and cognitive challenges. This chapter examines findings from four focus groups with 36 older people examining the importance of mobility and future changes in mobility and transport. Older people were generally sceptical of potential transport futures, though they welcome technologies that reduce physical difficulty in mobility, gave real-time information, and reduced issues with interchange. There were mixed feelings of automated vehicles, often dependent upon the individual's willingness to accept technology taking over their own skills and abilities, trust in the technology and concerns over future built environments.

Keywords Transport · Mobilities · Ageing · Gerontology · Needs · New technologies · Transport futures · Mobility-as-a-service · Automated vehicles Driverless vehicles

1 Introduction

This chapter examines older people's attitudes towards technologies that could potentially help them to stay mobile in later life. The introduction begins by looking at how much society is ageing, followed by the importance of mobility to us as we age, which goes beyond simple notions of connectivity to also include psychosocial aspects of mobility as is shown here through different theoretical perspectives. After the introduction the paper shows the methodology of how we asked older people their views on mobility futures and innovation, before presenting and discussing the findings.

C. Musselwhite (✉)
Centre for Innovative Ageing, College of Human and Health Sciences, Swansea University, Singleton Park Campus, Swansea, Wales SA2 8PP, UK
e-mail: c.b.a.musselwhite@swansea.ac.uk

B. Müller and G. Meyer (eds.), *Towards User-Centric Transport in Europe*,
Lecture Notes in Mobility, https://doi.org/10.1007/978-3-319-99756-8_7

1.1 Ageing Across the Globe

Countries across the globe are ageing, largely due to longevity and a reduction in birth rate. There are now 840 million people over 60 across the World, representing 11.7% of the population. In 1950, there were only 384.7 million people aged over 60, representing only 8.6% of the global population (UN 2015). Projections suggest there will be 2 billion people aged over 60, representing 21.2% of the global population by 2050 (UN 2013). Better health care across the life course mean people are living longer. Years spent in better health in later life are not increasing at the same rate as increases in age and, in the United Kingdom (UK), 40% of people aged over 65 and 2/3 of those aged over 85 years, have some form of longstanding illness that limits their lives (ONS 2015, 2017). These limitations and their impact vary, for example in the UK, 70% of over 70 year olds have a hearing impairment (Action on Hearing Loss 2017), 35% of over 75s have sight loss (Living with Sight Loss 2015) that impacts on their lives and around 16% of over 75s have a form of dementia (Alzheimer's Society 2017). Combined with gradual muscle weakness and deterioration seen as people age, this can result in difficulty in being mobile. Older people are more likely than any other age group to suffer mobility deprivation, in that they cannot access the places they want because they cannot physically get to them (Holley-Moore and Creighton 2015; Mackett 2017). Being mobile in later life is linked to quality of life (see Holley-Moore and Creighton 2015 for review). Research has shown that giving up driving is related to a decrease in wellbeing and an increase in depression and other related health problems, including feelings of stress and isolation and also increased mortality (Edwards et al. 2009; Fonda et al. 2001; Ling and Mannion 1995; Marottoli 2000; Marottoli et al. 1997; Mezuk and Rebok 2008; Musselwhite and Haddad 2010, 2017; Musselwhite and Shergold 2013; Peel et al. 2002; Ragland et al. 2005; Windsor et al. 2007; Zieglar and Schwannen 2011).

Research has examined how to mitigate the relationship between driver cessation and negative health and wellbeing. This can be done through providing better quality alternatives taking into account barriers and enablers of each. Public transport, for example, can be improved for older people by having concessionary or free fares, level access from kerb to bus and training drivers to understand older people's issues and to speak clearly or not to drive off before they have sat down (Broome et al. 2010). Improving the public realm to be more accessible (providing public toilets, benches, wide and well maintained pavements), legible (maps, spaces that make you feel you should be there) and desirable (use of quality materials, landscaping, arts and vegetation to make you want to be there) can help older people walk more often (Musselwhite 2017). Community transport in the UK, can provide a door to door alternative to the car. However, as Musselwhite and Haddad (2017) note that all the alternatives are not viewed as comfortable, practical or desirable as the car.

1.2 Theoretical Models of Ageing and Mobility

Musselwhite and Haddad (2010) propose a three-tier model of needs and motivations for travel in later life (Fig. 1). The model was developed through interviews and re-convened focus groups with drivers and ex-drivers aged over 65 discussing the importance of mobility and travel to their daily lives. The three levels are hierarchical reflecting in general when in conversation with participants each need was mentioned. At the base of the hierarchy, reflecting needs mentioned early on in the conversations, were practical or utilitarian needs. These include the need to get from A to B cheaply, reliably, safely and quickly. This is followed by a middle level of needs grouped under the term social or affective needs. These came out a little later on in the conversation and were to do with the affective or emotional needs that mobility satisfies, such as affording a sense of purpose, of independence, of identity and of control. Finally, the top level reflects needs mentioned towards the end of the conversations, labelled aesthetic needs, including travelling for its own sake, to see the world passing by and to view nature; in other words with little explicit purpose, often referred to in the literature as discretionary needs. Musselwhite and Haddad's (2010, 2017) work on this suggests the car satisfies all three levels of need and alternative modes only meet some of the levels of need. For example, walking satisfied the middle level of

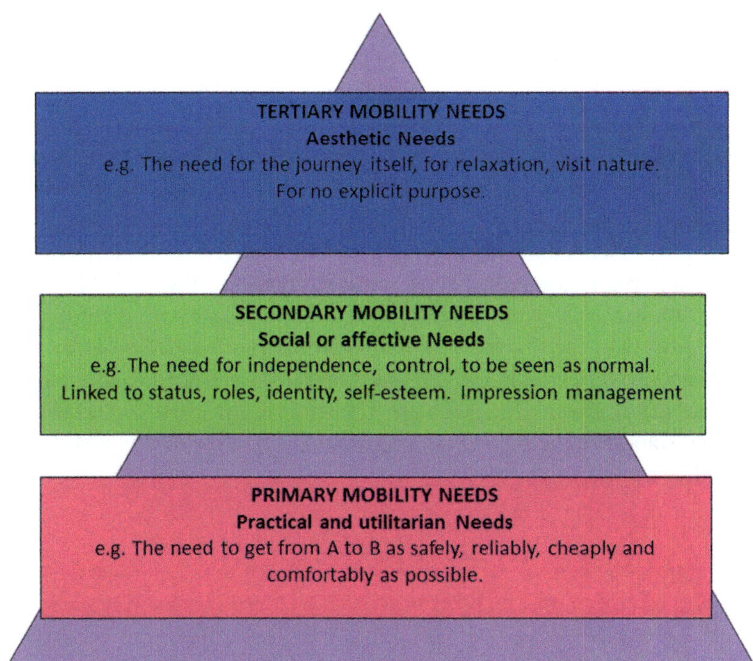

Fig. 1 A model of a hierarchy of travel needs in later life, showing the importance of utilitarian, social and aesthetic needs (after Musselwhite and Haddad 2010)

needs and, if the public realm is of sufficient accessibility and quality, the aesthetic needs too. However, practical or utilitarian needs are not met well with walking, which is physically demanding and takes time (relative to other modes). Buses, and even community transport, can meet practical or utilitarian needs. These can even meet aesthetic needs if the services run through to aesthetically rich places but the sense of freedom, independence and control is still lost.

Parkhurst et al. (2014) propose a model that suggests that mobility does not have to be replaced literally in later life. Advancements in technology have enabled mobility to be supported virtually, for example, through e-shopping or e-health. They also propose that mobility needs can be met through imaginative means, reminiscing about past travels or watching the world moving from a still position.

1.3 Moving Forwards: The Future of Mobility for Older People

Looking ahead, a number of exciting technology developments are promised to revolutionise the transport network. Improved digital technology, increased automation and the growing prevalence of the sharing economy are all promised changes in transport provision over the forthcoming years (Metz 2017). This paper examines older people's attitudes towards future technologies, examining whether they think they would make an impact on their travel behaviour and whether they would meet travel needs as set out by Musselwhite and Haddad (2010) and Parkhurst et al. (2014). It will examine changes in light of improved digital infrastructure in the transport world, including real-time transport information (to show live delays, departures and arrivals), smart ticketing (integrating ticketing between services), interfaces between technology and enhanced taxi services (e.g. taxi services enhanced through technology such as Uber, Lyft), and mobility as a service (provision of different transport services as one, bringing together taxis, car park charges, train tickets, bus fares as one payment through one portal), automated vehicles (cars with different levels of automatic elements of the driving task which are currently carried out by the driver themselves) and driverless cars (where the driver does not need to drive any part of the vehicle at all). It also investigated any transport innovations older people themselves have developed to help cope with mobility difficulties when meeting their day to day mobility needs.

2 Methodology

In order of examine older people' attitudes to future technologies that may enhance mobility, the study employed a qualitative design, to capture people's attitudes and perceptions. This chapter describes the methodology employed, including design, participants, procedure and tools and analysis, discussing the rationale for choice made.

2.1 *Design*

Four focus groups were carried out with 36 individuals over the age of 65 years to explore travel and mobility needs and behaviour in relation to new technologies. The focus groups were arranged into groups based on the participants' usual mobility mode: (1) regular drivers; (2) people who usually walk; (3) regular bus users and; (4) non-drivers who regularly rely on friends and family (who don't live with them). Focus groups were deliberately selected to help people to think more widely about an everyday element—mobility and transport use which might otherwise be hard to talk about individually, since it is so automated and taken for granted, and also to help generate new ideas and discuss new technologies around driving which might otherwise be hard individually.

2.2 *Participants*

Participants were sought through a research network of older people in South Wales, United Kingdom, answering an advert for people needed under the four different categories. People were placed into each category if they used that mode most often for their journeys (they did not need to be exclusive users of that mode). A total of 36 participants answered questions on future transport provision with an average age of 73.2 years, 25 were cohabiting with a partner and 11 lived alone (see Table 1). They were asked to self-report their health on a scale from 1 very poor to 9 very good. An average of 7 on the scale was found overall with the highest average, indicating best average health, among the people who walked and lowest among the people getting lifts from family and friends.

Table 1 Background of participants in the study by type of transport commonly used

	n	Age range (average)	Living arrangement	Health (self-score 1 = poor to 9 = good)
Focus group 1: Drivers	9	65–87 (73.9)	In couple = 8 On own = 1	6.7
Focus group 2: Bus users	9	65–86 (71.7)	In couple = 7 On own = 2	6.5
Focus group 3: Lifts from family and friends	8	72–92 (78)	In couple = 4 On own = 4	6
Focus group 4: Walkers	10	65–81 (70.2)	In couple = 6 On own = 4	8.2
Total	36	65–92 (73.2)	In couple = 25 On own = 11	7

2.3 Procedure and Tools

Focus groups lasted around an hour long and took place at a local community centre. Participants were free to talk around two key set themes, (1) current mobility and meanings of that mobility to the individual and their life and; (2) future mobility and how they felt it might change, using apprenticing and abstraction style questions.

Apprenticing (Robertson and Robertson 2012) was used for the current transport needs and current usage. This got the participants to talk through a particular day's worth of travel, taking the researcher through step by step in a high level of detail, so that researcher would be able to complete the journey themselves, then focused on a specific journey itself, "take me through a recent trip you went on step by step".

Abstraction (Robertson and Robertson 2012) techniques were used when discussing future transport provision and asks the participant what would happen to their everyday mobility if their experience was different. It involves both two processes: (1) counterfactual detail, asking the participants what they would do if their health or individual circumstance changed (for example if they were older, less mobile or less healthy) and; (2) future scenario testing. Here 5 short examples where given about future transport in order for discussions to start, (1) automated and driverless vehicles; (2) Mobility as a service; (3) Increase in sharing services; (4) using enhanced taxi services e.g. Uber/Lyft, which are enabled and connected through Information Communication Technology (ICT) through smart phones and apps, and; (5) real-time transport information and smart ticketing. Each scenario began with, "imagine a transport world sometime in the future" and then explained the scenario, for example driverless cars were described as, "where a car does not need a driver at all but you can access it from home and command it to take you anywhere for a price equivalent of what you pay for your transport at present". They were then asked, "how far do you think this is a desirable future?" which generated the most discussions, including examining how it might be improved or enhanced to maximize benefits, followed by what are the barriers were to implementation and whether they though the technology was feasible or not. Finally, participants were then asked to develop their own future scenarios.

2.4 Analysis

A typical thematic analysis was employed on the data. Data was recorded and then transcribed word for word (and individual people identified, if possible, by the transcription service). The researcher then highlighted key themes in each transcript and then collapsed themes further when all individual transcripts were analysed. Etic (stemming from themes derived from current theory, models and literature) and emic (stemming from analysis of the data itself) coding was then carried out on the data. Etic codes looked to place the data within categories of practical, psychosocial or aesthetic need based on Mussselwhite and Haddad's (2010) model and among the literal, potential, virtual and imaginative mobility categories based on Parkhurst et al. (2014) model. Additional challenges to the model found through emic style analysis were then added.

3 Findings

Findings about the importance of mobility are presented first as these were most often talked about. This is followed by a discussion on general attitudes towards new technologies and then finally a discussion linking the importance of mobility with attitudes towards new transport technologies concludes this section.

3.1 Importance of Mobility to Older People

The most important aspect of mobility for older people was how it kept people connected to family and friends and helped people access the shops and services they wanted. This came out in discussions early on and was frequently repeated, with people stressing how important it was.

> I need to get to the hospital for regular checks now with my heart. I get a taxi or a lift from friends for that as the bus isn't always convenient. (female, aged 80, bus user)

Almost always the car was seen as the easiest way of meeting this, even when people no longer drove and had infrequent, if any, access to a car,

> Well I get all my stuff in because I can drive God knows what I'd do without the car. The supermarket is so far away you see. (female, 78, car driver)

Older people noted how they could stay connected to events in the town, to groups, so that they felt part of a community or importantly could be part of democratic change for a community,

> If I couldn't walk, now I haven't got a car, I couldn't go to the local council meetings I need to go to. I was a councillor for many years, and I still help out. I get my voice heard still and I'm helping others like me have a say. (male, 80, walker)

People saw mobility as a way of warding off loneliness, of staying connected to people, stopping people being isolated, even in rural areas,

> Well it's beautiful out here but it is out the way, the car keeps me connected, without it I'd go spare, I wouldn't hardly see a soul. I go to my lunch club and play golf in the town but it's 40 miles away. (male, aged 78, car driver)

Older people frequently discussed how mobility was linked to their own health and wellbeing, both indirectly and as a direct result of the physical activity of mobility,

> "I am certain that it keeps me happy. I always feel better when I've been out, even if it is a visit to the shops'. (male, aged 80, bus user)
>
> Walking always makes me feel better. Fitter, more awake. (male, aged 75, walker)

Mobility is also clearly linked to Musselwhite and Haddad's(2010) second level of needs, keeping people feeling independent and in control,

> The freedom of the open road is still there you know. We were sold this in the 60s and I still feel it now! (male, car driver, aged 77)

And especially for females how it kept them connected to a family role,

> It means I can help the family. I can pick up grandkids. I can be grandma for them. (female, aged 80, car driver)

And for males a head of family style role,

> Well as I can drive, I can still take my family out, grandchildren and my children for a meal or something. I can help there you see. (male, car driver, aged 80)

Most of this level of need is satisfied by owning and driving a car, but walking can meet these needs if someone is physiologically fit enough and the built environment is of sufficient quality,

> I'm lucky to be fit enough to go walking, so I get out and about when I feel like it. There are nice walks to have round here and it isn't far to go to the shops and there are pavements all the way. (female, aged 75, walker)

Linked to this was the ability to go out when and where someone chose to do so, linked to potential mobility noted by Parkhurst et al. (2014). Just going for a drive or a walk was noted quite often by drivers and walkers,

> If I wake up and I want to go and see the mountains I just can with a car. I decide. (male, aged 80, car driver)

The ability to actually drive was linked to control and to status,

> Being able to drive still is important to me. It demonstrates I've still got my faculties. That is important to me. I'm still in charge of my own destiny with a car. (male, car driver, aged 77)

But mobility for its own sake was very rarely noted by public transport and community transport users. However, they did mention the social nature of their mobility, the interacting with others while being co-present with them,

> It's a social club in itself the bus, we all look forward to a natter and catching up once a week. (female, community transport user, aged 79)

3.2 *Non-Literal Mobility User-Innovation*

The focus of mobility in the interviews was almost exclusively around literal mobility, but older people did get their needs satisfied through non-literal modes as well. For example, it was quite common for older people to use virtual mobility through e-shopping,

> "Yeah I get stuff delivered now. It's so much easier and they'll bring it right up to your kitchen door and even help with some of the heavy things you know'. (female, aged 77, bus user)

The use of tele health and telemedicine wasn't used very much by the participants, who almost always travelled to the doctor's surgery, health centre or hospital. The only remote incidences of this were consultation with doctors over the telephone or using the internet to self-diagnose.

People stay connected to family and friends at distance usually using telephone, but a significant number noted the use of video and talking methods including Skype and Face Time,

> I have family in Australia I don't get to see yet I feel quite close as we Skype quite a bit. It's so lovely to see their little ones, my great-nieces and nephews. (female, aged 80, community transport user)

Linked to this was just seeing parts of the world they had once been connected to either through these video calls or in some cases through use of webcams,

> It's great to see the places I grew up in on those webcams. Some haven't changed much at all. I love using them! I don't suppose I shall go back there but I can still see it. (male, aged 80, walker)

3.3 *Attitudes Towards Future Transport Innovation*

3.3.1 Automated and Driverless Vehicles

Automated vehicles were the technology discussed by participants in most detail. This was largely because this technology has received a lot of press coverage recently in the United Kingdom. They were largely viewed very positively especially by those who had given-up driving or did not drive,

> Yeah I can see driverless cars being very popular, you don't have to drive, brilliant. I'm all up for that. I'd still have car then. It'd be easier for me to get around. (male, aged 80, bus user)

There was overall less support for driverless vehicles by current drivers, especially concerns about giving-up control and technological failure,

> I simply can't see technology being better than a human being in control. I just can't see how it can cope. It isn't for me this tech. (male, car driver, aged 75)
>
> I hear that human error is a major factor in driver crashes but I would hate, and I think many would agree with me, hate to give up control of the vehicle. I just couldn't trust the technology. Maybe I'm a control freak. (male, car driver, aged 77)

However, this was placated somewhat in the counterfactual discussions when older drivers projected ahead to having to give-up driving,

> Well if it came down to a choice where I couldn't drive anymore then yeah sure I'd rather have an automated vehicle driving me than taking a bus or burdening people for lifts. (male, aged 80, car driver)

The reasons being stated that it kept people in control of time and destinations of travel, linking to potential for travel theory again,

> well it would mean that at least I'm not beholden to a bus timetable. I'll still pick and choose when and where I go. (female, car driver, aged 72)

There was also the assumption that it wouldn't be shared and that individuals would own their automated vehicles,

> again it has the advantage over the bus of being your own vehicle, your own private space. (female, car driver, aged 78)

Those who walked were more likely to worry about interaction of driverless vehicles with themselves,

> I worry about them stopping for pedestrians. I hear they can but I wouldn't trust it, I wouldn't try it out! (female, aged 75 walker)

One individual even noted he could not see them being on local roads, perhaps leading to pedestrianised areas and areas for vehicles being segregated,

> You won't see them on residential roads or town centres; I can see them on motorways. I can't see them being on the same streets as pedestrians. Maybe they'll be behind fences. It isn't a future I want but I bet it is one the politicians and car companies want. (male, walker, aged 70)

People discussed how automated vehicles might act like taxis that were on call but there was little appetite for such a model, citing concerns over cost, waiting times and potentially sharing such vehicles with other people,

> I guess you would be waiting for ages for them to turn up and they'd be incredibly expensive. More expensive than taxis are now because of the tech needed. (male, aged 80, car driver)

> I wouldn't want driverless cars you have to share. I'd want my own to have to look after. And have my own space. (female, aged 74, car driver)

The people who were positive about automated vehicles saw how it could keep them connected to communities and individuals further away and this was seen as hugely positive,

> well it would be so easy to go and see my daughter and her family. I don't like driving so far to see them, but this would just be so easy. I'm all for it. (female, aged 75, lifts from family and friends)

> I guess it adds choice, as you get older you are restricted in what you can do, this can take you to the shops and the services you choose. (male, aged 70, car driver)

It was especially seen as being advantageous to keeping people connected to hobbies and leisure and this was frequently mentioned,

> Well, I'd go back and see the Swans <Swansea football team> again if I had a car driving for me. I wouldn't have given my season ticket up. That was a major factor, just couldn't get there by bus and too far to walk with my knees. (male, aged 80, bus user)

3.3.2 Mobility as a Service

Mobility as a service had not previously been heard of by all but two of the participants in two separate focus groups. The concept was readily and easily understood, but a reality could not be seen due to difficulty in joining up of services, of data and the costs associated with it,

> You mean they'd all have to talk to each other, buses, trains, taxis, car parks – that's isn't going to happen. There isn't political will for that or desire among the different parties. It won't happen. Not in my life time or my children's. (male, aged 72, car driver)

Using counterfactual discussions to suggest mobility as a service had come in, older people were fairly favourable of it, especially as a way of helping keeping track of costs or even driving down costs and affording choice.

> I can see the advantages, we would know what we were paying from door to door, providers would be encouraged to compete and offer better and cheaper services for us, potentially we'd have more choice but would it end up being so joined up? I'm skeptical. (male, car driver, aged 70)

3.3.3 Increased Shared Services

Older people really didn't like the idea of increased sharing of transport services. Many older people noted already that ageing was synonymous with more public based use of services and reduction of privacy compared to middle-age.

> Typical isn't it. Provision of transport for older people is always about sharing. Like we're not worth it as an individual. I wouldn't want to see it. I still value independence and freedom and sharing transport isn't for me. Not always anyway. (male, aged 71, car driver)

3.3.4 Enhanced Taxi Services

Four older people had used technological-led taxi service Uber, three in their local city and one when they visit friends in London. They all cited how much they thought it was an excellent concept. They especially liked the security of knowing who the driver was and being able to know the cost up front,

> When you get used to it, you end up preferring it as you know who the driver is going to be and you know the cost up front. For us pensioners, we haven't always got a lot of cash, that's really helpful and secure. (male, aged 78, lift from friends and family)

There was also a feeling of being youthful and associated high self-esteem of using such a service,

> I'm down with the kids! It was my kids who introduced me to it. I thought why not, I'll try it when I'm next in London. I haven't looked back. I use it every time I'm there. They're right proud of their trendy dad now my kids!! (male, aged 73, car driver)

Negative feelings attached towards such services emanated from poor business models and negative press associated with such companies and so some older people felt insecure and vulnerable to attack both personally and with their data,

> I'm worried about the safety. I don't think they're vetted like professional drivers are they? And the data? Not sure that I'm very safe with them. I'll stick to what I know. (female, aged 74, lifts from friends and family)

3.3.5 Increased Use of IT and Digital Technologies

People were most positive over increased use of IT in everyday transport services, especially in terms of provision of information, for example real-time bus information,

> Like the trains, now buses tell you when they're coming. It's great to know you haven't missed it or sometimes you have! But you know where you are. It's extra security and yes it seems quite reliable. (male, bus user, aged 80)

Smart ticketing was less of an issue as many who used buses would use a concessionary bus pass anyway. More joined up services and ticketing was acknowledged as a good step forwards though there were few examples they could give. Older people especially favoured help getting to and from a train station, a journey often completed by taxi,

> I mean I would get a bus, I do at home, but when I go to a less familiar place, to save the hassle I get in a taxi, its costly but I can't be doing with looking for a bus. Often I'll have

> luggage. It might be dark. Yes a taxi is easier. I would use bus if I knew more about it, timetables, but also where it goes from and to, what to expect, you know. (male, aged 75, bus user)

There was concern, however, that the increased use of IT, especially in information provision, was at the expense of members of staff who they felt most secure and trust with,

> Well my problem with it is that it is all becoming so automated there is no staff left. And when it goes wrong or we can't read it who can we turn to then? (female, bus user, aged 75)

3.4 User Transport Innovation

Older people quite often had their own solutions that aided with mobility issues surrounding old age. One older person took in parcels for the entire street, all of whom were usually out at work during the day, which was reciprocated with gifts of food, flowers and even being taken out for the day. But, more importantly, the daily contact of people coming to get their parcels was seen as keeping this older person connected to the people in the street and the wider community.

Another older person had taken the role for booking older people onto community transport and was a community transport user herself. Having used the service for three years, she had become expert at the routes and got to know the other riders and when a vacancy appeared decided to help. This keeps her connected closely to other passengers and the drivers, also giving a sense of purpose.

Many older people give lifts to other people in their local community and it was noted by one person that making the service more joined up and offering a "grey uber" style system would be the way forwards,

> I'm driving people in the local community around, people I know, but I know sometimes, because when I get there they say, that I could've taken a few others who've got a taxi there themselves. So something that joins us all up. Something like Uber for oldies? Grey Uber would be the way forwards? (male, aged 76, car driver)

Older people also projected forwards to think how other forms of mobility might change. A number of older people talked about the built environment being more supportive. One older person noted she had seen playgrounds for older people and bus stops with swings which she'd like to see introduced more often,

> too often older people are seen as needing support, sometimes we also like to play! (female, aged 77, bus user)

Another older lady supported two other older people in terms of doing their own line shopping for them, which kept her connected to them but also a sense of responsibility and purpose,

> "Since they <name> can't drive anymore, I offered to do them a shop, then when I didn't drive I do their shop online. They're very grateful and I find it quite purposeful, feeling quite useful. (female, bus user, aged 79)

4 Discussion

Mobility is clearly linked to self-reported health and wellbeing which agrees with previous research (Musselwhite and Haddad 2010; Schlag et al. 1996). The reasons for this are that being mobile provides activity at the end of the journey, helping people stay connected to people and access and use shops and services they see as important. It is also linked to status, role and independence and the act of mobility itself is seen as being positive and the notion of just getting out and about for its own sake was seen as vital escapism. As was found by Musselwhite and Haddad (2017), these needs were most clearly met for car drivers and all older people using alternative modes found some deficit. People who walked lamented poor pavements, fast traffic and inability to travel long distances. People who used public transport missed the potential of travel, the ability to go when and where they wanted, even if they didn't always want to go at a different time or place than the bus went. Those receiving lifts and using community transport missed the opportunity to go when and where they wanted and to be able to just go out for a journey itself. Hence, discussions around the future transport solutions often centred on technologies that mimicked journeys undertaken by driving a car, most noticeably around autonomous and driverless vehicles. Most of those who had given up driving or did not drive were very favourable towards such technology, feeling all of their mobility needs would be met in relation to Musselwhite and Haddad's (2010) hierarchy of mobility needs; ability to get from A to B when and where the individual felt they could, including travelling for its own sake, and ability to keep independence and enhance status and roles if they owned the vehicle themselves.

Those who drove were less favourable towards autonomous vehicles, though their support increased if a layer of counterfactual thought was added where the alternative would be to give-up driving altogether. It is the psychosocial needs from Musselwhite and Haddad's (2010) model that would be missed through autonomous vehicles for drivers, the ability to do the driving themselves and associated identity of being a driver and of being in control of the vehicle and lack of trust in automation. Interestingly, although there was much support for driverless cars there was not actually a huge amount of insight, often it was assumed that such vehicles would simply replace the driver and a like for like comparison was made. Naturally, therefore, the reduced effort needed to drive was very positive among those who had given up driving. There was little insight for example into how autonomous vehicles might change society and change interpersonal spaces. Older people still talked of how the journeys they made would be the same, just easier to perform. There was no notion that journeys involving collecting or picking up good or items or journeys taking other people would not be necessary any more, for example. There was no insight into changes in public realm that would accompany such vehicles, except one individual who noted that such vehicles may need dedicated infrastructure away from pedestrians. That if journeys were easier, there could be potential changes to the location of shops and services, was not discussed by participants. Overall, automated vehicles were equated very much as business as usual in terms of mobility.

Older people had trouble with the idea of sharing services. This was especially true again of those who were drivers and owned their own vehicles. There was very little appetite for moving away from ownership and having private space. Thus many of the benefits of automated vehicles and of better scheduling associated with taxis increasing capacity and making better use of the road space were not in evidence and it is suggested that it would require changes in policy to change habit and practice. There was in fact resistance among many older people in terms of not owning vehicles and sharing services. Thus, sharing mobility is probably more of a challenge than is currently acknowledged and emphasises the importance of the secondary psychosocial level of need in relation to mobility evidenced in Musselwhite and Haddad's (2010) model.

There was significant support for improved information provided through technological means, especially if that helped reduce physical and cognitive burden of mobility. Real-time travel information and accessing mobility through apps has some appeal. Mobility as a service was seen as a relatively useful concept but there was concern about the ability for it to work in reality. In relation to Musselwhite and Haddad's (2010) hierarchy of mobility needs, the accessibility to information which both improved the utilitarian elements of reliability and convenience and psychosocial level of trust in the services were met. If the technology was new and innovative an element of aesthetic needs may also be met in using such technology.

Transport innovation from individuals themselves is relatively low-tech. It often involves providing support in the form of social capital, supporting others in their mobility needs, either virtually (e-shopping, taking in parcels) or literally (giving-lifts). Yet, despite being low-tech in operation, these innovations tend to meet all three levels of mobility needs in Musselwhite and Haddad's (2010) hierarchy of needs; keeping people practically connected, emotionally connected and offering a pleasurable aesthetic service in the community. There are obviously opportunities for technology to help facilitate this style of social capital on mobility by joining up provision with need in real-time, however this is yet to happen in practice.

5 Conclusions

Future transport provision has the potential to help older people meet their mobility needs. Autonomous vehicles, as long as they operate like cars and that society does not change much, has the potential to fulfil all three level of Musselwhite and Haddad's (2010) mobility needs hierarchy, practical, psychosocial and aesthetic needs. However, psychosocial needs may be somewhat reduced if older people don't drive the vehicle themselves and certainly would be if the vehicle was shared and not owned by the individual. Increased information provision helps reduce practical barriers and improves the utilitarian aspect of public transport. However, it does little to change psychosocial or aesthetic needs, although the novelty of using modern technology can improve psychosocial aspects of "feeling young" and "up with the latest", though how long these feeling might stay is questionable. Improving the urban realm through more playful spaces was one of the more novel innovations showing the concept of mobility as a playful activity.

It has to be noted though, that despite using future scenario work in the interviews, people's visioning of the future is very much business as usual in terms of needs, mobility provision and place. It is suggested, future research needs to spend a longer period of time visioning different futures in order to generate future scenarios wider than this project has achieved. This may mean reconvened focus groups or creative methods of data collection in a workshop style taking place over a day or two to allow development of ideas.

Overall, older people shared in-depth stories and insights in terms of mobility and transport and such tacit knowledge would be beneficial to planners and could be fed into the design of transport systems, policy and practice. To conclude, it is suggested that more in-depth work with older people could help illuminate and develop transport innovation in a human centred way that will be beneficial to older people themselves.

References

Action for Hearing Loss (2017) Facts and figures. Available at https://www.actiononhearingloss.org.uk/about-us/our-research-and-evidence/facts-and-figures/. Last Accessed 14 Dec 2017

Broome K, Nalder E, Worrall L, Boldy D (2010) Age-friendly buses? A comparison of reported barriers and facilitators to bus use for younger and older adults. Australias J Ageing 29(1):33–38

Edwards JD, Perkins M, Ross LA, Reynolds SL (2009) Driving status and three year mortality among community-dwelling older adults. J Gerontol Series A Bio Sci Med Sci 64:300–305

Fonda SJ, Wallace RB, Herzog AR (2001) Changes in driving patterns and worsening depressive symptoms among older adults. J Gerontol Series B Psychol Sci Soc Sci 56(6):343–351

Holley-Moore G, Creighton H (2015) The future of transport in an ageing society. ILC, UK, London. Available at http://www.ilcuk.org.uk/index.php/publications/publication_details/the_future_of_transport_in_an_ageing_society. Last Accessed on 14 Dec 2017

Ling DJ, Mannion R (1995) Enhanced mobility and quality of life of older people: Assessment of economic and social benefits of dial-a-ride services. In: Proceedings of the seventh international conference on transport and mobility for older and disabled people, vol 1. DETR, London

Living with Sight Loss (2015) Updating the national picture RNIB and NatCen, 2015 Available at https://www.rnib.org.uk/sites/default/files/LivingwithSightLoss.pdf. Last Accessed 14 Dec 2017

Mackett R (2017) Older People's Travel and its Relationship to their Health and Wellbeing, In: Musselwhite CBA (ed) Transport, travel and later life (transport and sustainability, vol 10).Emerald Publishing Limited, pp 15–36

Marottoli RA, Mendes de Leon CF, Glass TA, Williams CS, Cooney LM, Berkman LF (2000) Consequences of driving cessation: decreased out-of-home activity levels. J Gerontol B Psychol Sci Soc Sci 55B(6):334–340

Marottoli RA, Mendes de Leon CF, Glass TA, Williams CS, Cooney LM Jr, Berkman LF, Tinetti ME (1997) Driving cessation and increased depressive symptoms: prospective evidence from the New Haven EPESE. J Am Geriatr Soc 45(2):202–206

Metz D (2017) Future transport technologies for an ageing society: practice and policy In: Musselwhite C (ed) Transport, travel and later life (transport and sustainability, vol 10). Emerald Publishing Limited, pp 207–220

Mezuk B, Rebok GW (2008) Social integration and social support among older adults following driving cessation. J Gerontology Soc Sci 63B:298–303

Musselwhite C (2017) Creating a convivial public realm for an ageing population. Being a pedestrian and the built environment. In: Musselwhite C (ed) Transport, travel and later life (transport and sustainability, vol 10). Emerald Publishing Limited, pp 129–137

Musselwhite, C, Haddad H (2017) The travel needs of older people and what happens when people give-up driving. In: Musselwhite C (ed) Transport, travel and later life (transport and sustainability, vol 10). Emerald Publishing Limited, pp 93–115

Musselwhite C, Haddad H (2010) Mobility, accessibility and quality of later life. Q Ageing Older Adults 11(1):25–37

Musselwhite CBA, Shergold I (2013) Examining the process of driving cessation in later life. Eur J Ageing 10(2):89–100

ONS (Office for National Statistics) (2015) Life Expectancy at Birth and at Age 65 by Local Areas in England and Wales: 2012 to 2014. Office for National Statistics, Titchfield, UK. https://www.ons.gov.uk/peoplepopulationandcommunity/birthsdeathsandmarriages/lifeexpectancies/bulletins/lifeexpectancyatbirthandatage65bylocalareasinenglandandwales/2015-11-04. Last Accessed 14 Dec 2017

ONS(Office for National Statistics) (2017) Health state life expectancies, UK: 2013 to 2015 (with April, 2017 correction). Office for National Statistics, Titchfield, UK. https://www.ons.gov.uk/peoplepopulationandcommunity/healthandsocialcare/healthandlifeexpectancies/bulletins/healthstatelifeexpectanciesuk/2013to2015/previous/v2. Last Accessed 14 Dec 2017

Parkhurst G, Galvin K, Musselwhite C, Phillips J, Shergold I, Todres L (2014) Beyond transport: understanding the role of mobilities in connecting rural elders in civic society In: Hennesey C, Means R, Burholt V, (Eds) Countryside connections: older people, community and place in rural Britain. Policy Press, Bristol pp 125–175

Peel N, Westmoreland J, Steinberg M (2002) Transport safety for older people: a study of their experiences, perceptions and management needs. Inj Control Safe Promotion 9:19–24

Ragland DR, Satariano WA, MacLeod KE (2005) Driving cessation and increased depressive symptoms. J Gerontol Series A Biol Sci Med Sci 60:399–403

Robertson S, Robertson J (2012) Mastering the Requirements Process. Addison-Wesley, Boston, MA

Schlag B, Schwenkhagen U, Trankle U (1996) Transportation for the elderly: Towards a user-friendly combination of private and public transport. IATSS Res 20(1):75–82

The Alzheimer's Society (2017) What is dementia? Fact sheet 400, https://www.alzheimers.org.uk/download/downloads/id/3416/what_is_dementia.pdf. Last Accessed 14 Dec 2017

United Nations (UN) (2015) World Population Ageing. United Nations, New York. Available at http://www.un.org/en/development/desa/population/publications/pdf/ageing/WPA2015_Report.pdf. Last 14 Dec 2017

Windsor TD, Anstey KJ, Butterworth P, Luszcz MA, Andrews GR (2007) The role of perceived control in explaining depressive symptoms associated with driving cessation in a longitudinal study. Gerontologist 47:215–223

Ziegler F, Schwanen T (2011) I like to go out to be energised by different people: an exploratory analysis of mobility and wellbeing in later life. Ageing Soc 31(5):758–781

Changing the Mindset: How Public Transport Can Become More User Centered

Ineke van der Werf

Abstract There is a growing demand for public transport. It provides sustainable mobility that tackles air pollution and climate change as well as adapts to the needs of the elderly, commuters and a growing urban population. However, in order to become a real alternative for the car and to attract other customer groups as well as retain existing passengers, public transport needs to be improved and become more user centered. A user-centric public transport system contains three aspects: a strong(er) focus on customer satisfaction is needed; it is necessary to research customer satisfaction thoroughly and have a closer look at different perceptions of service quality of various customer groups; and the involvement of users in the design, planning and implementation of public transport services is important. Some examples, best practices and outcome of various user-centric approaches in Europe and elsewhere are described in this article.

Keywords Sustainable mobility · Public transport · Users · Passengers
User-centric design · Participatory approach

1 The Necessity of More and Better Public Transport

Today, the population growth of cities and hence the importance of public transport is becoming evident. All over Europe voices are heard that urge us to tackle air pollution and climate change, as well as adapt to the needs of the elderly, commuters and growing urban population. Public transport is very important in providing sustainable mobility, to meet growing transport demand in and between cities and to improve the efficiency of the existing infrastructure. With more and better public transport, cities can cut traffic congestion, foster social inclusion and reduce pollution (Imre and Çelebi 2017). However, in order to become an alternative for the car and to attract

I. van der Werf (✉)
Reizigersvereniging Rover, Braillestraat 16, 3822 XB Amersfoort, The Netherlands
e-mail: gorby@planet.nl

B. Müller and G. Meyer (eds.), *Towards User-Centric Transport in Europe*,
Lecture Notes in Mobility, https://doi.org/10.1007/978-3-319-99756-8_8

other customer groups as well as retain existing passengers, public transport needs to be improved. But how do we do that?

'A passenger's choice of transportation modes depend on several factors such as travel and transfer times, total cost, cruise comfort, reliability and total experience of the trip' (Imre and Çelebi 2017). These rational considerations for travel behavior, which can be supplemented with action radius, flexibility and the capacity to carry goods, are not the only considerations for people to travel and choose certain transport means. Social factors (for instance status) and emotions (such as stress) also play a role. However, traditional transport research and planning mostly deny these intrinsic values. Travel behavior is dependent on other mobility choices of people as well, like the purchase of a private car, choice of a place to live and lifestyle choices for work, household and leisure. Travel behavior depends 'on the activities in which individuals want to participate at their destination(s) or while travelling and the options they have to fulfil these needs.' The motivation arises from needs and the presence of opportunities of an individual, like the supply of transport alternatives and distance to destinations. It also depends on individual abilities: the available time, money, skills and capacity for certain travel choices (Dijst et al. 2013).

People might choose for public transport if it brings them profit: in time, money or image for instance. It can be faster and cheaper than the car and you are able to do some work or be active on social media. Ease is also important: it should not be too difficult to use. Thus, information is important, it should be easy to pay for and easy to find your way and vehicles should be accessible, Thirdly, joy or pleasure is also a factor: public transport should be fun! You can sit and relax, meet interesting people and enjoy the scenery. However, safety and reliability come first (Maartens 2017).

1.1 The Necessity of User Centered Public Transport

Understanding the needs of potential public transport users and the loyalty of existing public transport users is important in order to be able to improve public transport and make it grow. 'It can help public transport managers and marketers design effective strategies to meet passengers' needs and retain existing passengers as well as attract new ones from other modes' (Lai and Chen 2011). Also, a deeper understanding of passengers' attitudes towards public transport and the perception of comfort in public transport services is needed. Only then it is possible to design transport services in such a way that it accommodates the service levels required by customers, which in itself is a prerequisite for public transport usage to increase (Imre and Çelebi 2017).

According to Mark Wardman, addressing convenience is important for improving the well-being of public transport users as well as attract non-users to use public transport. Comfort, convenience and reliability are undervalued in transport research and evaluation. Mostly its focus is on speed and price, resulting in cost-effective improvements being overlooked such as opportunities for modal integration. This would involve improving walking and cycling conditions as well as enhance connections with other modes of transport. By overlooking these potential improvements, the attractiveness of public transport is reduced in relation to the use of the private car, 'contributing to a cycle of increased automobile dependency and sprawl and reduced transit ridership and revenue' (Wardman 2014). More information is needed on what people want and what they are willing to pay for.

1.2 User Centered Public Transport

Public transport users traditionally have a weak position towards service providers as compared to the consumers' situation on other markets. This is on the one hand due to the fact that public transport is dominated by monopoly conditions. Therefore, customers cannot express dissatisfaction with the service on offer by simply switching to another operator, like shoppers can do if they are not satisfied with the service of products sold in a certain store. They can choose to shop somewhere else, while public transport users can only choose another mode of transport, e.g. the car, if they are not satisfied with the services of a certain transport company. But this is not always possible and certainly not desirable. Also, there is a strong political influence on public transport, so passengers may seek to influence the services provided by political lobbying. However, this is a time-consuming and indirect approach. And not without difficulties. To attract political interest, they have to compete with the car as well (Schiefelbusch and Schmithals 2005).

Three aspects are important for public transport to become more user-centric. First of all, a strong(er) focus of public transport companies on customer satisfaction is needed. Therefore, the role of customer service departments should be increasing. (Although companies at the same time should bear in mind that service is not a department but a mentality!) Secondly, it is necessary to research customer satisfaction thoroughly and have a closer look at different customer groups. Are they living in the city or in the region? Are they male or female? What is the difference in age groups? What is their incentive to use public transport; e.g. do they use public transport to go to work or to visit their family and friends? Do they travel for pleasure and recreation or do they use public transport as part of a business trip? And how do these different types of public transport users perceive different attributes of service quality in public transport? (Mouwen 2017) Finally, the involvement of users is important to create attitudinal loyalty. Consumer watchdogs and other customer-focused organisations can play a vital role in this. 'They help ensure passengers have a voice and serve as the link between passengers, transport operators and service-regulating authorities' (McHugh 2017a). At the same time, it may also be useful to involve variable public transport users directly in the planning process.

2 The Formal Involvement of Users in Public Transport Planning

According to Schiefelbusch and Schmithals (2005), there are four levels where user interests can be identified: the political or strategic level, the planning level, the operational or provisional level and the practical level. At the political level, 'user interests can be expressed through the general process of political decision-taking as well as through consultation and participation procedures which can be formal or informal in character.' Instruments are political lobbying, formal consultation and informal participation exercises. Usually, this type of advocacy for user interests is done by the consumer watchdogs or customer-focused organisations mentioned above (See Sect. 2.1).

On the planning level, public transport services are designed in detail. It involves issues such as the routing of services, timetables, ticketing conditions, design of vehicles and location of stops as well as the development of new service concepts. A customer-oriented service requires that the users' expectations are known by the planners and that they are given attention in the planning procedure. This can be done by dialogue, for instance with the interest groups of Sect. 2.1 or by consulting specific passenger groups or users (see Sect. 3.1). Another possibility of 'getting to know the customer' is doing market research (see Sect. 4) or analyse customer feedback (e.g. as collected by the customer service department).

On the operational or provisional level, the concepts of the planning level are implemented. At this level, users need a reliable service and detailed information about the service. In other words, they like to experience a high-quality product. However, on the practical level things can go wrong. Therefore, user-oriented public transport operators need to have a backup plan: they should deliver suitable alternatives for their customers, provide adequate redress and handle complaints in an uncomplicated way. Hence the importance of a solid and qualitively good customer service department within public transport companies.

2.1 Interest Groups as Voice of Passengers

In several European countries, different consumer 'watchdogs' or interest groups are active in order to protect the right of passengers. Transport Focus and London TravelWatch in the UK, Passenger Pulsen in Denmark, TreinTramBus in Belgium, Reizigersvereniging Rover in the Netherlands, FNAUT in France and PRO BAHN and VCD in Germany are just several of these formal consumer organisations. They all somehow receive funding from their respective governments and/or from their members (public transport users), but nevertheless work independently to represent the voice of passengers. All over Europe (21 countries), there are 37 public transport users' organisations. They are associated in EPF, the European Passengers' Federation (www.epf.eu). They all represent the voice of passengers and try to make sure these are embedded in their countries' decision-making procedures.

2.2 Rocov's and Locov in the Netherlands

In the Netherlands, since 2001 regional governments are responsible for public transport in their region, which is subject to tendering. By law (Wet Personenvervoer 2000), it is mandatory that users are involved in the planning and monitoring process of regional public transport (bus, metro, trams and regional train services as well as public transport on water ways). These users are usually represented in a board by customer organisations such as Rover (Reizigers Openbaar Vervoer), a consumer organization of public transport users. Other customer organisations in these boards represent cyclists, students, the elderly, chronically ill or handicapped persons, inhabitants of small villages and even car users. Several times a year these boards of users (Rocov's or Regionaal Overleg Consumenten Openbaar Vervoer) meet with representatives of the regional governments and public transport companies and discuss all kinds of subjects related to public transport services. Some of the subjects are mandatory to advice by these Rocov's, among others tariffs and timetables (Maartens 2017).

On the national level such a board also exists, called Locov (Landelijk Overleg Consumenten Openbaar Vervoer, www.locov.nl), in which representatives of the consumer groups mentioned above discuss all relevant subjects with representatives of the national railway company and maintenance provider as well as representatives of the Ministry involved. Usually, also the Minister and Parliament are very much interested in the advices of this board. It is seen as an independent instrument in which users (organisations) can give their advice on important matters. Newspapers may publish about the advices given as well. On the long run, Locov can be a powerfull instrument for consumer organisations to act as a watchdog for the rights and needs of users.

2.3 LondonTravelWatch and Transport Focus in the UK

The London Transport Users Committee (LTUC) was established in 2000 and renamed London TravelWatch in 2005. The Committee has the right to be heard on all aspects relevant to transport users. According to Wikipedia 'it is the transport watchdog for services provided by Transport for London, which includes facilities for pedestrians, cyclists, and motorists. It is also the watchdog for National Rail companies in the London area and significant parts of the South East and deals with individual complaints by passengers against the relevant transport operators.' However, it takes up a complaint only after the operator has already dealt with it and if the complainant remains dissatisfied. In the Netherlands, such a (more informal) body of appeal also exists and is called OVloket (www.ovloket.nl). London TravelWatch has 7 members and a small staff (a secretariat providing support, research and day-to-day activities) who have regular contacts with operators, politicians, the media and other groups with similar interests. The members are appointed for 4 years, 'to ensure a regular inflow of new ideas'. The secretariat prepares the Committees meetings, collects material and conducts or commissions analyses and research on particular issues, based on the Committee's activities (Schiefelbus and Schmithals 2005).

As a watchdog for passengers and road users, Transport Focus wants to provide evidence to inspire change. As Chief Executive Anthony Smith of Transport Focus explains, "we carry out satisfaction surveys and do research on fares and ticketing in order to help those making decisions about transport to make better decisions" (McHugh 2017a). With offices in Manchester and London, the organisation ensures that public transport operators and government authorities are well informed about passengers' opinions on public transport services and about their priorities for future improvements. "Our mission is to get the best deal for passengers and road users. With a strong emphasis on evidence-based campaigning and research, we ensure that we know what is happening on the ground" (www.transportfocus.org.uk). The organisation specializes in large scale surveys, such as the National Rail Passenger Survey, where over 50,000 rail passengers a year give their views about their rail journeys. There is also a Bus Passenger Survey and Tram Passenger Survey. At the moment, they are developing a survey for road users. Furthermore, in-depth research into transport user experiences and needs for the future, results in lists of top priorities.

2.4 Passenger Pulsen in Denmark

Danish watchdog Passenger Pulsen—as part of the Danish Consumer Council—operates through fact-finding and by investigating consumer satisfaction with regard to passenger rights. They conduct national surveys of passenger satisfaction regarding the use of bus, metro and train, through a web based survey group (passenger panel), with the aim of examining the users of public transport according to their habits and satisfaction. They also have a team of 200 volunteer Passenger agents who conduct and document their experiences with public transport all over the country (passengerpulsen.taenk.dk). Once a year, they use their fact-finding to publish a customer service report, ranking the Danish bus and train companies by level of customer service. As a positive incentive, they hand out a prize (McHugh 2017a).

3 User-Centric Design

Possible objectives to let users participate in planning could be to mobilize creativity or to resolve conflicts and obtain agreement. 'The knowledge, experiences and ideas of users hold a creative potential which is normally not tapped by internal planning procedures' (Schiefelbusch and Schmithals 2005). While planning implies the allocation of scarce resources, participation can be used to develop a reasonable compromise and improve the users' understanding for the choices made. Hence, the acceptance of final decisions are greater than for those imposed from 'above'. And although participation of user groups usually does not include the actual decision-taking, it does recognise users as everyday-experts and their wishes and knowledge could directly flow into the design process. Such a form of direct customer orientation can make staff members learn to think and act in a more customer-oriented way. This is very important but not easy: 'Designing for usability is not about making something that's cool in the designer's book; it is about making something that works for the user (and that the user *may* find cool). That requires a lot of knowledge about the user group, about methods, about design techniques. And it requires the attitude of putting the user centre stage, while acknowledging that you are not the user' (Kuijk 2010).

Many organisations pay lip service to usability and user-centric design but seem at a loss how to achieve it. Some companies may find it difficult to change their developing processes while others do not recognize the benefits of involving users in the development process. However, active user involvement can be judged to be the number one criterion on how to be successful (Gulliksen et al. 2003).

User centered design is a development method that guarantees that a product or service will be easy to use. The principles and activities that underlie user centered design are:

1. The design is based upon an explicit understanding of users, tasks and environments.
2. Users are involved throughout design and development.
3. The design is driven and refined by user-centric evaluation.
4. The process is iterative.
5. The design addresses the whole user experience.
6. The design team includes multidisciplinary skills and perspectives.

'In other words, user-centric design is based upon an explicit understanding of users, tasks, and environments; is driven and refined by user-centered evaluation; and addresses the whole user experience. The process involves users throughout the design and development process and it is iterative. And finally, the team includes multidisciplinary skills and perspectives' (www.userfocus.co.uk). When taking into account user experiences, companies should focus on having a deep understanding of users, what they need, what they value, their abilities, and also their limitations. According to Peter Morville (semanticstudios.com), custom friendly services should be *useful* (fulfill a need), *usable* (easy to use), *desirable* (image, identity, brand, and other design elements are used to evoke emotion and appreciation), *findable* (easy to find), *accessible* (inclusive design or design for all) and *credible* (trustworthy).

3.1 User Involvement Throughout the Planning Process

Schiefelbusch identifies a range of typical subjects that arise more or less regularly in transport planning in which users may be involved somehow (Table 1).

Table 1 Main topics of public transport planning and user involvement

Physical design	Service availability	Commercial planning and controlling
Infrastructure design (routes)	Development of new concepts	Fares and ticketing conditions
Infrastructure design (stations)	Timetables and routes	Marketing activities
(Internal) vehicle design	Additional services (e.g. assistance for disabled users)	Formulating tender documents and assessment of submissions
	Temporary measures and changes to services	Monitoring of service quality

Source Schiefelbusch and Schmithals (2005)

Several methods and procedures of user participation are possible, depending on the subject and its character. For users it is easier to express their interests the closer the issue at stake is to their everyday experience. Less known or new topics or developments require more complex and possibly more long-term forms of participation than well-established ones. Detailed design asks for product clinics and usability tests, while surveys, focus groups and simulation studies can be used when the definition of a product is at stake (Schiefelbusch and Schmithals 2005).

Some examples of the involvement of users in the design cycle can be given here.

On the subject of fares and ticketing conditions, Transport Focus wants to involve passengers in the early stages of the development process. 'Passengers are increasingly coming to expect services to be delivered to them in smarter ways as technology becomes a bigger part of everyday life' (www.transportfocus.org.uk). The introduction of smarter ticketing in public transport could make life easier and cheaper for passengers. For this to happen though, it is essential that these smarter ticketing schemes are well designed and properly implemented. In order to make sure that products are designed for *ease of use* rather than what is *convenient to administer*, Transport Focus is working on a wide-ranging smart-ticketing research programme on behalf of the Department of Transport. Their first study in 2016 reveals that passengers are typically open towards proposed smartcard options as long as it reflects 8 core principles—value for money, convenient, simple, flexible, secure, tailored, leading edge, trustworthy—and if the necessary actions make sense to them (Transport Focus 2016).

In the Netherlands, the national train company Nederlandse Spoorwegen (NS) in 2016 and 2017 tested the internal vehicle design and the choice for train seats of a new train (Intercity New Generation or ICNG) together with a group of train passengers. They were selected from a passenger panel, though not randomly. They asked for tall and short people, thin and fat people, students and elderly people, commuters and incidental passengers, etc. This group was invited to test seats of different companies and write a report on it. Together with comments from personnel and cleaners, the company was able to choose a train seat that hopefully is acceptable and comfortable for the average passenger. A few months later, the same group of passengers (Trein

Test Team) was invited to give comments on a mockup of a train unit of the ICNG (Werf 2017).

In the summer of 2002, potential customers were involved in the planning phase of a new idea for a demand-responsive shared-taxi service in Berlin, CharterCAB. Residents and citizens' groups contributed their knowledge about the region and the wishes and needs for mobility services. This feedback helped to define specific target groups and to develop a customized service, which was tested by some 100 residents before it became official. Experiences and results of the tests were evaluated in workshops where staff of the operating company together with some of the participants refined routing, a fare system and a service and marketing strategy. After that, the service operated on a trial basis for another 15 months. During this time, user feedback was still part of the process. It is a good example of short-term participation of users, resulting in a reduction of development time, the saving of costs and providing the company a competitive edge in the market (Schiefelbusch and Schmithals 2005).

3.2 Recommendations for a Participatory Approach

Several elements of a successful participatory approach can be mentioned here. First of all, *power*: make sure there are some formally guaranteed possibilities for users to participate in the planning process. Secondly, *flexibility*: the range of instruments at hand should cover all planning issues (ranging from detailed reading of proposals to open forum discussions and from small-scale problems to bigger ones). Thirdly, *timing*: participation should take place at the right time, i.e. leave those consulted enough time to assess the question. Also, the receiving party should have enough time left to do something with the advice given. Fourth, *forming of opinion*: different people (users, passengers) have different needs and wishes, so different opinions are unavoidable. But, so are compromises. Nevertheless should there be room to make differences of opinion transparent. Last (but not least), *representation*: different users (age, abilities, habits, motives for travelling, etc.) should be represented in the planning process. Also, a balance must be found between a rotary schedule in order to introduce new ideas at a regular basis (or avoid becoming 'too professional') and the need to build up competencies in order to follow developments over a longer period. (Schiefelbusch and Schmithals 2005).

4 Getting to Know the Customer

Another way of 'involving' users is doing research on (specific) customer groups or perform a questionnaire survey in order to collect empirical data which can be used to improve public transport in a specific region or for certain customer groups. It can also be an inspiration for new ideas and for innovative design which users approve of and will be comfortable with. Several examples of such research and results can be presented here.

For instance, in 2008 in Taiwan a survey was done to find out why actual ridership of a specific public transport system was lower than expected. Around 800 passengers filled in a questionnaire which examined service quality, perceived value, overall satisfaction, public transport involvement, behavioral intentions as well as the respondents' demographic and travel behavior. More than 60% of the respondents were female. The study revealed that several service attributes influenced passenger behavioral intentions. At the same time, in order to satisfy passengers and create customer loyalty, providing passenger-value-oriented quality services is even more crucial. Thus, push (offering quality) and pull strategies (marketing: eye-catching advertisements, celebrity endorsements) are equally important. 'When people are satisfied with the public transport, their involvement towards it will increase' (Lai and Chen 2011).

Within 7 communities in Scotland in 2010 travel behavior research was used to plan smarter and more sustainable travel (i.e. walking, cycling and public transport). In total over 12,000 households were surveyed door-to-door, more than 3000 people were approached by telephone, 14 focus groups were established (150 persons) and 70 in-depth interviews were undertaken in order to investigate whether people could be encouraged to make sustainable travel choices. The research showed that, in order to change travel behavior effectively, marketing and promoting techniques need to be appropriately targeted. Therefore, the local communities involved need to know the aspirations of their inhabitants. For 'the people who might benefit most from some interventions can be the least likely to pay attention to new information or initiatives'. Some keys to success of changing travel behavior are: accelerate change where travel behavior is already changing (less car use, more walking and cycling and use of public transport), target people who are both willing and able to reduce car use, work with broader lifestyle changes (health issues, families with growing children), alter the perception of travel costs, raise expectations of bus services and improve the (perceptions of) safety of walking, cycling and bus use (Halden et al. 2010).

In 2014, both the European Union and the Civitas Wiki Consortium presented a report on gender equality and mobility within Europe. Both reports use several studies and literature as well as a huge amount of empirical data from different European countries as a foundation for their analyses. Civitas focusses on urban mobility. 'Achieving the target of sustainability in urban mobility also means considering the needs of different users and thereby offering equal levels of accessibility to transport to all different groups.' Both the Consortium and the European Union argue that, although there still is a lack of knowledge of gender issues and while gender mobility data and statistics are scarce, the gap between genders is evident and has effects on mobility patterns: 'women travel differently than men in relation to transport modes used, distance travelled, the daily number of trips and their pattern and they also travel for different purposes.' The Consortium comes up with some gender-sensitive policy recommendations that the European Union supports as well: women should participate in decision-making; the accessibility, safety and comfort of transport modes should be improved; the planning of transport services should be gender-sensitive: affordable and flexible fares for multi-trips, promotion of dedicated services, marketing, better provision of transport information, especially for non-rush hour trips. 'An important driver in this process could be the fact that women more likely than men support or accept sustainability and green economy policies as they appear to be more sensitive to environmental risks and more prepared to behavioral changes.' Moreover, better transport for women benefits all (Civitas Wiki Consortium 2014; European Union 2014).

In the Netherlands, researchers of the Vrije Universiteit Amsterdam did research on customer satisfaction in public transport in 2015. They used data of a yearly survey of public transport users in the Netherlands (Klantenbarometer 2015), a sample of more than 50,000 male and female passengers of different age groups (10–17, 18–27, 28–40, 40–65, >65) from rural as well as urbanized areas and cities. They found out that there is a difference between the appreciation of public transport services of passengers in cities and passengers in rural areas. In cities, customer satisfaction in general is significant lower compared to rural areas. This outcome seems to be related to the fact that in cities, passengers may find it more difficult to find a place to sit (public transport is more crowded) and (therefore?) they give lower rates when asked about the friendliness of staff and are less satisfied with the fare price. On the other hand, public transport users in cities are more satisfied compared to rural public transport users with the frequency provided and the information when a service is disrupted. So, customer research is important for public transport companies in order to find out what people want and what differences can be found between different users in different areas. Furthermore, it is important to know how different customers perceive several service attributes differently. In that way, measures taken to improve public transport will be more user centered and thus more effective (Mouwen 2017).

In 2016, user interest group PRO BAHN organised passenger workshops in order to find out about their needs and wishes for the timetable, vehicle facilities and quality of the services provided by local trains and buses in the region of Augsburg in Germany. They used social media to 'recrute' the passengers, which attracted also a lot of young people. The outcome was a report, which they successfully used to lobby for improving public transport from the passengers' point of view.

5 Putting Passengers First

According to Alain Flausch, Secretary General of UITP, "placing customers first is vital if we want to achieve our goal to double the public transport modal share by 2025". Although customer expectations are rapidly evolving and ever-rising, "delivering the right basic experience to customers should be a priority; efforts to surprise and delight them should come later" (Flausch 2017).

Although there has been more focus on passengers in the past 5-10 years, Lone Fruerskov Andersen of Passenger Pulsen thinks there is room for improvement. "There is an increasing awareness that the aim is not only to drive a bus or a train—it is to transport passengers. But they should be looking more into the plans with the eyes of passengers—to take them by the hand and see *how would this fit into my world?*" (McHugh 2017a).

In the end, the arrival of technology-driven mobility services is changing the way transport is operated, viewed and consumed and provides opportunities for public transport to better meet the demands and experiences of consumers. In Kansas City for instance, Ride KC offers an on-demand service for its city's public transport users. "Buses typically run on a fixed route going in circles, regardless of demand. We take a different approach, using the technology available to us. Instead of the user having to catch the bus, in essence the bus is catching the user in adapting to these demand patterns", says Matt George of transport provider Bridj (McHugh 2017b).

Such public-private partnerships evidently will grow in the future, where innovative technology meets long-established systems to better provide the best service for customers and really become user-centric.

Acknowledgements The author wishes to thank the consortium of the Mobility4EU project for the intense and fruitful collaborations and the European Commission for the funding of the Mobility4EU project in the framework of the Horizon 2020 program (EC Contract No. 690732).

References

Civitas Wiki Consortium (2014) Smart choices for cities. Gender equality and mobility: mind the gap! Policy note

Dijst M, Rietveld P, Steg L (2013) Individual needs, opportunities and travel behavior: a multidisciplinary perspective based on psychology, economics and geography. In: Wee van B, Annema J, Banister D (eds) The transport system and transport policy pp 19–50

European Union (2014) She moves. Women's issues in transportation

Flausch A (2017) Putting passengers first. Public Trans Int 66–1:3

Gulliksen J et al (2003, November–December) Key principles for user-centred systems design. Beh Info Technol 22(6):397–409

Halden D et al (2010) Changing the mindset using travel behavior research to plan smarter travel in Scotland. Association for European Transport and contributors

Imre S, Çelebi D (2017) Measuring comfort in public transport: a case study for Istanbul. Trans Res Procedia 25C:2445–2453

Kuijk van J (2010) Recommendations for usability in practice. www.designforusability.org

Lai W, Chen Ch (2011) Behavioral intentions of public transit passengers—the roles of service quality, perceived value, satisfaction and involvement. Trans Policy 18:318–325

Maartens M (2017) Handboek reizigersinvloed. Wat je moet weten om het regionale openbaar vervoer te verbeteren, Rover

McHugh K (2017a) Watching out for passengers. Public Trans Int 66–1:22–24

McHugh K (2017b) Customer needs and integrated new mobility services. Public Trans Int 66–1:32–35

Mouwen A (2017) Reizigerstevredenheid in stedelijke gebieden. Tijdschrift Vervoerswetenschap 53–1:3–18

Schiefelbusch M, Schmithals J (2005) Customer participation in public transport planning: conceptual issues and experiences. Association for European Transport and contributors

Transport Focus (2016) Using smartcards on rail in the south east of England: what do passengers want?
van der Werf I (2017) Lekker zitten is heel belangrijk. De Reiziger 40–1:22–24
Wardman M (2014) Valuing convenience in public transport. Roundtable summary and conclusions, OECD Discussion Paper 2014-2

Part III
Improving Urban Mobility

Mobility Planning to Improve Air Quality

Lluís Alegre Valls

Abstract Mobility using vehicles with internal combustion engines is one of the major sources of pollutant emissions in urban areas the world over. This article reviews some relevant actions of cities to reduce such emissions and proposes strategies to improve their effectiveness. One particularly important challenge is the electrification of buses, requiring an infrastructure plan to ensure the fleet can operate. The metropolitan region of Barcelona has mobility planning and also a plan to improve air quality. This article examines their compatibility and proposes a joint action to make the most of synergies and achieve the targets in terms of reducing pollution. The urgent need to improve air quality also demands commitment on the part of all parties involved to act together and agree future actions to be carried out.

Keywords Mobility · Health · Public transport · Emobility · Electrification
Pollutants · Climate change · Regional planning

Abbreviations

ZeEUS	Zero Emission Urban Bus System
pdM	Mobility Master Plan
PIRVEC	Strategic plan for the deployment of charging infrastructure for the electric vehicle in Catalonia
P&R	Park and Ride
DKV	Insurance policies
LEZ	Low Emission Zones
WHO	The World Health Organisation
CREAL	Centre for Research in Environmental Epidemiology

L. Alegre Valls (✉)
The Metropolitan Transport Authority of Barcelona (ATM), Muntaner Street 315-321, zip code 08021 Barcelona, Spain
e-mail: lalegre@atm.cat

B. Müller and G. Meyer (eds.), *Towards User-Centric Transport in Europe*,
Lecture Notes in Mobility, https://doi.org/10.1007/978-3-319-99756-8_9

1 The Environmental Impact of Daily Mobility

The relationship between motorised vehicles and air quality has been examined quite extensively over the past few years from the perspective of vehicle emissions (Ajuntament de Barcelona i Barcelona Regional 2010; EEA European Environment Agency 2012; European Environment Agency, EEA 2014; Institut Cerdà 2014). However, when it comes to establishing the relationship between mobility and immission values (the emissions that actually affect people), other factors are also involved, such as the climate and physical environment, making it difficult to establish a direct correlation between mobility and pollution levels (Leicester City Council 2007). This is because, since the air does not respect the municipality's boundaries, pollutant emissions from vehicles spread everywhere.

Regarding mobility in Europe's metropolitan regions, peak travel appears in city centres. Every day, hundreds of thousands of people go to and from the city centre, with private vehicles or public transport, mostly using highly pollutant combustion engines. Figure 1 illustrates this factor for the example of Barcelona. Every day more than 1 million commuters go to Barcelona half in private cars and half in public transport (ATM 2016).

In the Barcelona Region the mobility is largely work-related. On the one hand, the number of jobs available in many municipalities is lower than the number of

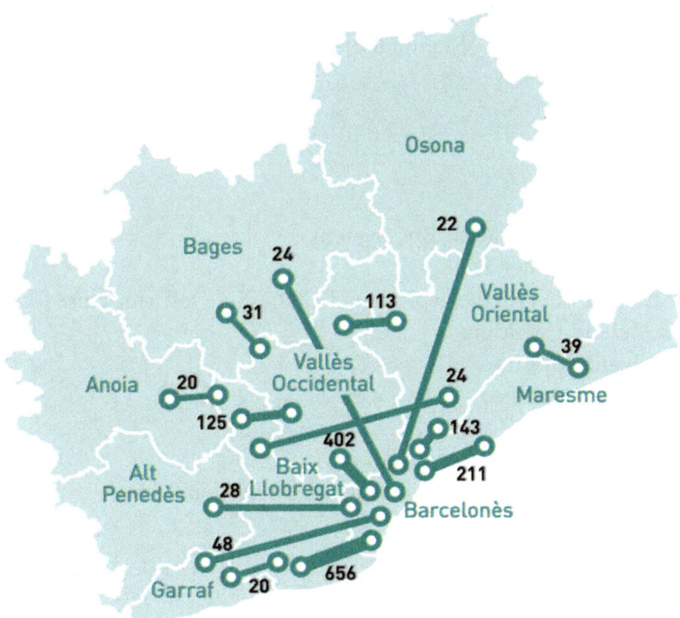

Fig. 1 Main daily journeys between counties in the metropolitan area of Barcelona. (In thousands) (ATM 2016)

people of working age. On the other, those jobs available are not always suitable for many of the city residents' professional profiles. This results in highly concentrated movement to and from city centres (Institut d'estudis territorials 2012).

Motorised vehicles therefore come from areas outside the city into the centre. This mobility is in addition to the mobility of the urban area itself, which is intensive in energy consumption and in emissions of harmful pollutants for human health and greenhouse gases (Aphekom 2008; Ostro et al. 2011; Sevillano and Balsells 2011; Science for Environment Policy 2013; Science for Environment Policy 2013; Science for Environment Policy 2014). There is also another factor related to environmental quality, namely the noise associated with the large number of combustion vehicles being used. Motorbikes and cars produce noise far above the levels permitted by law, affecting people's quality of life, both in urban and interurban areas (Practitioner handbook for Local Noise Actions Plans 2007).

To ensure mobility free from pollutant emissions, it is therefore necessary to reduce the circulation of combustion vehicles by implementing different strategies. For example, encouraging a technological shift towards electric vehicles, increasing the average vehicle occupation rate, restricting the circulation of the most pollutant vehicles and, especially, by encouraging a modal shift towards mobility on foot, by bicycle and public transport (Barcelona 2015; Tackling air pollution in Berlin 2010).

2 The Contribution of Public Transport to Improving Air Quality

Public transport guarantees people's right to universal access irrespective of their social status. It is therefore a fundamental factor for the well-being of our society. Ensuring the most disadvantaged districts have public transport allows their residents to enjoy the right to travel to those places where there are more chances of finding work. But it is also an essential tool to reduce the environmental impact of mobility and improve urban air quality for all the city's inhabitants (EEA European Environment Agency 2012). If a journey cannot be made on foot or by bicycle, we therefore need to ensure it can be made by public transport because this has clear advantages over private vehicles.

In the metropolitan region of Barcelona, the energy consumed in kilos of oil equivalent per person transported in one thousand kilometres of a railway is, on average, 8.5 kep/10^3 km, 2.7 times less than the average consumption of a bus and 6 times less than cars (Institut Cerdà 2014). These figures are even more compelling in terms of pollutant emissions. A train's nitrogen oxide emissions per person transported over one km of distance are, on average, 20 times less than a bus and 50 times less than a diesel car (ATM 2018b).

Such values can even double at peak times, when there are traffic jams and public transport vehicles are also more heavily occupied. Speed-emissions curves are parabolic (see Fig. 2) and emission factors increase rapidly at very low speeds. This is one of the reasons for ensuring a minimum speed for buses during peak times

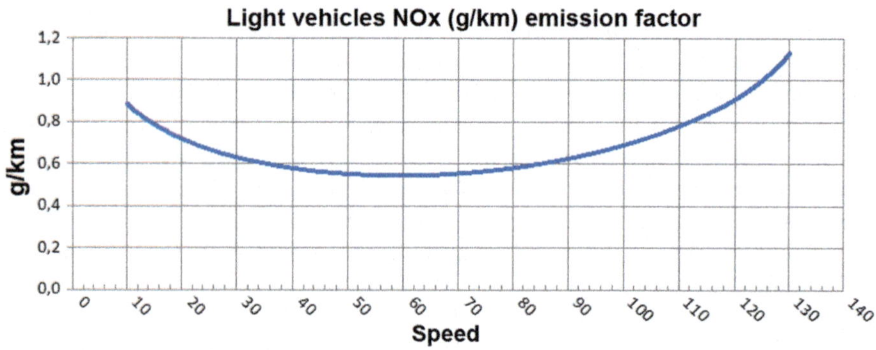

Fig. 2 NOX emission factor for vehicles (vans, cars) (Institut Cerdà. Seguiment de l'evolució de la mobilitat i les emissions de gasos d'efecte hivernacle i contaminants a la Regió Metropolitana de Barcelona 2014)

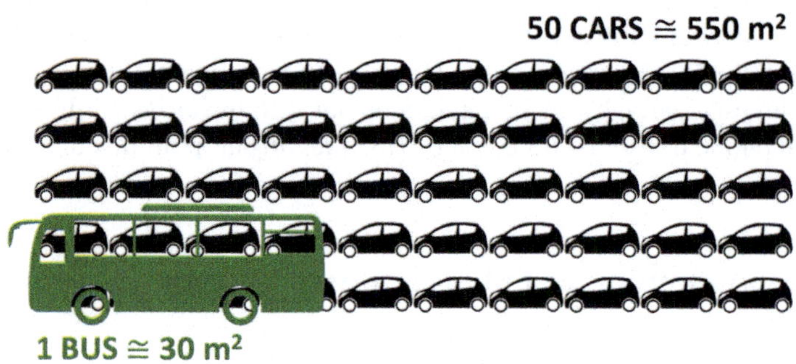

Fig. 3 Comparison of the area taken up by buses and cars (image by Jordi Martin)

by helping them to use bus lanes and giving them priority at green lights. Public transport also takes up less room than cars, freeing up space on streets, squares and roads that can be used to widen pavements, add bike lanes or increase urban green and biodiversity (see Fig. 3).

3 The New Challenge Posed by Electrification of Public Transport: The Bus Fleet

Public transport has great potential to support vehicle market transformations as it has already done in the past (European Commission 2013). Bus fleets have included hybrid and gas vehicles for several years now and existing vehicles have even been adapted by modifying their diesel engines to make them operate as hybrids. This pioneering approach is once again coming to the fore with the first trials of electric buses in the fleets of many European cities.

The European Commission has published "Electrification of the Transport System, Studies and Reports" (Directorate-General for Research and Innovation 2017), detailing the main trends and developments in the electrification of transport and proposing a strategic implementation plan. It also funds projects to implement electric vehicles, as in the case of the ZeEUS project (ZeEUS Zero Emission Urban Bus System 2016), with the aim of placing this technology at the heart of urban bus transport networks.

Bus fleet electrification represents a huge challenge for the transport sector. Firstly, because high initial investment is required to set up the electricity infrastructure and secondly, because electric buses are not in widespread production and the purchasing cost of electric vehicles is more expensive than conventional diesel or hybrid models. It should also be noted that bus lines can become less flexible since they require fixed charging stations that can be used by vehicles while these are in service. The customary solution for these charging stations is to incorporate a pantograph that charges the vehicle via the roof. Nevertheless, other solutions are also being examined, such as inductive charging, which has the advantage of less visual impact on the environment although the initial investment is much higher at present. Both technologies are illustrated in Fig. 4.

In any case, the electrification of the public transport bus fleets requires at least a regional plan that guarantees a sufficient electricity supply from renewable sources, that sets up synergies between charging stations for different transport modes so that they can be shared, making the infrastructure and supply more efficient, as well as ensuring any new consumption requirements resulting from this radical change in the transport system can be covered. Such planning must be incorporated within all sector plans concerning mobility infrastructures and within metropolitan and urban mobility plans.

In the case of the metropolitan region of Barcelona, the current Mobility Master Plan 2012–2018, (pdM) (ATM 2018c) already includes several measures related to promoting electric vehicles, very much in line with the recent strategic plan to deploy charging infrastructures for electric vehicles in Catalonia 2016–2019 (PIRVEC) (ICAEN. PIRVEC 2016) and the European Clean Bus Deployment Initiative. At present, a large number of European cities are electrifying their urban bus networks but, very soon, electric buses will also be used for metropolitan networks. In the medium term, regional bus networks will also operate with electric vehicles and plans need to include a future when even private coaches are electric.

The electrification of the bus network will considerably reduce the energy consumed per vehicle-km driven, bringing this down to the same level as the railway network in terms of its contribution to improving air quality and reducing greenhouse gases.

Fig. 4 Buses with pantograph charging system and with inductive charging system. (Photos and images. Source TMB and author. Images arranged by Jordi Martin)

4 Mobility Planning of the Conurbation of Barcelona—Good Practice Example

The Metropolitan Transport Authority of Barcelona is the institution in charge of drawing up the Mobility Master Plan for the metropolitan region. This Plan covers all transport modes, both for passengers and goods, in its strategic planning. The aim of the Plan is to encourage journeys using non-motorised modes and public transport, in line with the principles and goals contained in Catalonia's mobility legislation. The Plan establishes 9 core areas of action and 75 measures, many of which are related to improving air quality and people's health (ATM 2018d).

Energy is one of the areas most directly related to pollutant emissions and air quality and the regional strategy endorses actions that help to incorporate electric charging stations throughout the territory, proposing localizations such as P&R and network connection points with the largest circulation of vehicles, if possible with a format similar to traditional petrol stations. It also promotes the incorporation of a smart electricity network to handle the challenges of efficiently managing energy supply and consumption by vehicles, for example, the difference in peak demand. If the 2.1 million cars in the Metropolitan Region of Barcelona were efficiently charged, it would result in a 5000-MW reduction in the peak demand (Red eléctrica 2018).

In terms of air quality and public health, one of the core areas of the strategy and proposals is to widen the scope of those actions carried out to improve environmental quality beyond municipalities that do not comply with the maximum limits of pollutant emissions established in European legislation for particles and nitrogen oxides. The Catalan Government and the Barcelona Municipality drew up plans to improve air quality in Barcelona 2015–2017. The aim of all these plans is to establish air quality targets in areas that generate mobility towards the most polluted zones and set actions to achieve them.

This creates positive synergy in terms of achieving the goals set. On the one hand, areas with a high concentration of pollutants, with colour dark green in the Fig. 5, come under restrictions for private vehicles that affect the vehicles entering from outside (see Fig. 5). On the other hand, those municipalities creating traffic towards zones where emissions need to be reduced also provide support by promoting public transport that connects with high capacity railway networks and by making it easier to park and ride. This effect is further reinforced if other measures, such as encouraging car sharing and ridesharing, working from home or eliminating the circulation of the most pollutant vehicles, are carried out throughout the region, ensuring that external mobility towards the central conurbation is also involved in reducing pollutant emissions.

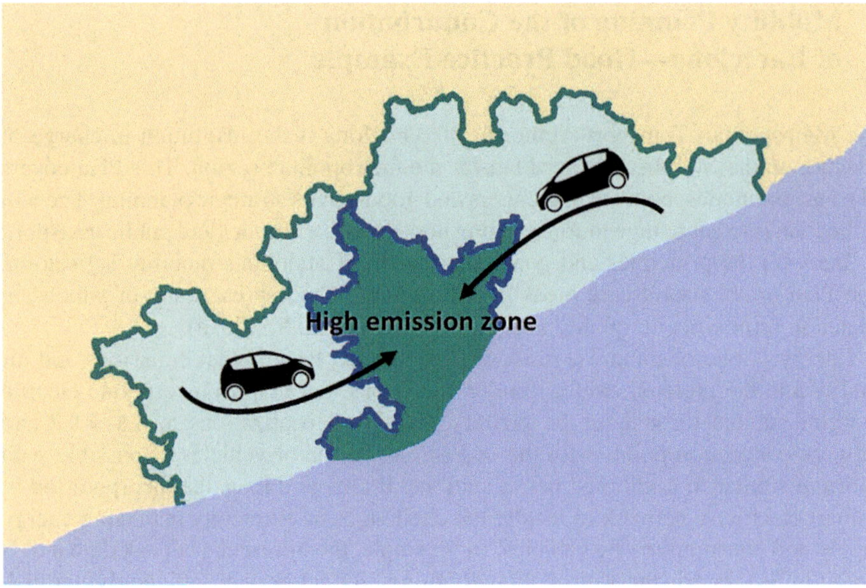

Fig. 5 Scope established by the action plan to improve air quality in the metropolitan region of Barcelona (image by Jordi Martin)

5 Mobility Measures to Improve Public Transport and Reduce Emissions

Regional and urban mobility plans play a crucial role in improving air quality and some have a huge impact and that's why many European networks foster them (EMTA European Metropolitan Transport Authirities 2018; POLIS 2018). The institutions of the Barcelona ATM are engaged in some of them and they foster measures to reduce the environmental mobility impacts.

In the area of planning, the Catalan government approved the Decree 344/2006 that establishes that studies assessing the mobility generated by town planning and large commercial and other development plans must now include an environmental assessment and also carry out actions to correct traffic emissions. Over the last ten years, more than 600 new town planning and other specific implementations have been carried out in the metropolitan region, with an area of 12,200 ha (1.2 times the municipality of Barcelona) (ATM 2018a). These mobility plans contain measures to prioritise transport modes on foot, by bicycle and public transport, as well as installing charging stations for electric vehicles, all with the aim of guaranteeing greater environmental quality in urban areas.

In the area of energy, the fleet of vehicles needs to be transformed, incorporating increasingly efficient cars and vans free from fossil fuels, with strategies to encourage both private individuals and companies to choose vehicles that minimise consumption

Fig. 6 New T-verda (T-green) (image by ATM Barcelona)

and emissions of pollutants and greenhouse gases during their journeys. These new habits and behaviours can be consolidated by incentives as discounts on tolls, easier parking and reductions in some taxes and actions that reward those who commit to this change by means of a "clean fleet" symbol.

A report by the medical insurance DKV (Ecodes Ecología y Desarrollo and Observatorio DKV de Salud y Medio Ambiente en España 2010) has stated that traffic needs to be reduced significantly, by more than 30%, for any improvements in health to be noticed in Spain. To achieve this goal, mobility management must promote a change in transport modal from private vehicle to the rest. Consequently, the metropolitan region of Barcelona encourages owners of older vehicles, once these are withdrawn from circulation, not to buy another car by giving them, in exchange, a free integrated public transport ticket that is valid for 3 years, called T-verda (T-green) (Fig. 6). Another action producing good results is the creation of a green metered zone. In this case, street parking places can only be occupied by residents and not by vehicles from outside the district, thereby reducing journeys by private vehicle for work.

In the area of regional public bus transport, fast buses are required with good connections with the rest of the networks, without forgetting the fundamental role of citizen information. This network needs to be accompanied by infrastructures and services that guarantee reliable journey times and punctuality.

Agreements are particularly required in the area of goods transport since this sector uses mostly private vehicles. For example, the aim of the logistics committee for the metropolitan region of Barcelona is to achieve a more efficient and sustainable logistics system and thereby reduce pollutant emissions. It also promotes a change in technologies towards cleaner vehicles and measures to increase the load transported per kilometre driven, as well as moving goods transport outside the peak time when people travel.

Finally, regional planning must encourage innovation and new technologies to be included in all modes of transport and throughout the metropolitan region, so that these are not limited to the most densely populated part of the city. Not only do new supports need to be promoted for contactless or mobile-based public transport tickets but this must also be extended to all journeys made by passengers. Consequently, over the coming years other modes of mobility need to be incorporated, enabling modal interchange between all types of mobility and with public transport.

6 Challenges of the Future

As already noted, mobility planning and plans to improve air quality share goals in terms of reducing emissions of pollutants and greenhouse gases associated with the consumption of fossil fuels (as well as noise pollution), thereby also improving air quality and protecting people's health. For example, the Mobility Master Plan (pdM) for the metropolitan region of Barcelona in agreement with the air quality plan of the region established environmental operational indicators regarding energy consumption and pollutants for the next 6 years, as the Table 1 shows.

However, in-depth knowledge of the trend in pollutant immissions is also required. This is why data from the measurement stations must be monitored on a daily basis. Over the last few years, in many European cities the number of measurement points exceeding maximum particulate levels has tended to fall, but not for nitrogen oxide emissions, especially NO_2. This is largely due to diesel vehicles not complying with the values established by the European Commission (Science for Environment Policy 2013; Science for Environment Policy. New generation diesel cars are likely to exceed emissions standards on the road. 10 January 2013).

Table 1 Operational indicators of the Mobility Master Plan 2013–2018 (ATM 2018b)

Operational indicators of the Mobility Master Plan 2013–2018	2012	2018 pdM	pdM forecast (%)
Reduce the consumption and energy intensity of transport	*List*		
Total energy consumption (thousands toe/year)	1801	1642	−8.80
Reduce mobility's contribution to climate change			
CO_2 emissions (thousand tonnes/year)	5106	4442	−13.00
Reduce mobility's impact on the air			
PM10 emissions (tonnes/year)	1754	1351	−11.30
NOx emissions (tonnes/year)	5397	4641	−4.60

Therefore, since the non-compliance of pollutant emission limits in many European cities is closely related to the emissions produced by motorised vehicles, hundreds of cities are creating low emission zones (LEZ). In these zones, cities impose restrictions on the circulation of pollutant vehicles and other measures to manage mobility in order to improve air quality for their inhabitants. The website "Urban Access Regulations" lists the cities carrying out low emission zone initiatives (http://urbanaccessregulations.eu).

The complexity of carrying out, within a short period of time, the actions required to ensure pollutant emissions are reduced, has led the local governments of the metropolitan region of Barcelona and the Catalan government to sign a political commitment to improve air quality (Catalan Goverment 2017). In this document, each institution undertakes to carry out specific actions with dates for their execution and which, in many cases, must be implemented by all of them together. Because of this, the initiative has set up a working group that coordinates the actions to be carried out in cases of high environmental contamination. In the case of public transport, a protocol has been signed and a whole procedure created that guarantees the necessary resources are provided within a maximum period of 48 h, whenever it's necessary to establish exceptional restrictions on private vehicles.

The World Health Organisation (WHO) proposes stricter pollutant immission limits than those currently established by the European Commission. A scientific study carried out by CREAL (Laura Pérez, Jordi Sunyer and Nino Künzli. Centre de Recerca en Epide-miologia Ambiental (CREAL), Institut Municipal d'Investigació Mèdica (IMIM), Centro de Investigación Biomédica en Red de Epidemiología y Salud Pública (CIBERESP), Institut Català de Recerca i Estudis Avan çats (ICREA) 2009) states that there would be 3500 fewer premature deaths in 57 cities the size of Barcelona if current pollution levels were reduced to those established by WHO. For this reason, all the actions presented in this article are merely the beginning of a quest to enjoy cleaner, healthier air in cities and their surrounding areas. Over the coming years, more ambitious targets will be set that must be incorporated in regional, metropolitan and urban mobility plans.

Significant changes will have to be made to public transport for it to make a more decisive contribution to improving air quality. In addition to issues regarding energy and the adaptation of supply to growing demand, other fundamental aspects of planning will be efficiently combining different modes of transport in a single journey, more flexible timetables and routes, applying transport system tariffs and, especially, encouraging greater involvement to raise awareness of the collective advantages of improving air quality, with citizens taking decisions aimed at achieving more efficient and emission-free journeys. Tackling such challenges decisively and together will surely result in cleaner air for future generations.

References

References in Tet

Aphekom. Improving Knowledge and Communication for Decision Making on Air Pollution and Health in Europe. Summary report of the Aphekom project 2008–2011
ATM (2016) Daily mobility survey. Enquesta de mobilitat en dia feiner EMEF
ATM (2018a) Barcelona https://www.atm.cat/web/ca/eamg.php
ATM (2018b) Master Plan for Mobility (pdM). https://doc.atm.cat/en/_dir_pdm2013-2018/Executive-summary/files/assets/basic-html/page-1.html
ATM (2018c) Master Plan for Mobility (pdM). https://doc.atm.cat/en/_dir_pdm2013-2018/Executive-summary/files/assets/basic-html/page-33.html
ATM (2018d) Master Plan for Mobility (pdM). https://doc.atm.cat/en/_dir_pdm2013-2018/Executive-summary/files/assets/basic-html/page-25.html
Ajuntament de Barcelona i Barcelona Regional. Barcelona city council. Avaluació de la reducció d'emissions de NOX i PM10 dels vehicles que circulen per la ciutat de Barcelona, en base a la caracterització del parc mòbil de la ciutat. Novembre 2010
Barcelona council Plan to improve air quality in Barcelona 2015–2018 https://ajuntament.barcelona.cat/qualitataire/sites/default/files/pdfs/PMQAB_EN_2014.pdf (2018)
Catalan Goverment http://mediambient.gencat.cat/es/detalls/Noticies/cimera-qualitat-aire (2017)
Directorate-General for Research and Innovation. Smart Green and Integrated Transport. European Commission Directorate General for Research and Innovation. Electrification of the Transport System. Studies and reports (2017)
Ecodes Ecología y Desarrollo, DKV Seguros Médicos. Observatorio DKV de Salud y Medio Ambiente en España 2010. Estado de la cuestión: Contaminación atmosférica y salud. Núm. 2 (2010)
EEA European Environment Agency. The contribution of transport to air quality TERM 2012: transport indicators tracking progress towards environmental targets in Europe. Report No 10/2012
EMTA European Metropolitan Transport Authirities http://www.emta.com/ (2018)
European Commission. Communication from the Commission to the European Parliament, the Council, The European Economic and Social Committee and the Committee of the Regions: Clean Power for Transport: A European alternative fuels strategy (2013)
European Environment Agency, EEA. Adaptation of transport to climate change in Europe. Challenges and options across transport modes and stakeholders. Report. No. 8/2014
ICAEN. PIRVEC. http://icaen.gencat.cat/ca/plans_programes/pirvec/ (2016)
Institut Cerdà. Seguiment de l'evolució de la mobilitat i les emissions de gasos d'efecte hivernacle i contaminants a la Regió Metropolitana de Barcelona 2014. Master Plan for Mobility (pdM).Website: https://doc.atm.cat/ca/_dir_pdm_estudis/emissions_gasos_qualitat_aire_2014/files/assets/basic-html/page-1.html
Institut d'estudis territorials. Els elements territorials de la regió metropolitana de Barcelona i la seva relació amb la mobilitat (2012)
Laura Pérez, Jordi Sunyer and Nino Künzli. Centre de Recerca en Epide-miologia Ambiental (CREAL), Institut Municipal d'Investigació Mèdica (IMIM), Centro de Investigación Biomédica en Red de Epidemiología y Salud Pública (CIBERESP), Institut Català de Recerca i Estudis Avan çats (ICREA). Estimating the health and economic benefits associated with reducing air pollution in the Barcelona metropolitan area (Spain) (2009)
Leicester City Council, North East South West Interreg IIIC, Cite air. Air Quality Management Guidebook (2007)

Ostro B, Tobias A, Querol X, Alastruey A, Amato F, Pey J, Pérez N, Sunyer J (2011) Research: the effects of particulate matter sources on daily mortality: a case-crossover study of Barcelona, Spain. Centre for Research in Environmental Epidemiology, Barcelona, Spain. Institute of Environmental Assessment and Water Research, Spanish ResearchCouncil, Barcelona, Spain. Environmental Health Perspectives, vol 119, Number 12. December 2011

Red eléctrica http://www.ree.es/sites/all/SimuladorVE/simulador.php (2018)

POLIS Cities and Regions for Sustainable Transport https://www.polisnetwork.eu/ (2018)

Practitioner handbook for Local Noise Actions Plans. Silence Project. EU. Sixth framework programme (2007)

Science for Environment Policy. Diesel cars' climate impacts not as beneficial as believed, scientists conclude. 5 December 2013, Issue 353

Science for Environment Policy. Health impacts of air pollution: the evidence reviewed. 8 May 2013, Issue 327

Science for Environment Policy. Living close to heavy traffic strongly linked to heart disease deaths. 30 May 2013, Issue 330

Science for Environment Policy. New generation diesel cars are likely to exceed emissions standards on the road. 10 January 2013, Issue 312

Science for Environment Policy. PM2.5 air pollution strongly linked to increased risk of heart attacks. 1 May 2014, Issue 370

Sevillano EG, Balsells F (2011) La contaminación de los diésel resta 13 meses de vida a los barceloneses. Barcelona es una de las ciudades más perjudicadas por la polución, según un estudio europeo - Los coches generan asma al 25% de los menores de edad. Madrid, Barcelona. Marzo

Tackling air pollution in Berlin: the low emission zone & other transport related measures. Martin Lutz, Senate Department for Health, Environment and Consumer Protection Directorate III, Environment Policy. Berlin Senatsverwaltung für Gesundheit, Umwelt und Verbraucherschutz (2010)

Website: http://urbanaccessregulations.eu (2018)

ZeEUS Zero Emission Urban Bus System (2016) ZeEUS eBus Report An overview of electric buses in Europe

Car Sharing as an Instrument for Urban Development

Conceptual Framework and Simulation on Interactive CityScope Tables

Jörg Rainer Noennig, Lukas Schaber, Jochen Schiewe and Gesa Ziemer

Abstract Starting from the assumption that development of urban areas is linked to mobility and access to transportation, the paper investigates car-sharing services as an active instrument for urban development. It hypothesizes that purposeful implementation of car-sharing invigorates "sleeping" urban areas, and potentially turns them into places of high urban and real estate value. For testing this hypothesis, a custom made simulation software was designed for application on the interactive modeling table CityScope. The tool enables the dynamic analysis of urban accessibility in the form of heatmaps and lines out with a special "Impact Mapping" feature the change of accessibility through car-sharing. Using demographic and mobility data of the city of Hamburg, the paper demonstrates how the accessibility of cities can be analyzed with the tool, and how appropriate areas for car-sharing services can be identified.

Keywords City science · Urban analysis · Urban data · Impact tool
Visualisation · Urban upgrading · Value creation · Accessbility analysis
CityScope · GIS · Spatial reference system · Dijkstra-algorithm
Interactive model · Use cases · Hamburg · Population density · Supply quality
Intelligent transportation systems · ITS

J. R. Noennig (✉) · G. Ziemer
CityScienceLab, HafenCity University Hamburg, Hamburg, Germany
e-mail: joerg.noennig@hcu-hamburg.de

G. Ziemer
e-mail: gesa.ziemer@hcu-hamburg.de

L. Schaber · J. Schiewe
g2lab, HafenCity University Hamburg, Hamburg, Germany
e-mail: lukas.schaber@hcu-hamburg.de

J. Schiewe
e-mail: jochen.schiewe@hcu-hamburg.de

B. Müller and G. Meyer (eds.), *Towards User-Centric Transport in Europe*,
Lecture Notes in Mobility, https://doi.org/10.1007/978-3-319-99756-8_10

1 Introduction

1.1 Urban Challenge

The presented work was conducted at HafenCity University Hamburg (HCU), an academic institution dedicated to studies of the metropolitan and built environment. Over a period of one year, researchers from two labs at HCU—the CityScienceLab and the g2lab Lab for Geoinformatics and Geovisualization—carried out focused research, thesis projects, expert talks and workshops on the topic of urban mobility.

Taking an instrumental approach, this research positions itself at the intersection of mobility research, spatial analysis, and interactive information technology. Relating to recent developments especially in the field of city science (M. Batty et al. at UC London, K. Larson et al. at MIT Media Lab) the research aims to provide application-oriented instruments in support of data-based urban development.

The investigations centered around the question how urban mobility enables access to social, commercial, cultural and other amenities. Good accessibility is a key factor for livable cities and well-functioning urban districts. For densely populated urban areas the priority task is to make the variety of amenities efficiently reachable to as many as possible citizens: "A location has good accessibility if many attractive activity locations can be reached with little effort" (Schwarze 2015). Not all urban areas, however, have sufficient transportation access and mobility infrastructure. This paper therefore such addresses "sleeping" urban areas—places that are hard to reach by the existing means of transportation, and therefore remain inaccessible and underdeveloped. In any city, supposedly, such blind spots of development exist, in many cases even in close proximity to city centers. Although they may have large potential for urban development e.g. as sites for housing, commerce, or creative industries, they remain inactive assets as along as sufficient accessibility is not established (Salchow 2017).

1.2 Research Challenges

Mobility and accessibility have long since been understood as key ingredients of well-working cities, however, research on urban mobility is faced by a two-fold disruption today. First, new forms of mobility e.g. autonomous vehicles, e-mobility, or drone and robot transportation raise a multitude of unexpected challenges, while—second—new digital simulation and modeling tools enable new approaches in urban analysis, planning, and design that go far beyond the established repertoire of planning and management media, such as cartographic maps, zoning plans, or physical volumetric models.

The effects of new mobility concepts e.g. car-sharing on urban development and planning are far from being well-understood. Mobility providers as well as urbanists struggle hard to grasp the consequences of the swift technological developments. On that background, impact models that clarify the relationship between new mobility and urban development have emerged as a major demand. To address this desideratum, a data-based approach integrating data from urban environments and mobility systems within a comprehensive modeling and simulation environment seems promising. This paper presents such a prototypical impact tool and approaches the issue both on conceptual as well as on technical level. Building upon accessibility research conducted by El-Geneidy and Levinson (2006), Krizek et al. (2007), it presents a new theoretical framework prepared by Salchow (2017) and a new technological solution created by Schaber (2017).

1.3 Aims

Being part of the overall research area of City Science, i.e. data-based urban analysis and development, this paper contributes especially to the discussion how dynamic models can help estimating the impact of new mobilities on cities (Schürmann and Spiekermann 2011). Although digital tools are at hand for simulating respective effects (one of them will be presented in this paper), their practical utilization and applicability for urban development and strategic scenario analysis is still unclear. Hence, a general aim of the research was to provide an easy-to-use analytic instrument for planners and decision makers. While the overall scope of this research addresses all modes of new mobility, the present study focusses on the urban effects of car-sharing, and how these can be effectively investigated with a new interactive tool.

Two specific aims guided the research:

(1) An analytic method was sought to carry out dynamic calculations of accessibility resulting from the different modes of transportation, and to assess the impact on urban development. This method should enable the identification of places with low accessibility, and quantify the change of accessibility resulting from the provision of shared mobility services.
(2) The method should be supplied by a low-threshold modelling and visualization tool. For this purpose, the interactive modeling table CityScope suggested itself as a well-established technological platform and proven tool for expert decision support and participative planning (Noyman et al. 2017; Schiewe et al. 2017). This study is to validate the usability of the CityScope also for research issues in urban mobility and accessibility.

2 Conceptual Framework and Hypothesis

2.1 Hypothesis

The viewpoint that this paper takes is an urbanist one; it argues from a perspective of urban development and urban value creation. As mobility creates accessibility to and within urban areas, and thus qualifies them for further development, the starting assumption is that new forms of mobility will also lead to higher level accessibility. This in turn shall result in upgrades of the development potential and real estate value of the urban area at stake. Value creation through urban and spatial development is directly depending on mobility designs that adequately synthesize the available means in respect to the qualities of place and environment (Huber and Spiekermann 2014). As all mobility innovations of the recent past, from car-sharing to drone flight, may be regarded such "urban upgraders", a basic hypothesis can be derived: neglected and underdeveloped ("sleeping") urban areas can be purposefully activated by the deployment of new mobility services.

As this paper deliberately focuses on car-sharing, the working hypothesis may be further specified: car-sharing is a potential tool for urban development and upgrading. Thus the paper advocates a shift of perspective: it views car-sharing not only as a new business model for mobility providers (whose concern is mainly the modes of implementation and marketing), but as an active instrument for the evolvement of cities and neighbourhoods. What is the added public value generated from car-sharing services in underdeveloped urban areas? Can car-sharing give effective leverage for urban (re)vitalisation?

2.2 Research Questions

The discussion of the stated hypotheses leads to two key questions: (1) How to measure the urban impact and added value through car-sharing? (2) How to identify those urban areas that suggest themselves for upgrading through car-sharing?

The estimation of the impact of complex urban interventions e.g. new mobility systems remains a difficult task. With different viewpoints, fields like Science and Technology Studies (societal), Policy Management (political) and Technology Forecast (technological perspective) have endeavored in providing reliable insights and theories (Schürmann and Spiekermann 2011). Although reliable models have been established for scenario planning, impact measurement and assessment, there is still a lack of tools and processes that can be handed over to planners and developers. Applicable instruments are needed to support their decisions about where and how to start development programs, implement new infrastructures, and build new work and living areas (Schiewe et al. 2017).

Certainly not all urban areas can benefit on similar level from new mobility services. Many of them may be already sufficiently supplied, especially downtown areas and city centers where car-sharing would not add substantially more accessibility. Hence, a reliable procedure is needed that indicates the urban areas in which car-sharing makes creates a genuine difference in terms of reached places, costs efforts, and travel comfort (El-Geneidy and Levinson 2006; Krizek et al. 2007).

3 Technical Approach and Implementation

Above mentioned hypothesis and research questions were pursued with a custom made simulation tool "AccessFinder" that is based on a new numerical accessibility model. In brief, the model allows a free definition of multiple input parameters especially in regards to mobility options, and a dynamic calculation of the resulting accessibility levels of places (Schaber 2017). The algorithm computes data-input re. population, employment, routing networks, public transportation into cartographic visual output (accessibility heatmaps) that are projected onto the digital modeling tables CityScope, in order to be interacted. A first application of the "AccessFinder" tool was prepared for the city of Hamburg.

3.1 Interactive Tool CityScope

To carry out systematic analysis of urban accessibility and study their impact on urban development, city scientists at the HafenCity University have developed the new digital tool "AccessFinder" based on the technology platform known as CityScope (Schaber 2017) (Fig. 1). The CityScope is an interactive mapping device developed by the MIT Media Labs CityScience/Changing Places groups over the past years, whose core technology are real time projections of cartographic information onto camera-equipped modelling tables. On these, in turn, programmed Lego blocks are used as tangible user interfaces ("marker stones"), enabling direct interaction and manipulation (Noyman et al. 2017; Hadhrawi and Larson 2016).

In previous projects like (e.g. FindingPlaces 2017) the CityScope has proved being an effective tool that enables collaboration and exchange among expert group as well as in large stakeholder workshops. The CityScope provides a technological basis not only for effective social interaction but also for representing complex information in

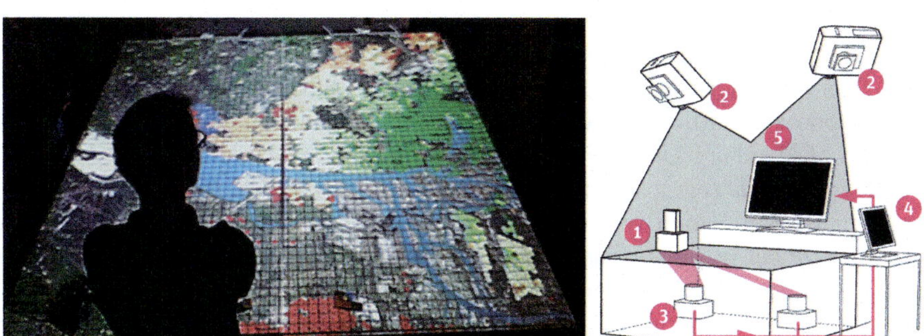

Fig. 1 CityScope interactive mapping table (left); Technical components (right): 1—Marker Stone, 2—Overhead projectors, 3—Camera, 4—Control station, 5—Data display

an easy-to-comprehend manner, thus strongly supports decision making processes (Noyman et al. 2017; Schiewe et al. 2017).

3.2 Data and Technical Background

The overall system is based on a WebGIS that ingests the needed data (map layers, social data, traffic data, routing network etc.) through a geoserver from a PostGIS database (extended with a pgRouting) using parametrized SQL queries. Websockets enable real-time communication between all applications. The key data resources are briefly described below.

3.2.1 Spatial Reference System

Due to data availability, this study limited it's area of investigation to Hamburg city, while maintaining that routing network as well as activity parameters stretch beyond the city limits, in order to prevent outer city areas being miscalculated with only low accessibility. Apart from administrative zones (wards, boroughs, postal areas etc.) cellular grids were used for the micro analyses and computation (Schürmann and Spiekermann 2011). To each cell a knot in a routing network was assigned, being the starting point for the calculation of accessibility (Schwarze 2015).

To calculate and generate the envisioned accessibility heatmaps, the Hamburg area was re-structured into a grid of ~16,000 hexagon cells with edge length 133 meters, which were to be linked with a routing network. In similar manner, population and employment data were aggregated into the same hexagonal geometry.

3.2.2 Data

Population data from the Northern Authority of Statistics amounted to more than 400,000 inhabited cells with a total population of ~9 Mio. people. To reduce data load, cells were aggregated into 13,000 model points, having a population in the core area of approx. 2.6 Mio. people. *Workplace data* from the institute of transportation research of Hamburg University of Technology (TUHH) came as point data with multiple attributes (e.g. name of employer, address, overall number of employees) totaling ~1.5 Mio. employees in the core area. Aggregated into the hexagon structure, the number of points was reduced from 55,000 to 8000. *Transportation data* from Hamburger Verkehrsverbund (Public Transportation Agency) and from Hamburg University of Technology (TUHH) comprised all lines and stops of public transportation in the city, plus mobility data derived from the accessibility software VISUM. *Routing networks* from the basis of the algorithmic calculation of travels between target points (e.g. from home to work, from work to social facilities) (Noyman et al. 2017; Schiewe et al. 2017). In the networks, the most economic route must be found according to an optimization of cost attributes (e.g. time, distance, or combined parameters). However, the number of freely available street networks is strongly limited; for this study only from the open-source project OpenStreetMap a routable street net could be retrieved. (Schäfer et al. 2003; Obe and Hsu 2015).

3.2.3 Software

The research involved following softwares: For accessibility analysis, the geoinformation systems ArcGIS 10.3.1 and QGIS 2.16.2 were used. As data bank systems for geodata, PostgreSQL, PostGIS, as well as pgRouting were used, the latter offering multiple functionalities for routing and network analysis. The programming of the WebGIS was carried out in languages HTML, CSS und JavaScript. For the interactive web maps, the open-source JavaScript-library OpenLayers was used. As geoserver, the open-source product Mapserver was chosen, which shares spatial data and publishes them as a geoservice (Studer 2016).

The capability of the software, before all, is to enable interactive accessibility analysis without having the core parameters already defined. Instead, users can freely adjust search parameters, upon which respective accessibility analyses and maps will be generated in real-time. This has strongly determined the software design in terms of (a) user interaction and functionality, (b) processing speed and computation performance.

3.3 Model and Algorithm

As its key operation, the "Access Finder" calculates and visualizes the accessibility of urban cells as a function of the different modes of mobility: public transportation, car mobility, car-sharing, bicycle riding, and pedestrian walk (Meschik 2008). The basis for the accessibility calculation is a matrix that combines activities (before all: housing, working) and means of transportation, whose travel efforts are determined with the routing network (El-Geneidy and Levinson 2006). The derivative algorithmic model reflects the following parameters as input: mobility mode, travel time, target places and activities.

The formal modeling of accessibility is based on the Dijkstra-algorithm which calculates the cost efforts that are necessary to reach all knots within a routing network from a given starting knot. In practical application: one selects one specific cell (or knot) of interest on the urban map, and investigates the accessibility of other cells (knots) within a selected time frame (e.g. 30 min) and a selected form of mobility (e.g. public transport + bicycle).

Assuming that substantial urban value is created if citizens can easily reach their work place from their place of residence, the computation focused on the accessibility of workplaces from a given housing area, respectively the accessibility of housing areas from given workplace. Recognizing the overall goal of this study—the assess-

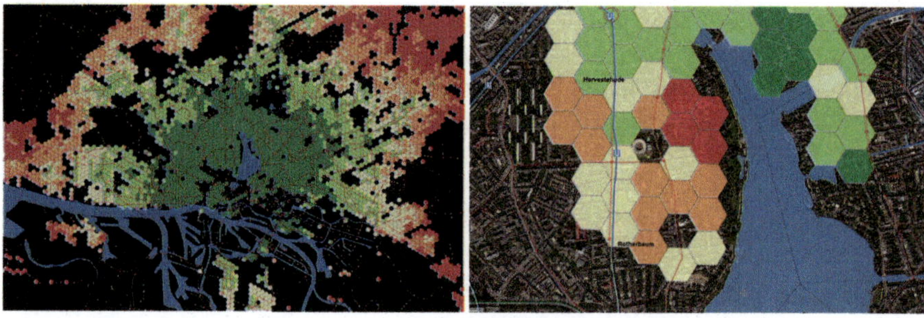

Fig. 2 Output: Accessibility maps—on city level (left)/on district level (right)

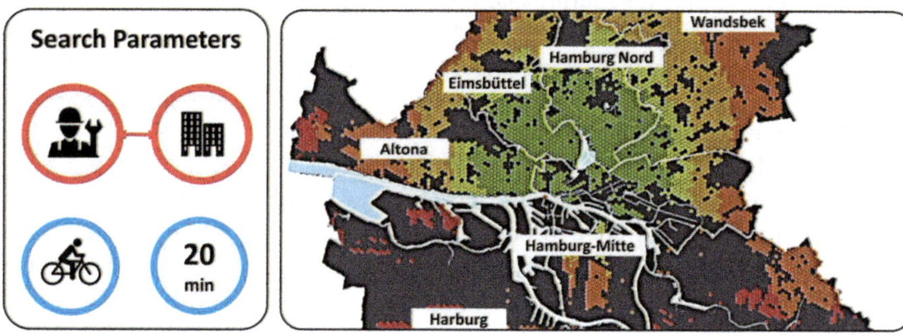

Fig. 3 Sample analysis: for a user request "Bicycle Accessibility of workplaces from residential areas within 20 min" (left), a real time heatmap will be generated (right)

ment of the potential urban impact of car-sharing—these accessibilities based on car-sharing were modelled with special attention, e.g. increased extension of range and accessibility were assumed, and time consumption for finding and catching a shared car regarded (Schäfer et al. 2003). Output of the computation are accessibility heatmaps that cover the complete metropolitan area of Hamburg. For each cell in the hexagonal spatial grid, accessibility is calculated in two-fold manner: (1) as catchment of the individual cell i.e. its accessibility from other cells, (2) as outreach of the cell, i.e. accessibility to other cells. Resulting heatmaps indicate areas of high accessibility in green colours, low-access areas in red (Fig. 2).

With the procedure mentioned, a multitude of accessibility maps can be generated, according to the setting of input and context parameters, e.g. "Car accessibility of medical services from housing areas in 10 min", or "Pedestrian accessibility restaurants and pubs from working areas in 10 min" etc. (Fig. 3).

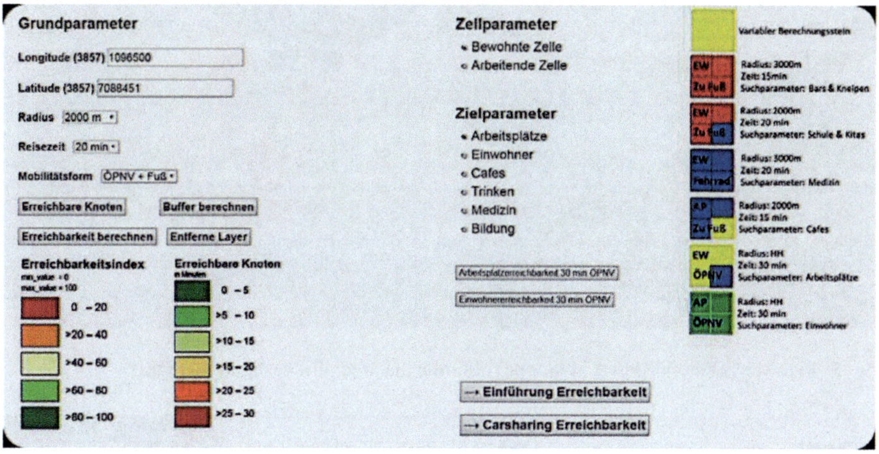

Fig. 4 Control monitor for setting the basic parameters

3.4 Dynamic Interaction

Cartographic representation of accessibility is common state-of-the-art and not especially innovative. Hence a new quality was sought for the "AccessFinder" by conceiving the tool as a dynamic one, "dynamic" standing for the free manipulation of the parameters. Addressing current demands in mobility research, the envisioned tool should replace static accessibility maps with dynamic analyses that users can interact with—that is: maps that are calculated and visualized in real time.

3.4.1 Mobility and Accessibility Parameter Setting

A first key idea for enhancing the dynamic usage and interactivity of the tool was to allow users to freely set and change the input and context parameters for the accessibility analysis. Thus not only the different kinds of places and activities (workplace, housing, social infrastructures) can be selected as target locations, but also different modes of mobility can be chosen and mixed to investigate cross-modal accessibility (e.g. bicycle + train, bicycle + car sharing etc.) The setting of these parameters is carried out with a control monitor (Fig. 4).

3.4.2 Search Area Definition

A second idea was to let users freely define the search area for which accessibility was to be analysed. As the CityScope technology supports tangible user interfaces (e.g. tagged Lego bricks whose positions on the interactive tables are recognised by the system), the freehand selection of focus areas was developed as a key interaction

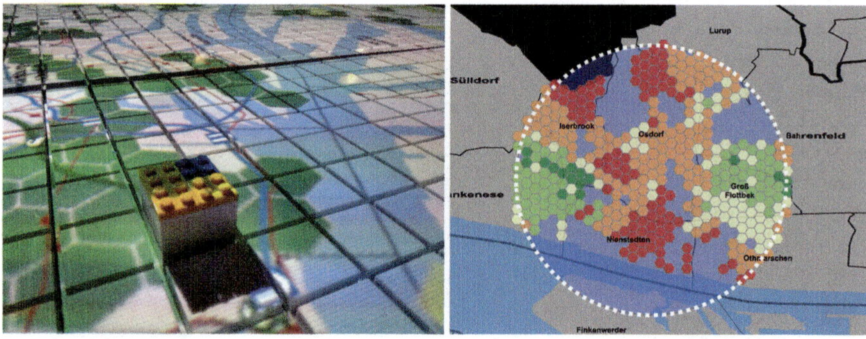

Fig. 5 Fig. Marker stone (left: CityScope) delimits the area of investigation (right)

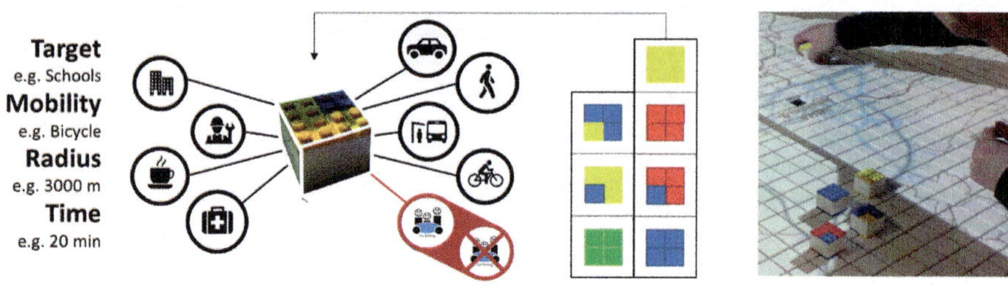

Fig. 6 Programming the marker stones with default parameters, e.g. with/without car-sharing (left); "hands-on" research on the CityScope table (right)

feature. By simply moving the Lego marker on the camera-monitored CityScope surface, users can directly define and investigate areas of their interest (Fig. 5). By determining the range of area to be analyzed (e.g. 1 km), not only a spatial focus of attention is defined but also computational efforts are reduced to a necessary minimum as the system's live calculation skips the accessibility of other cells outside of the area of interest. This substantially contributes to the dynamics of the interaction, as it enables fast generation of accessibility maps according to parameters the user just defined before.

In order to enable still easier and faster accessibility analysis with the tool, a set of "programmed markers" was developed that bear the most common accessibility queries. Thus the parameter setting via the control monitor can be leapfrogged, and a standard request like "Pedestrian Accessibility of Schools and Kindergarten within 20 min in a radius of 2 km" be investigated by simply placing on the CityScope the respective marker which "contains" that query (Fig. 6).

3.4.3 Car-Sharing Impact Map

The central question of this study being the impact of car-sharing on the accessibility of urban areas, the "AccessFinder" also makes the areal accessibility *with* versus

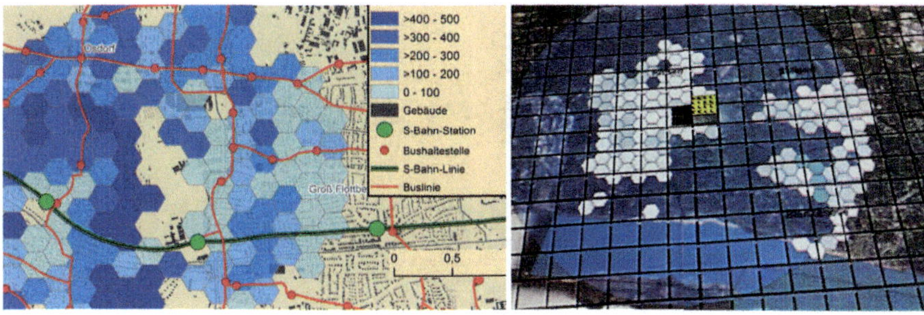

Fig. 7 Car Sharing Impact: Mapping the difference of 30 min accessibility to workplace (left); "Marker" on the CityScope reducing the areal range of impact analysis

without car-sharing immediately clear and comparable (Schürmann and Spiekermann 2011). For this purpose the "Car-Sharing Impact Map" was conceived as a key function of the tool and process (Fig. 7).

The effect of car-sharing offers (free floating model) in areas with low accessibility is investigated by calculating the accessibility gains with individual car mobility ontop the other existing mobility options, while acknowledging time losses (10% of average travel time) for searching the vehicle.

This feature calculates the difference of accessibility with and without car-sharing for each cell in the search area, and projects the differential results on the city map. Areas with a high level of difference (colour-coded dark blue) indicate places suitable for the implementation for car-sharing. In them, new mobility service may provide substantial improvement of the existing situation. In areas with a low level of difference (light blue or white), car-sharing would bring little benefits, even if the accessibility of the place may be high already (as the case is in downtown areas).

4 Results and Discussion

4.1 Use Case 1: Dense yet Insufficiently Accessible Areas

First investigations with the "AccessFinder" tool have outlined two typical urban use cases for the purposeful deployment of car-sharing. Use case 1 addresses densely populated areas with yet low accessibility (Fig. 8). For this application, two maps need to be generated and compared with the tool: population density on the one hand, local accessibility on the other. In areas with high density but low accessibility, the deployment of car-sharing services suggest successful operations of fleets etc. due to sufficient local demand. Such a case is found in the densely populated HafenCity area, Europe's largest urban development project under construction, which in terms of infrastructure is not yet fully developed though.

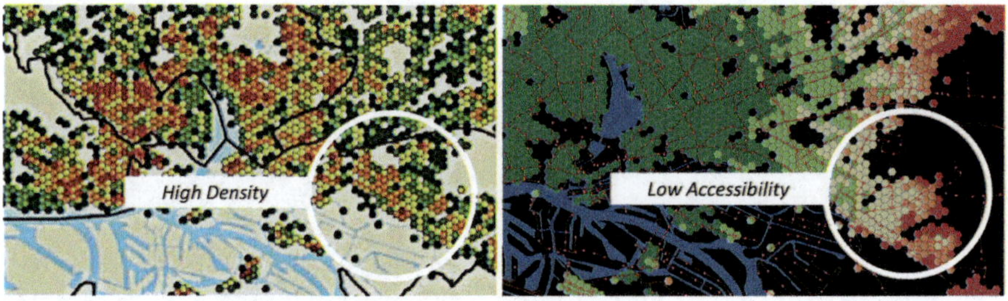

Fig. 8 Dense yet insufficiently accessible areas (left: population; right: accessibility)

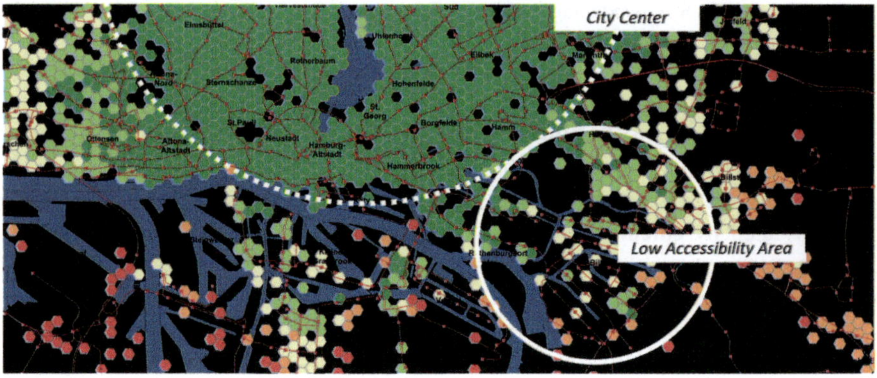

Fig. 9 Central yet insufficiently accessible areas

4.2 Use Case 2: Central yet not Sufficiently Accessible Areas

A second use case for the purposeful implementation of car-sharing is given in urban areas with high centrality (i.e. proximity to downtown locations, activities, infrastructures etc.) but low-level accessibility, as indicated by the "AccessFinder"-tool. In such "sleeping areas", the purposeful deployment of car-sharing services suggests urban value upgrade and activation. By making central areas instantly more accessible, users with comparatively low demands in regards to spatial quality but high demand re. access and proximity to urban amenities can be attracted e.g. new urban industries in search of workspaces. In Hamburg, such cases can be found in the districts like Rothenburgsort or Bergedorf (Fig. 9).

Especially the second case implies urban business and development models that negotiate and balance the implementation costs for new car-sharing services with the gains resulting from urban upgrade, especially rising real estate value. In this balance, also social and demographic benefits need to be incorporated that may result from an mobility-invigorated urban area.

5 Conclusions

This paper showed how the custom made tool and methodology "AccessFinder" can support the identification of urban areas suitable for an implementation of car-sharing. Beyond the presented use cases, the tool provides further opportunities for data-based urban research and scenario analysis. It can be also applied for the assessment of supply quality and coverage of public transportation, retail facilities, or social infrastructures.

Technically, the "AccessFinder" demonstrates the feasibility of dynamic computation and visualization of urban accessibility on the CityScope. Going beyond static map representations, the interactive tool enables "hands-on" expert discussion and decision support through a set of distinct features: real-time analysis and mapping, variability of input parameters, free definition of search area, and spatial scalability.

The conceptual and technical research at CityScienceLab and g2lab will be further pursued in the context of Hamburg's Intelligent Transportation Systems (ITS) strategy. Future investigation will pay special attention to hardware issues and the problem of computation time. In practical application, the "AccessFinder" will support urban strategists by helping to discover high-potential sites in the urban fabric that can be activated by new modes of mobility.

References

El-Geneidy A, Levinson D (2006) Development of accessibility measures. "Access-to-Destinations"-Forschungsbericht Nr. 1. University of Minnesota, Minneapolis

Hadhrawi M, Larson K (2016) Illuminating LEGOs with digital information to create urban data observatory and intervention simulator. In: Proceedings of the 2016 ACM conference companion publication on designing interactive systems, pp 105–108 http://dl.acm.org/citation.cfm?id=2909400

Huber F, Spiekermann K (2014) Deutschland in Europa, Ergebnisse des Programms ESPON 2013, Heft 5: Erreichbarkeit und räumliche Entwicklung. Bundesinstitut, für Bau-, Stadt- und Raumforschung (BBSR) im Bundesamt für Bauwesen und Raumordnung (BBR), Bonn

Krizek K, El-Geneidy AM, Iacono M, Horning J (2007) Refining methods for calculating non-auto travel times. "Access to Destinations"-Forschungsbericht Nr. 2. University of Minnesota, Minneapolis

Meschik M (2008) Planungshandbuch Radverkehr. Springer, Wien

Noyman A, Holtz T, Kröger J, Noennig JR, Larson K (2017) FindingPlaces: HCI Platform for Public Participation in Refugees' Accommodation Process, KES2017 International conference on knowledge based and intelligent information and engineering systems, Procedia Computer Science, KES2017, Elsevier Procedia Computer Science

Obe R, Hsu L (2015) PostGIS in action, 2nd edn. Manning Publications Co., Shelter Island, NY, USA

Salchow M (2017) Car-Sharing als neues Instrument der Stadtentwicklung – Impakt-Modellierung von Stadt- und Mobilitätsentwicklung: Verbesserung der Erreichbarkeit von Wohnund Arbeitsorten durch Car-Sharing-Angebote (Thesisprojekt HafenCity Universität Hamburg 2017)

Schaber L (2017) Entwicklung und Visualisierung dynamischer und interaktiver Impact-Modelle auf CityScopes (Thesisprojekt HafenCity Universität Hamburg 2017)

Schäfer R, Gühnemann A, Thiessenhusen K (2003) Neue Ansätze im Verkehrsmonitoring durch Floating Car Daten. 19. Verkehrswissenschaftliche Tage "Mobilität und Verkehrsmanagement in einer vernetzten Welt", Dresden. http://elib.dlr.de/6612/2/VWT_Dresden_Schaefer.pdf (Stand 02.09.2017)

Schiewe J, Kröger J, Zobel K, Mensing T (2017) Aufbau, Nutzung und Evaluation eines interaktiven Stadtmodells für die partizipative Flächensuche für Flüchtlingsunterkünfte. gis.Science, Die Zeitschrift für Geoinformatik, Ausgabe 2/2017. VDE Verlag, Berlin, Offenbach

Schürmann C, Spiekermann K (2011) Räumliche Wirkungen von Verkehrsprojekten. Ex post Analysen im stadtregionalen Kontext. In: BBSR-Online-Publikation 02/2011. Bundesinstitut für Bau-, Stadt- und Raumforschung (BBSR) im Bundesamt für Bauwesen und Raumordnung (BBR), Bonn

Schwarze B (2015) Eine Methode zum Messen von Naherreichbarkeiten in Kommunen. MV-Verlag, Münster

Studer T (2016) Relationale Datenbanken – Von den theoretischen Grundlage zu Anwendungen mit PostgreSQL. Springer-Vieweg, Berlin, Heidelberg

Active Mobility: Bringing Together Transport Planning, Urban Planning, and Public Health

Caroline Koszowski, Regine Gerike, Stefan Hubrich, Thomas Götschi, Maria Pohle and Rico Wittwer

Abstract Active mobility is related to various positive effects and is promoted in urban planning, transport planning, and in public health. The goals of these three disciplines differ in many respects but have a strong overlap in the ambition to foster active mobility. Until now, efforts for strengthening active mobility have typically not been combined, but rather promoted separately within each discipline. This paper presents a review of research on determinants and impacts of active mobility and of policy measures for supporting active mobility, including the three disciplines of transport planning, urban planning, and public health. The paper further shows the different perspectives and ambitions of the three disciplines and, simultaneously, the substantial synergies that can be gained from an interdisciplinary collaboration in research and practice.

Keywords Active mobility · Walking · Cycling · Public health · Planning

---

C. Koszowski (✉) · R. Gerike · S. Hubrich · M. Pohle · R. Wittwer
Chair of Integrated Transport Planning and Traffic Engineering, "Friedrich List" Faculty of Transportation and Traffic Sciences, Technische Universität Dresden, 01062 Dresden, Germany
e-mail: caroline.koszowski@tu-dresden.de

R. Gerike
e-mail: regine.gerike@tu-dresden.de

S. Hubrich
e-mail: stefan.hubrich@tu-dresden.de

R. Wittwer
e-mail: rico.wittwer@tu-dresden.de

T. Götschi
University of Zurich, Physical Activity and Health, CH-8001 Zurich, Switzerland
e-mail: thomas.goetschi@uzh.ch

M. Pohle
Fraunhofer Institute for Transportation and Infrastructure Systems, 01069 Dresden, Germany
e-mail: maria.pohle@ivi.fraunhofer.de

B. Müller and G. Meyer (eds.), *Towards User-Centric Transport in Europe*,
Lecture Notes in Mobility, https://doi.org/10.1007/978-3-319-99756-8_11

1 Introduction

As a core objective in public health strategies, the promotion of active mobility aids in increasing physical activity levels within the daily routine. Active mobility is defined as utilizing walking and cycling for single trips or within a trip in combination with public transport (Gerike et al. 2016). The World Health Organization (World Health Organisation (WHO) 2017) recommends at least 150 min of moderate-intensity physical activity per week for adults and at least 60 min of daily moderate to vigorous physical activity for children. In 2010, 20% of adult men, 27% of adult women, and 78% of boys and 84% of girls (between 11 and 17 years of age) did not fulfil these recommendations globally (World Health Organisation (WHO) 2017). This results in the increased risk for non-communicable diseases and reduces life-expectancy (World Health Organisation (WHO) 2011). The WHO (2011) lists the promotion of active mobility as one core strategy to overcome these problems of insufficient physical activity.

Active mobility also supports transport planning ambitions. Walking and cycling are space efficient; these modes of transport are flexible; they cause low individual and societal costs; and, in combination with public transport, they can cover almost all mobility needs. Increased active mobility can thus help to mitigate the adverse effects caused by motorized private vehicles, especially in urban areas. Common transport-related and environmental problems include safety, congestion, climate change, air-pollution, noise and land consumption. In 2013, 22% of all fatalities in EU road transport were pedestrians and 8% were cyclists (European Commission (EC) 2015). Compared to their share of overall traffic volume these numbers are high, and in recent years their decrease in fatalities is slower than the total fatality development.

The transport sector is responsible for major parts of the overall greenhouse gas emissions with no substantial reductions thus far (European Environment Agency (EEA) 2017a). Reductions of greenhouse gas emissions in transport might be more expensive as compared to other sectors, but effective climate protection will not be possible without this sector. Emissions from transport need to lower by around two thirds by 2050 (base year 1990) in order to meet the long-term 60% greenhouse gas emission reduction target set in the 2011 White Paper on transport of the European Commission (European Environment Agency (EEA) 2017a).

In 2010, around 420,000 people died prematurely from air pollution in the European Union (European Commission (EC) 2014), and significant proportions of the urban population in the EU-28 are exposed to air pollutant concentrations above the EU limit or target values and even more in relation to the more stringent WHO air quality values set for the protection of human health. The critical pollutants are particulate matter ($PM_{2.5}$, PM_{10}), ozone (O_3), nitrogen dioxide (NO_2), and benzo[*a*]pyrene (BaP) (European Environment Agency (EEA) 2017b). More than 100 million people in the 33 member countries of the European Environment Agency (EEA) are affected by harmful noise levels above 55 decibels [dB]. Out of these, 32 million people are exposed to very high noise levels above 65 dB (European Environmental Agency (EEA) 2017c).

Additionally, from the urban and city planning perspective, increased levels of active mobility provide a promising outlook, as this increase allows for less space-consuming transport systems with lower speeds. This opens various opportunities for designing more attractive, inclusive, and livable cities. Some shopkeepers and representatives of local industries insist on having parking spaces nearby in order to increase their accessibility and attractiveness for customers; however, many examples worldwide show that destinations in areas with attractive public spaces are very successful and that success even increases when space is re-allocated from the car to active modes (Gehl 2010). Gehl (2010) demonstrates the reinforcing cycle of attractive public spaces: People are drawn to the area; this calls on political support for assistance in increasing and improving the space for these individuals, consequently leading to the presence of even more people in the streets and public spaces.

The ministers responsible for urban development in the member states of the European Union have committed to "create attractive, user-oriented public spaces and [to] achieve a high standard in terms of the living environment" (European Union (EU) 2007). In 2016, the EU Ministers agreed on the "Pact of Amsterdam" which establishes the so-called Urban Agenda for the EU (European Union (EU) 2016). This agenda focusses on an operational framework to encourage improved involvement of urban authorities within the EU policy processes. It presents an initial list of priority themes including the general topic "Urban Mobility" and, more specifically, "Soft Mobility", referring to walking, cycling, and public spaces. More and more cities recognize the importance of people in public realms for developing attractive, economically successful, democratic and inclusive cities. There is a growing understanding that cities must be designed to invite pedestrian traffic and city life in order to be successful.

The ambitions as well as the strategies for increasing active mobility differ substantially between the disciplines of transport planning, urban planning and public health, but, at the same time, they have a strong overlap in the objective to foster active mobility. However, despite this common interest in active mobility, efforts thus far have been primarily individual rather than collaborative. Substantial synergies could be harnessed by better coordinating activities, as well as by combining approaches for promoting active mobility from the various disciplines, and by pooling financial and personnel resources. Increased walking and cycling volumes yield various environmental, social, and economic benefits. These key factors contribute to the functioning of cities and support sustainable urban development (Gerike and Koszowski 2017).

This paper presents an overview of determinants and impacts of active mobility, as well as of policy measures for supporting active mobility, including the three disciplines of transport planning, urban planning, and public health. The paper is composed of three main parts: First, determinants of active mobility are presented based on a conceptual framework that combines findings from the literature in all three disciplines. Second, the impacts of active mobility and physical activity are introduced in order to show the substantial societal gains that result from high levels of active mobility. Third, key strategies for promoting active mobility are outlined. The different foci of each discipline become apparent. This shows again the great potential that exists in combining the efforts in all the disciplines for promoting active mobility more efficiently and successfully. In the last section, the findings are summarized, and an outlook is given on perspectives for active mobility.

2 Conceptual Framework of Active Mobility

Research on active mobility has been exponentially growing in the last years, and, as a part of this research, various conceptual frameworks of active mobility have been developed (see Götschi et al. 2017) for a systematic review). Frameworks exist separately for walking and cycling as well as for overall active mobility. They are rooted more firmly in transport and public health and less in urban planning. Frameworks from transport research often put active mobility into the context of the overall travel behavior and do not consider non-transport physical activity; frameworks from public health often put active mobility into the context of the overall physical activity and do not consider travel with motorized modes. Few approaches include the overall picture of determinants, behavior, and impacts of active mobility; most focus on specific parts of the whole system. The built environment and transport systems, as well as the psychological, socio-demographic, and socio-economic characteristics as important determinants of active mobility obtain about equal attention. Most frameworks describe active mobility in generic terms and only a few offer a more detailed perspective, including, for example, modes and purposes of the individual trips. Some address the impacts of active travel; these cover health, safety, and environmental outcomes (e.g., carbon emissions). Most of the frameworks are static, only a few include feedback loops, policies or changes over time. Ogilvie et al. (2011, 2012) provide a framework that allows for the qualification and evaluation of changes in active mobility and physical activity resulting from measures which promote active mobility.

Socio-ecological frameworks are often used in public health related research (Giles-Corti and Donovan 2002; Sallis et al. 2006, 2008), distinguishing the layers of environmental (ecological) and individual (socio-demographic, socio-economic, socio-psychological) determinants of active mobility and physical activity. The theory of planned behavior developed by Aizen (1991), see also Bamberg and Schmidt (2003), Haustein and Hunecke (2007) has been widely applied. It states that socio-psychological variables such as attitudes and norms influence the intention to choose a specific (travel) activity as a mediator variable, which, in turn, influences the actual behavior. Frameworks from the transport literature mainly focus on mode choice (see De Witte et al. 2013 for a comprehensive review.). Schneider (2013) propose a five-step mode choice process, covering (1) awareness and availability of choice options, (2) safety and security, (3) costs and convenience, (4) enjoyment, and (5) habits. Socio-demographic variables are included as moderators in these steps.

Singleton and Clifton (2015), see also Alfonzo (2005) conceptualize travel decisions based on a hierarchy of travel needs which was developed using Maslow's theory of human motivation (Maslow 1954). The bottom of the hierarchy assesses feasibility, followed by accessibility of relevant destinations, safety, comfort and delight. The distinction between reasoned influences on behavior (such as perceptions, preferences, and attitudes) and unreasoned influences (such as habits and impulsiveness) as important determinants of active mobility are stressed by van Acker et al. (2010). Kroesen et al. (2017) study the relationship between attitudes and travel behavior; they demonstrate that this relationship is bidirectional, however, the influence of travel behavior on attitudes is more dominant. Persons with dissonant (i.e., non-aligned) attitude-behavior patterns are less stable compared to persons with well aligned attitudes and behavior. These persons tend to rather adjust their attitudes to their behavior than vice versa. Pikora et al. (2003) examine particular physical environmental factors as part of the social ecological model by reviewing studies from public health research, urban planning, and transport-related research. Their frameworks include the influencing factors for physical activity in the context of transport and recreation. Götschi et al. (2017), based on their literature review as part of the PASTA-Project (Physical Activity through Sustainable Transport Approaches, see Gerike et al. 2016), develop a comprehensive and holistic framework of active mobility behavior. They use a detailed, multi-layered model to differentiate the socio-spatial levels, where socially, physically, and individually related factors are located. The interaction of these elements leads to the choice of travel and, thus, behavior.

Figure 1 presents the framework developed specifically for this contribution which is based on Götschi et al. (2017). The built environment in Fig. 1 is distinguished by the type of area as well as by the characteristics of public space and the transport system. The whole transport system, including all modes and all components as active mobility, can only be understood in the context of overall travel behavior. For example, when commuting distances are too long, walking or cycling are not attractive options for these trips. Public space comprises not only the density and connectivity of networks for active mobility but also the design of each individual part of these networks (e.g., the design of intersections, streets, or squares). The type of area includes the spatial structure as, for example, the type, size, density, and spatial arrangement of buildings and additionally land-use as the specific usage of buildings and open spaces (e.g., for dwellings or work places). Most aspects of these three parts of the built environment can be characterized by their quality, attractiveness, availability, and accessibility. In Fig. 1, the built environment spreads across the city/regional and neighborhood layers and represents, together with the natural factors such as climate and topography, the supply side of the framework.

The social context of active mobility is conceptualized by a multi-layer approach which considers the (neighborhood) community, the peers, the household, and finally the individual itself. Objective characteristics of each individual such as socio-demographic variables, the accessibility of destinations, and the availability of mobility tools (e.g., car ownership, membership in a carsharing scheme, possession of a public transport pass) are interpreted by each individual into their subjective perception of their mobility options. These, together with the socio-psychological variables

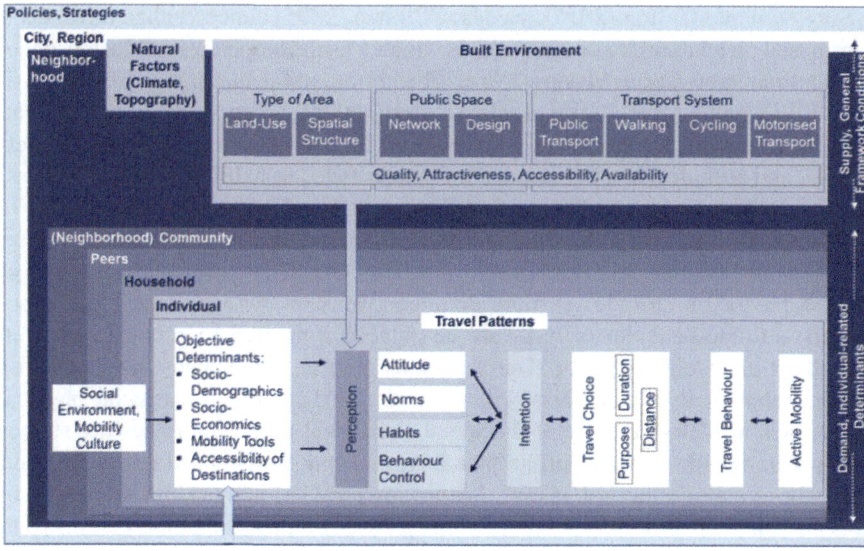

Fig. 1 Conceptual framework of active mobility

shape the intention to choose specific travel activities and, finally, the behavior itself. The causal relationship between these different determinants is a two-way process, as shown by Kroesen et al. (2017): Attitudes and objective determinants affect travel behavior but travel behavior, in turn, also affects attitudes, the choice of residential location, mobility tools, etc.

The outer box "Policies, Strategies" brings temporal dynamics into the static framework. Policies might directly influence travel behavior when, for example, new walking and cycling infrastructure is developed. Policies might also indirectly influence travel behavior when these change attitudes and mindsets that then in a second step change travel behavior. Policies include all aspects of activities for promoting active mobility as described below but also the governance structures, institutions, processes, finance issues, etc.

3 Determinants of Active Mobility

In this section, we review the most pertinent determinants of active mobility behavior as identified by research in the past years, in the context of the conceptual domains described above.

3.1 Built Environment

3.1.1 Density of Spatial Structures, Diversity of Land-Use

Land-use planning determines the shape of transport networks, the attractiveness of public space, and the distances between relevant generators and attractors of travel. It is therefore one core determinant of active mobility (Stead and Marshall 2001; Banister 2007; Ewing and Cervero 2010; Buehler et al. 2016). The following requirements for spatial structures that support active mobility are listed in the literature (sometimes also called the "5 Ds"): density, destination accessibility, design, distance to public transport, and diversity (Parkin and Koorey 2009; Gerike and Parkin 2015; Kang 2015). Stead and Marshall (2001) describe the relationship between density and diversity of spatial structures on mode choice as follows: A higher density of development and of population lead to (i) a higher amount of personal contacts and possible activities, (ii) a higher amount of commercial facilities and services in the neighborhood and, consequently, to (iii) a reduction of distances. Dense developments (iv) are thus beneficial for the supply of public transport as well as for active mobility. In addition to the density of spatial structures, the diversity or mix of land-uses contributes to the vitality of the public spaces and to a pedestrian- and cycling-friendly environment. Kang (2015) has observed increased daily walking times in areas with mixed-use developments, whereas mono-functional residential and industrial land-uses show opposite effects (see also Cervero and Duncan 2003; Bentley et al. 2010).

3.1.2 Connectivity and Density of Transport Networks

The density and connectivity of transport networks are closely linked to the spatial structures since these determine the size of blocks and hence the transport grid meshes. Dense, well-connected, safe, and comfortable infrastructures are important determinants for mode choice in general, including all modes as alternatives; they are particularly important for walking and cycling. Pedestrians and cyclists are the most vulnerable road users; they cover all age groups from infants to elderly people. They require muscle power and are thus highly detour-sensitive modes.

The public transport infrastructure and services are also decisive determinants of active mobility. Most public transport trips include at least two walking legs (access to the first public transport leg, egress from the last public transport leg). Walking is not suitable for all travel needs on its own, but in combination with public transport it can serve almost all travel needs. The infrastructure for the motorized individual modes (mainly car and motorcycles) is also of high relevance for active mobility: Attractive infrastructures for the motorized modes encourage car use for all purposes and distances, even for very short distances, and thus hinders active mobility. Less attractive infrastructures for the motorized modes with, for example, lower speed and/or less parking spaces support active mobility—on its own or in combination with public transport.

Hooper et al. (2015) measure the connectivity of the road network and the accessibility of centers within a specific radius of 1600 m in the neighborhood by using a Geographic Information System [GIS] with the following result: The higher the road connectivity and the higher the accessibility to commercial land-use within the neighborhood, the higher the odds of walking. In addition, Cervero et al. (2009) find that inhabitants of neighborhoods with highly connected road networks tend to have higher active mobility levels of a minimum of 30 min per day on average. A high connectivity is mainly related to fine-grained road networks with many intersections [nodes] and requires many route options to get from each starting point to the respective destination. The smaller the distances are between the intersections, the higher the share of pedestrians and cyclists in the local modal split (Guo et al. 2010; Lee et al. 2013; Lamíquiz and López-Domínguez 2015; Kaplan et al. 2016).

For cycling specifically, various studies confirm the importance of the density and connectivity of the network of cycling paths (Dill and Carr 2003; Christiansen et al. 2014; Carver et al. 2015; Damant-Sirois and El-Geneidy 2015). The accessibility of cycling paths particularly near to the residence and workplace as well as near to other destinations has a positive influence on cycling (Braun et al. 2016). Lowry et al. (2016) introduce aspects of infrastructure quality and traffic density into their connectivity measure, emphasizing the role of the weakest network links (i.e., highest stress) in limiting accessibility.

3.1.3 Accessibility

Land-use characteristics (density, diversity of spatial structures) and transport networks (density, connectivity, quality of transport networks) together determine the accessibility of destinations. For achieving high levels of active mobility, good accessibility is of special importance for the following destinations: workplaces (Lamíquiz and López-Domínguez 2015), shopping facilities (Hooper et al. 2015; Guo et al. 2010; Borst et al. 2008; Haybatollahi et al. 2015; Carlson et al. 2016), gastronomy (Lamíquiz and López-Domínguez 2015; Borst et al. 2008; Haybatollahi et al. 2015), public transport (Borst et al. 2008; Carlson et al. 2016; Hoehner et al. 2005), residential buildings (Kang 2015), and educational institutions (Su et al. 2013). Giles-Corti et al. (2013), see also Hooper et al. (2015), Borst et al. (2008) stress the importance of good accessibility to parks and other green spaces for fostering recreational active mobility.

3.1.4 Availability of Infrastructure, Design of Public Space

In addition to the quantity of destinations and the characteristics of the transport networks, the design and the quality of public spaces must be considered. Kang (2015) and Kamargianni (2015) find that the width of sidewalks and a higher share of wide sidewalks positively impact walking. Test persons accepted even longer distances because of wider sidewalks (Guo 2009). On the contrary, a poor sidewalk quality has a hindering effect. The violation of the "design for all"-principles (e.g., by shortcomings in the surface or in the street layout but also damages such as cracks or uplifts in the sidewalk surface) is directly related to lowered pedestrian activities and fewer minutes of physical activity—especially for recreational purposes (Hoehner et al. 2005; Giles-Corti et al. 2013). Findings regarding the qualitative design of the bicycle network are similar to those of walking: The higher the comfort of the bike paths and also of parking facilities, the higher the regular use of the bicycle for transport purposes (Braun et al. 2016; Hunt and Abraham 2007; Handy et al. 2010; Bühler 2012; Heinen et al. 2013).

Attractive public places are also decisive for active mobility. Gehl (2010) emphasizes the importance of the following characteristics of public spaces for achieving high walking levels:

- Provision of opportunities to sit and stand/stay with high quantity and quality, (see also Ball et al. 2001; Van Cauwenberg et al. 2014)
- Good lighting and security (see also Carlson et al. 2016; Bird et al. 2010)
- Active frontages in the buildings adjacent to the street as a linking element between public space and the insides of the buildings
- Attractive frontages, including, but not limited to, interesting and inviting eye-level facades
- Clearly defined areas for various activities on the sidewalks (e.g., walking, standing, resting, gastronomy, and other commercial facilities) (Forschungsgesellschaft für Straßen- und Verkehrswesen (FGSV) 2011).

3.1.5 Scale of Buildings

The scale of buildings is another important factor for active mobility (Gehl 2010; Cervero 1996). Neighborhoods with medium-high buildings and appropriate proportions within the public space are convenient for human scale and adequate for the horizontal sensory apparatus (Gehl 2010). Gehl (2010) stresses the importance for adequate proportions between buildings and public spaces: the so-called human dimension of the built environment. Spatial structures and the transport systems need to be built in a matter that attracts people and activities, including not only the necessary engagements but also optional, spontaneous, and social activities. Inviting buildings, historical elements, and local attractions also support active mobility as well as other outdoor activities (Carlson et al. 2016; Ball et al. 2001; Van Cauwenberg et al. 2014; Carnegie et al. 2002; Borst et al. 2009).

3.1.6 Road Traffic, Traffic Safety

Road traffic and traffic safety are other aspects which influence active mobility; both are prerequisite and motivating factors (Gerike and Parkin 2015; Forschungsgesellschaft für Straßen- und Verkehrswesen (FGSV) 2014; Götschi et al. 2016).

Road sections with a major link function and high traffic volumes of motorized vehicles do not show high pedestrian volumes (Borst et al. 2009); this which is due to a low objective and subjective [perceived] safety on these roads. Additionally, Kaplan et al. (2016) state that an increased density of road traffic as well as a high share of heavy goods vehicles (HGV) decreases active mobility. In particular, the two variables "density of traffic" and "share of HGV" have a negative impact on children on their way to school (Kaplan et al. 2016). For children, the choice of walking and cycling is negatively related with the exposure of main streets in their neighborhoods (Helbich et al. 2016).

3.1.7 Traffic Calming

The presence of traffic calming elements as well as a high number of traffic-calmed roads have a positive influence on active mobility, particularly when there is a high amount of traffic-calmed roads within a radius of 500 m for pedestrians and 1500 m for cyclists around the home location (Kaplan et al. 2016). Again, regarding children as a part of the most vulnerable road users, traffic-calmed roads increase the odds of walking and cycling (Guliani et al. 2015).

3.2 Socio-Demographic, Socio-Economic and Socio-Psychological Determinants

Socio-demographic and socio-economic characteristics are important indicators for the propensity to walk and cycle (Gerike and Parkin 2015; Handy et al. 2014; Wittwer 2014).

In Germany, the overall number of trips is nearly constant for men in almost all age groups across their life cycle (Ahrens et al. 2010); however, for females, age is a key factor for explaining trip rates which vary substantially across their lifetime. Up to five trips per day are made between the ages of twenty and forty; afterwards trip rates decrease steadily (Ahrens et al. 2010). This often refers to the phases of family formation and family living. Compared to females, males have a stronger preference for the bicycle, rather than walking (Cervero et al. 2009; Carnegie et al. 2002; Downward and Rasciute 2015). Females have more walking trips when it comes to physical activity (Deka 2013; Hansen and Nielsen 2014).

The age distribution of cycling is more even in high-cycling countries, such as the Netherlands, than in low-cycling countries where cycling is dominated by the young (and male) (Götschi et al. 2015). Pupils generally have higher levels of active mobility compared to adult persons. In particular, in the age group of nine- to seventeen-year-olds, children and adolescents show a high affinity towards the active modes (Harms et al. 2014). Harms et al. (2014) especially find a reduction of cycling trips after the age of seventeen for both male and female adolescents. This decline in cycling trips could be associated with the obtainment of a driver's license.

Parents have a special responsibility to shape their children's travel behavior. The more parents travel by bicycle, the higher the odds of their children going by bicycle (Ghekiere et al. 2016). The mobility behavior of children is affected adversely by the car-oriented mobility culture of their parents (Guliani et al. 2015; Van Goeverden and De Boer 2013). Furthermore, the early personal contact of children to the streetscape promotes active mobility. For example, this can be manifested by the permission to walk independently to school, to play outside or to cycle (Deka 2013; Ghekiere et al. 2016; Carver et al. 2014). Other familiar persons and peers can have the same impact, as far as active mobility is concerned (Ball et al. 2007).

Seniors show an increase in walking trips as they get older and a concomitant decline of the overall trip rate per day and of the daily travel time by bicycle (Bühler et al. 2011; Prins et al. 2014). The availability of commercial facilities in the neighborhood, physical fitness, and personal health are decisive factors for active mobility (Prins et al. 2014; Marquet and Miralles-Guasch 2015a). Thus, healthy persons and persons with no functional limitations are more often active than persons with health restrictions (Handy et al. 2010).

Education is another important determinant of active mobility. The odds of walking (Ball et al. 2001) and cycling (Christiansen et al. 2014; Braun et al. 2016; Lee et al. 2016) rise with increasing levels of education. The impact of household size and structure on (active) mobility is evident as household members need to coordinate their travel activities as well as the usage of joint mobility tools, such as the car or a public transport season ticket. Guo et al. 2010, see also Kitamura et al. (1997) find a higher propensity to use motorized vehicles for utilitarian purposes as the size of the household increases.

The relation between the income at the person or at the household level in regards to active mobility is often investigated with inconclusive findings:

- Low-income neighborhoods have the highest amount of short walking trips (Marquet and Miralles-Guasch 2015b). Pucher et al. (2011) find that people with low income use the bicycle for work trips or other utilitarian purposes more often than for recreation.
- There are two opposite findings about the effect of high income on cycling (Harms et al. 2014). High-income person-groups are able to spend more money on bicycles. This has a positive effect on the usage frequencies of bicycles. Mainly this group of persons cycle for recreation purposes. In opposite, the same group has a higher car ownership rate, which decreases the bicycle usage mainly for utilitarian purposes (Harms et al. 2014; Pucher et al. 2011; Kotval-Karamchandani and Vojnovic 2015).
- The higher the parental income, the less pupils walk to school (Kaplan et al. 2016; Beck and Greenspan 2007). In contrast, McDonald (2007) reports higher odds of walking for pupils from high income families.

There is a clear consensus in the literature regarding both the ownership of a driving license and the availability of motorized vehicles. Both factors inhibit active mobility and encourage people to adopt a car-oriented travel behavior. The higher the number of motorized vehicles in one household, the higher the share of inactive mobility (Dill and Carr 2003; Damant-Sirois and El-Geneidy 2015; Bühler 2012; Downward and Rasciute 2015; Carver et al. 2014; Wittwer and Hubrich 2016). The same effect has been shown for the ownership of driving licenses (Aditjandra et al. 2013; Clark and Scott 2013). Accordingly, the steady usage of bicycles is related to the availability of bicycles in the household (Cervero and Duncan 2003; Guo et al. 2010; Bühler 2012; Heinen et al. 2013; Böcker 2014).

Socio-cultural indicators (e.g., origin and cultural background) are influencing factors. For instance, Cervero and Duncan (2003) find in their study for the San Francisco Bay Area (USA) that African Americans are undertaking more walking trips (for all trip purposes) than Caucasian or Asian Americans. Also, the cycling rates increased the fastest among African Americans, Hispanics, and Asian Americans in the United States during the 2000s (Harms et al. 2014; Pucher et al. 2011). In contrast, Harms et al. (2014) describe a lower propensity toward bicycle usage for people with a migrant background in countries with already high cycling rates, such as in the Netherlands. Instead, they use public transport and private motorized vehicles.

Factors such as perceptions, preferences, attitudes, habits, perceived controls on behavior (e.g., linked with the ability to carry luggage or to ride a bicycle), and social and personal norms are found to determine individual mode choice in general and active mode shares specifically. The theory of planned behavior (Ajzen 1991) is often applied in this context: This theory states that attitudes, habits, perceived behavioral control, and social and personal norms influence the intention to use certain transport modes; consequently, this intention influences the actual behavior. Affirmative opinions about active mobility are positively correlated with active mobility levels in a bidirectional, causal relationship (Kroesen et al. 2017). Motivating factors for cycling include status and lifestyle promotion, mental and physical relaxation, comfort, time savings, privacy and security (Heinen et al. 2011). Heinen (2011) lists the following individually perceived barriers for cycling: time loss compared to motorized transport modes, too-long distances to relevant destinations, bad weather, lack of parking facilities for bicycles, physical overload (e.g., caused by differences in altitude or insufficient fitness and sweating). Inhibiting aspects also refer to perceived security such as crime and fear of strangers (Lee et al. 2013; Foster et al. 2014). Factors like flexibility and environmental awareness (Damant-Sirois and El-Geneidy 2015; Handy et al. 2010; Böcker 2014; Heinen et al. 2011; de Geus et al. 2008) are identified only in a few studies. In all cases they have a positive influence on cycling trips.

Furthermore, "pleasant" neighborly relations in addition to participation in neighborly events set within a kind and trustful neighborhood increase the odds of walking (Van Cauwenberg et al. 2014; Clark and Scott 2013; Foster et al. 2014). Much less subjective motivational factors are identified for walking as compared to cycling. Ball et al. (2007) describe emotions such as fun and joy while walking as motivational factors. In particular, the purpose of each trip influences the decision for active mobility; the share of active mobility for leisure and for shopping purposes are higher than for others (Lindström et al. 2003; Ghani et al. 2016).

4 Impacts of Active Mobility

The main positive impacts of active mobility are improved health and enhanced quality of life as a consequence of the physical activity inherent to walking and cycling. Mueller et al. (2015) show, in their meta-analysis of health-impact assessments from Europe, North America, Australia, and New Zealand, the dominance of these effects. Improved health and quality of life generate 50–98% of the health effects from more frequent or prolonged walking and cycling trips resulting from measures for promoting active mobility.

Consistent moderate to vigorous physical activity reduces the risks of several non-communicable diseases such as cardiovascular diseases, type 2 diabetes, cancer, dementia, depression, and reduced life-expectancy (World Health Organisation (WHO) 2011). Insufficient physical activity is responsible for 25% of breast and colorectal cancer, 27% of diabetes and about 30% of ischemic heart disease (World Health Organisation (WHO) 2009). The most substantial health improvements are achieved for persons who are currently physically inactive and begin physical activities. Behavioral changes from increased walking and cycling for transport are more stable compared to sports activities—the latter of which are more often abandoned after a short time (Warburton et al. 2006).

Negative impacts of active mobility result from increased collision risks and from higher exposure to air pollution. Some studies also consider potential harms from increased noise exposure (Mueller et al. 2017). Pedestrians and cyclists have a raised risk per travelled distance of getting involved in an accident compared to car users, and, in addition, the accident consequences are on average more severe (Elvik 2009). Crash risks are lower where levels of active mobility are higher—a phenomenon often attributed to the so-called "safety in numbers" effect: This effect implies increased safety from the more prominent visibility of cyclists (or pedestrians) and resulting effects on driver awareness and behavior (Jacobsen 2003). While the non-linear relationship of crash risk and volumes has been confirmed by many, the term "safety-in-numbers" has been criticized for the implied direction of causality. It is very likely that safer infrastructure and traffic conditions (i.e., "safety") play the dominant role in increasing active mobility volumes (i.e., "numbers"), rather than vice versa (Götschi et al. 2016; Elvik 2009; Bhatia and Wier 2011). The number of underreported cases is high for collisions involving pedestrians and cyclists. von Below (2016) has observed, for single collisions with only cyclists, a share of unrecorded cases up to 96%. For collisions involving cyclists and cars, this share is up to 47%. This means that half of these collisions are not reported to the police and do not show up at all in the official collision statistics.

Panis et al. (2010) find five to nine times higher inhalation rates of air pollutants for cyclists while cycling compared to car occupants on the same routes. This leads to a significantly higher exposure to air pollution for cyclists than for car occupants. The resulting negative health effects from this exposure are, however, low compared to the abovementioned positive effects from improved health and quality of life resulting from the physical activity (Mueller et al. 2015). Tainio et al. (2016) show that cycling and walking are beneficial for the individual health even if engaged in excessively.

Environmental noise is, following air pollution, the second most significant environmental health risk in Europe (World Health Organisation (WHO) 2011). However, there is a lack of scientific evidence linking potentially increased exposure to noise while walking or cycling to negative health effects. Noise mapping is required every five years for all cities with more than 100,000 residents (European Commission (EC) Environmental Noise Directive 2002/29/EC 2002; Gerike et al. 2012). Unfortunately, these noise maps are not specifically focused on active mobility and not detailed enough to learn more about the noise exposure during walking or cycling compared to other transport modes.

The Health Economic Assessment Tool (HEAT, www.heatwalkingcycling.org) provided by the WHO is a user-friendly tool that allows for the quantification and monetization of the effects from measures for promoting active mobility including physical activity benefits, air pollution risks, crash risks and carbon emissions from modal shifts between active mobility and the motorized modes. Municipalities and other interested stakeholders can compute the expected impacts of planned measures for promoting active mobility (ex-ante) or evaluate the results from already implemented measures (ex-post) with the help of this tool.

5 Policies and Strategies for Promoting Active Mobility

Figure 2 visualizes how the three disciplines of urban planning, transport planning, and public health could and should collaborate for fostering active mobility. Each of these three disciplines pursue different specific objectives, but they all have a strong interest in achieving higher active mobility levels. This common objective is represented in the center of the figure in the middle triangle (see Fig. 2). The three disciplines are arranged at each triangle-apex in order to show that these are individual disciplines with substantial differences in their objectives, power, resources, competencies, and governance structures as well as in their policies for achieving the objectives. Acknowledging these differences is one core success factor for a fruitful collaboration.

In the figure, measures for promoting active mobility are highlighted in the outer circle representing the strategic level which encompasses an operative level in the form of a white triangle. On the strategic level, mid- and long-term strategic plans and concepts are developed. This is for urban planning the urban development plan that coordinates activities for all spatial elements including all land-use types. Urban development plans are highly important for transport planning since they determine the location, quantity, and quality of origins and destinations that need to be subsequently connected by transport planning. In addition, urban planning determines the location and widths of roads and, at times, can also contain prepared specifications of the street layout. Sustainable Transport Mobility Plans (SUMPs) are produced in transport planning. Following the distinction of the strategic and operational levels,

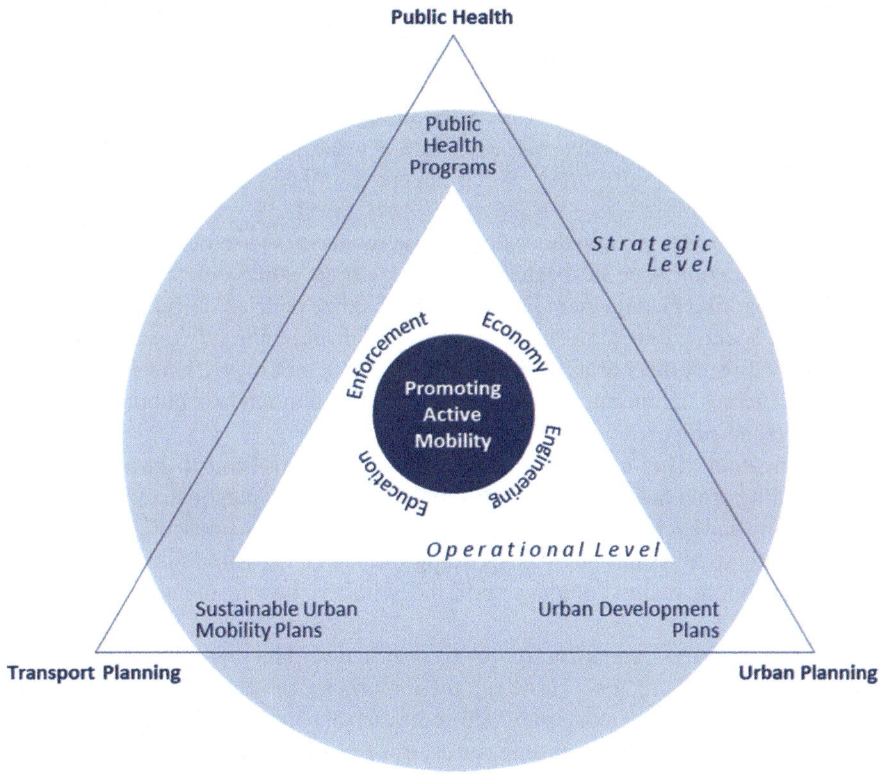

Fig. 2 Policies and strategies for fostering active mobility

these plans provide the strategies for developing future transport systems including all modes (GIZ 2015). Public health programs formulate health objectives and develop concepts and strategies for achieving those objectives. These strategic plans are each located at a triangle-apex and are specifically developed within the respective disciplines—this development should remain individual since each of the attributed strategic plans includes many objectives and strategies beyond active mobility. These are important for reaching the objectives of each discipline. The sections on active mobility provide for the coordination between the disciplines; this would create substantial synergies and improve the opportunities for fostering active mobility as a joint effort.

The operative level of measures for promoting active mobility is put into the white triangle based on the so-called "4 Es" as a proven classification of measures for promoting active mobility and also for transport planning in general (see e.g. Gerike and Parkin 2015; Winters et al. 2011):

E1 Engineering: Measures in this category address the built environment in urban planning and the transport supply in transport planning. They are supported by the processes of transport planning and traffic engineering. Examples are the strategic development of spatial structures, infrastructure networks, traffic signalization, layout of streets and intersections, (bicycle) parking facilities and the integration of transport modes.

E2 Enforcement: This category includes all legal issues such as land-use classifications, speed limits and rights-of-way. There is need for improvement in a number of jurisdictions, especially those which appear to "blame the victim" rather than "protect the vulnerable".

E3 Economy: Measures in this category use monetary instruments for incentivizing or discouraging specific behaviors. For example, some cities and companies implement parking management schemes, or subsidize pedelecs or bike-sharing systems. More economic incentives are possible that would indirectly encourage active mobility by directly discouraging car use: for example, pricing parking or congestion charging schemes. Economic measures are effective but often lack public or political acceptance and support.

E4 Education: This category includes all measures related to knowledge and understanding provided through information, campaigning, personalized travel planning, training and social marketing. For maximum effect, these measures will usually target specific user groups such as speeding offenders, school children learning to navigate the public realm, company staff, the elderly or new residents moving into a city.

In Fig. 2, the "4 Es" are not assigned to specific disciplines as all three disciplines are applied by all of the "Es". However, the importance of each of the "4 Es" differs between the disciplines. Transport planning and urban planning efforts are dominated by engineering and enforcement measures; the English-speaking countries Anglo-Saxon countries also have a high affinity for economic measures. Education measures are applied but with fewer resources and engagement than the other categories. Public health measures, on the contrary, are often focused on education. This difference in the importance of the different types of measures can be illustrated with the help of a "3L" classification of physical activity measures used in the public health field (Heath et al. 2012). This classification distinguishes three domains as follows:

L1: Campaigns and informational approaches: This category comprises community-wide campaigns, mass media campaigns, and decision prompts encouraging the use of stairs versus lifts and escalators.

L2: Behavioral and social approaches: These measures are meant to increase social support for physical activity within communities, specific neighborhoods, and worksites. The main difference to the first category is that measures in this second category target specific institutions that are classified in five different settings: school, workplace, community, clinical or health care, and other. Strategies for schools can encompass, for example, physical education, classroom activities or after-school sports, and the promotion of active transport.

L3: Policy and environmental approaches: These interventions create or enhance access to places for physical activity with outreach activities and include mainly measures from the "Engineering" category as described above.

The "3 Ls" thus put more weight into campaigning compared to the "4 Es". They assign the first two out of the three categories to community-wide versus institution-specific measures, classified altogether as "Education" in the "4 Es".

Packages of measures are needed for successfully achieving and maintaining high active mobility levels (Winters et al. 2011). Certain measures are highly effective but not well received; other measures are well received but are not very effective. Combining both into packages allows for their successful implementation while at the same time actually achieving behavioral changes and progress according to the specific circumstances and objectives.

6 Perspectives for Active Mobility

The above review of determinants and impacts of active mobility as well as of strategies for its promotion shows the substantial societal benefits that can be derived from high active mobility levels. Three disciplines are of particular importance for promoting active mobility: First, public health activities aim at increasing physical activity—this does not include, for example, traffic management or road safety improvements; second, transport planning efforts focus on providing efficient transport systems and leave, for example, the WHO standards of physical activity mainly unaddressed; third, the goals of urban planning are directed toward economically successful and attractive cities—urban planners focus less on the public health levels in the population. Thus, each of the three disciplines differ substantially in their ambitions and also in the policies used to achieve their respective goals while simultaneously promoting active mobility. Combining their expertise, resources, and policies has an enormous potential for fostering active mobility which would inevitably result in numerous societal benefits.

Acknowledgements Funding for the research project "Active Mobility: Improved quality of life in urban agglomerations" has been received from the German Environment Agency—UBA, and the research benefited from the European Union project "PASTA—Physical Activity Through Sustainable Transport Approaches".

References

Aditjandra PT, Mulley C, Nelson JD (2013) The influence of neighbourhood design on travel behaviour: empirical evidence from North East England. Trans Policy 26:54–65

Ahrens G-A, Hubrich S, Ließke F, Wittwer R (2010) Zuwachs des städtischen Autoverkehrs gestoppt!? Aktuelle Ergebnisse der Haushaltsbefragung "Mobilität in Städten – SrV 2008". Straßenverkehrstechnik 54(12):769–777

Ajzen I (1991) The theory of planned behaviour. Organ Behav Hum Decis Proc 50:179–211

Alfonzo MA (2005) To walk or not to walk? The hierarchy of walking needs. Environ Behav 37(6):808–836

Ball K, Bauman A, Leslie E, Owen N (2001) Perceived environmental aesthetics and convenience and company are associated with walking for exercise among Australian adults. Prev Med 33:434–440

Ball K, Timperio A, Salmon J, Giles-Corti B, Roberts R, Crawford D (2007) Personal, social and environmental determinants of educational inequalities in walking: a multilevel source. Epidemiol Community Health 61:108–114

Bamberg S, Schmidt P (2003) Incentives, morality, or habit? Predicting students' car use for university routes with the models of Ajzen, Schwartz, and Triandis. Environ Behav 35(2):264–285

Banister D (2007) Cities, urban form and sprawl: a European perspective by David Banister (United Kingdom). Organisation for Economic co-operation and development (OECD), European conference of Ministers of Transport (ECMT): transport, urban form and economic growth, Round table 137

Beck LF, Greenspan AI (2007) Why don't more children walk to school. J Safety Res 39:449–452

Bentley R, Jolley D, Kavanagh AM (2010) Local environments as determinants of walking in Melbourne, Australia. Soc Sci Med 70:1806–1815

Bhatia R, Wier M (2011) "Safety in Numbers" re-examined: Can we make valid or practical inferences from available evidence? Accid Anal Prev 43(1):235–240

Bird SR, Radermacher H, Sims J, Feldman S, Browning C, Thomas S (2010) Factors affecting walking activity of older people from culturally diverse groups: an Australian experience. J Sci Med Sport 13:417–423

Böcker L (2014) Climate, weather and daily mobility, transport mode choices and travel experiences in the Randstad Holland, Utrecht University

Borst HC, Miedema HME, de Vries SI, Graham JMA, van Dongen JEF (2008) Relationships between street characteristics and perceived attractiveness. J Environ Psychol 28:353–361

Borst HC, de Vries SI, Graham JMA, van Dongen JEF, Bakker I, Miedama HME (2009) Influence of environmental street characteristics on walking route choice of elderly people. J Environ Psychol 29:477–484

Braun LM, Rodriguez DA, Cole-Hunter T, Ambros A, Donaire-Gonzalez D, Jerrett M, Mendez MA, Nieuwenhuijsen MJ, de Nazelle A (2016) Short-term planning and policy interventions to promote cycling in urban centers: findings from a commute mode choice analysis in Barcelona, Spain. Transp Res Part A 89:164–183

Buehler R, Pucher J, Gerike R, Goetschi T (2016) Reducing car dependence in the heart of Europe: lessons from Germany, Austria, and Switzerland. Trans Rev 36:1–25

Bühler R (2012) Determinants of bicycle commuting in the Washington, DC region: the role of bicycle parking, cyclist showers, and free car parking at work. Transp Res Part D 17:525–531

Bühler R, Pucher J, Merom D, Bauman A (2011) Active travel in Germany and the U.S. contributions of daily walking and cycling to physical activity. Am J Prev Med 41(3):241–250

Carlson SA, Paul P, Watson KB, Schmid TL, Fulton JE (2016) How reported usefulness modifies the association between neighborhood supports and walking behavior. Prev Med 91:76–81

Carnegie MA, Bauman A, Marshall AL, Mohsin M, Westley-Wise V, Booth ML (2002) Perceptions of the physical environment, stage of change for physical activity and walking among Australian adults. Res Q Exerc Sport 73(2):146–155

Carver A, Panter JR, Jones AP, van Sluijs AMF (2014) Independent mobility on the journey to school: a joint cross-sectional and prospective exploration of social and physical environmental influences. J Trans Health 1:25–32

Carver A, Tomperio AF, Crawford DA (2015) Bicycles gathering dust rather than raising dust—prevalence and predictors of cycling among Australian school children. J Sci Med Sport 18:540–544

Cervero R (1996) Mixed land-uses and commuting: evidence from the American housing survey. Transp Res Part A 30(5):361–377

Cervero R, Duncan M (2003) Walking, bicycling, and urban landscapes: evidence from the San Francisco Bay area. Am J Public Health 93(9):1478–1483

Cervero R, Sarmiento OL, Jacoby E, Gomez LF, Neiman A (2009) Influences of built environments on walking and cycling: lessons from Bogotá. Int J Sustain Transp 3:203–226
Christiansen LB, Madsena T, Schipperijn J, Ersbøll AK, Troelsen J (2014) Variations in active transport behavior among different neighborhoods and across adult life stage. J Transp Health 1:316–325
Clark AF, Scott DM (2013) Does the social environment influence active travel? An investigation of walking in Hamilton, Canada. J Trans Geogr 31:278–285
Damant-Sirois G, El-Geneidy AM (2015) Who cycles more? Determining cycling frequency through a segmentation approach in Montreal, Canada. Transp Res Part A 77:113–125
de Geus B, de Bourdeaudhuij I, Jannesm C, Meeusen R (2008) Psychological and environmental factors associated with cycling for transport among a working population. Health Educ Res 23(4):679–708
De Witte A, Hollevoet J, Dobruszkes F, Hubert M, Macharis C (2013) Linking modal choice to motility: a comprehensive review. Transp Res Part A 49:329–341
Deka D (2013) An explanation of the relationship between adults' work trip mode and children's school trip mode through the Heckman approach. J Transp Geogr 31:54–63
Dill J, Carr T (2003) Bicycle commuting and facilities in major U.S. cities: if you build them, commuters will use them. Transp Res Rec 1828:116–123
Downward P, Rasciute S (2015) Assessing the impact of the national cycle network and physical activity lifestyle on cycling behaviour in England. Transp Res Part A 78:425–437
Elvik R (2009) The non-linearity of risk and the promotion of environmentally sustainable transport. Accid Anal Prev 41(4):849–855
European Commission (EC) (2014) Cleaner air for all (fact sheet). http://ec.europa.eu/environment/pubs/pdf/factsheets/air/en.pdf. 17 Dec 2017
European Commission (EC) (2015) Road safety in the European Union. https://ec.europa.eu/transport/sites/transport/files/road_safety/pdf/vademecum_2015.pdf. 17 Dec 2017
European Commission (EC) Environmental Noise Directive (2002/29/EC)
European Environment Agency (EEA) (2017a) Greenhouse gas emissions from transport. https://www.eea.europa.eu/data-and-maps/indicators/transport-emissions-of-greenhouse-gases/transport-emissions-of-greenhouse-gases-10. 17 Dec 2017
European Environment Agency (EEA) (2017b) Exceedance of air quality standards in urban areas. https://www.eea.europa.eu/data-and-maps/indicators/exceedance-of-air-quality-limit-3/assessment-3. 17 Dec 2017
European Environmental Agency (EEA) (2017c) Road traffic remains biggest source of noise pollution in Europe
European Union (EU) (2007) Leipzig Charter on sustainable European cities. Informal ministerial meeting on urban development and territorial cohesion
European Union (EU) (2016) The pact of Amsterdam: Urban Agenda for the EU. Informal meeting of EU ministers responsible for urban matters
Ewing R, Cervero R (2010) Travel and the built environment. J Am Planning Assoc 76(3):265–294
Forschungsgesellschaft für Straßen- und Verkehrswesen (FGSV) (2011) Empfehlungen zur Straßenraumgestaltung innerhalb bebauter Gebiete
Forschungsgesellschaft für Straßen- und Verkehrswesen (FGSV) (2014) Hinweise zur Nahmobilität – Strategien zur Stärkung des nichtmotorisierten Verkehrs auf Quartiers- und Ortsteilebene
Foster S, Villanueva K, Wood L, Christian H, Giles-Corti B (2014) The impact of parents' fear for strangers and perceptions of informal social control on children's independent mobility. Health Place 26:60–68
Gehl J (2010) Cities for people. Island Press
Gerike R, Koszowski C (2017) Sustainable urban transportation. Encycl Sustain Technol 379–391
Gerike R, Parkin J (2015) Cycling futures. From research into practice. Ashgate
Gerike R, Becker T, Friedemann J, Hülsmann F, Heidegger F (2012) Mapping external noise costs to the transport users—conceptual issues and empirical results. In: Proceedings of the Euronoise 9th European conference on noise control, Prague, 10–13 June 2012

Gerike R, de Nazelle A, Nieuwenhuijsen M, Panis L I, Anaya E Avila-Palencia I, Boschetti F; Brand C, Cole-Hunter T, Dons E, Eriksson U, Gaupp-Berghausen M, KahlmeierS, Laeremans M, Mueller N, Orjuela JP, Racioppi F, Raser E, Rojas-Rueda D, Schweizer C, Standaert A, Uhlmann T, Wegener S, Götschi T (2016) Physical Activity through Sustainable Transport Approaches (PASTA): a study protocol for a multicenter project. BMJ Open 6(1)

Ghani F, Rachele JN, Washington S, Rurrell G (2016) Gender and age differences in walking for transport and recreation: are the relationships the same in all neighborhoods. Prev Med Rep 4:75–80

Ghekiere A, van Cauwenberg J, Carver A, Mertens L, de Geus B, Clarys P, Cardon G, de Bourdeaudhuij I, Deforche B (2016) Psychosocial factors associated with children's cycling for transport: a cross-sectional moderation study. Prev Med 86:141–146

Giles-Corti B, Donovan R (2002) The relative influence of individual, social and physical environment determinants of physical activity. Soc Sci Med 54:1793–1817

Giles-Corti B, Bull F, Knuiman M, McCormack G, van Niel K, Timperio A, Christian H, Foster S, Divitini M, Middleton N, Boruff B (2013) The influence of urban design on neighbourhood walking following residential relocation: longitudinal results from the RESIDE study. Soc Sci Med 77:20–30

GIZ – Deutsche Gesellschaft für Internationale Zusammenarbeit: Recommendations for Mobility Master Planning, 2015

Götschi T, Tainio M, Maizlish N, Schwanen T, Goodman A, Woodcock J (2015) Contrasts in active transport behaviour across four countries: how do they translate into public health benefits? Prev Med 74:42–48

Götschi T, Garrard J, Giles-Corti B (2016) Cycling as a part of daily life: a review of health perspectives. Trans Rev 36(1):45–71

Götschi T, de Nazelle A, Brand C, Gerike R (2017) Towards a comprehensive conceptual framework of active. Travel behavior: a review and synthesis of published frameworks. Curr Environ Health Rep 4:86–295

Guliani A, Mitra R, Buliung RN, Larsen K, Faulkner GEJ (2015) Gender-based differences in school travel mode choice behaviour: examining the relationship between the neighbourhood environment and perceived traffic safety. J Trans Health 2:502–511

Guo Z (2009) Does the pedestrian environment affect the utility of walking? A case of path choice in downtown Boston. Transp Res Part D 14:343–352

Guo JY, Bhat CR, Copperman RB (2010) Effect of the built environment on motorized and non-motorized trip making: substitutive, complementary or synergistic. Trans Res Rec 1–11:2007

Handy SL, Xing Y, Buehler TJ (2010) Factors associated with bicycle ownership and use: a study of six small U.S. cities. Transportation 37:967–985

Handy S, van Wee B, Kroesen M (2014) Promoting cycling for transport: research needs and challenges. Trans Rev 34(1):4–24

Hansen KB, Nielsen TAS (2014) Exploring characteristics and motives of long distance commuter cyclists. Transp Policy 35:57–63

Harms L, Bertoloni L, te Brömmelstroet M (2014) Spatial and social variations in cycling patterns in a mature cycling country exploring differences and trends. J Trans Health 1:232–242

Haustein S, Hunecke M (2007) Reduced use of environmentally friendly modes of transportation caused by perceived mobility necessities: an extension of the theory of planned behavior. J Appl Soc Psychol 37(8):1856–1883

Haybatollahi M, Czepkiewicz M, Laatikainen T, Kyttä M (2015) Neighbourhood preferences, active travel behaviour and built environment: an exploratory study. Transp Res Part F 29:57–69

Heath GW, Parra DC, Sarmiento OL, Andersen LB, Owen N, Goenka S, Montes F, Brownson RC (2012) Evidence-based intervention in physical activity: lessons from around the world. Lancet 380(9838):272–281

Heinen E (2011) Bicycle commuting. Delft University of Technology (TU Delft), Delft Centre for Sustainable Urban Areas, 43

Heinen E, Maat K, van Wee B (2011) The role of attitudes toward characteristics of bicycle commuting on the choice to cycle to work over various distances. Transp Res Part D 16:102–109
Heinen E, Maat K, van Wee B (2013) The effect of work-related factors on the bicycle commute mode choice in the Netherlands. Transportation 40:23–43
Helbich M, van Emmichoven MJZ, Dijst MJ, Kwan M-P, Pierik FH, de Vries SI (2016) Natural and built environment exposures on children's active school travel: a Dutch global positioning system-based cross-sectional study. Health Place 39:101–109
Hoehner CM, Brennan Ramirez LK, Elliot MB, Handy SL, Brownson RC (2005) Perceived and objective environmental measures and physical activity among urban adults. Am J Prev Med 28:105–116
Hooper P, Knuiman M, Foster S, Giles-Corti B (2015) The building blocks of "Liveable Neighbourhood": identifying the key performance indicators for walking of an operational planning policy in Perth, Western Australia. Health Place 36:173–183
Hunt JD, Abraham JE (2007) Influences on bicycle use. Transportation 34:453–470
Jacobsen P (2003) Safety in numbers: more walkers and bicyclists, safer walking and bicycling. Inj Prev 9(3):205–209
Kamargianni M (2015) Investigating next generation's cycling ridership to promote sustainable mobility in different types of cities. Res Transp Econ 53:45–55
Kang C-D (2015) The effects of spatial accessibility and centrality to land use on walking in Seoul, Korea. Cities 46:94–103
Kaplan S, Sick Nielsen TA, Prato CG (2016) Walking, cycling and the urban form: a Heckman selection model of active travel mode and distance by young adolescence. Transp Res Part D 44:55–95
Kitamura R, Mokhtarian PL, Laidet L (1997) A micro-analysis of land use and travel in five neighborhoods in the San Francisco Bay Area. Transportation 24:125–158
Kotval-Karamchandani Z, Vojnovic I (2015) The socio-economics of travel behavior and environmental burdens: a Detroit, Michigan regional context. Transp Res Part D 41:477–491
Kroesen M, Handy S, Chorus C (2017) Do attitudes case behavior or vice versa? An alternative conceptualization of the attitude-behavior relationship in travel behavior modelling. Transp Res Part A 101:190–202
Lamíquiz J, López-Domínguez J (2015) Effects of built environment on walking at the neighbourhood scale. A new role for street networks by modelling their configurational accessibility? Transp Res Part A 74:148–163
Lee JS, Zegras PC, Ben-Joseph E (2013) Safely active mobility for urban baby boomers: the role of neighborhood design. Accid Anal Prev 61:153–166
Lee C, Yoon J, Zhu X (2016) From sedentary to active school commute: multi-level factors associated with travel mode shifts. Prev Med 95:28–36
Lindström M, Moghaddassi M, Merlo J (2003) Social capital and leisure time physical activity: a population based multilevel analysis in Malmö, Sweden. J Epidemiol Community Health 57:23–28
Lowry MB, Furth P, Hadden-Loh T (2016) Prioritizing new bicycle facilities to improve low-stress network connectivity. Transp Res Part A Policy Pract 86:124–140
Marquet O, Miralles-Guasch C (2015a) Neighbourhood vitality and physical activity among the elderly: the role of walkable environments on active ageing in Barcelona, Spain. Soc Sci Med 135:24–30
Marquet O, Miralles-Guasch C (2015b) The walkable city and the importance of the proximity environments for Barcelona's everyday mobility. Cities 42:258–266
Maslow AH (1954) Motivation and personality. New York, Harper & Row
McDonald NC (2007) Travel and the social environment: evidence from Alameda County, California. Transp Res Part D 12:53–63
Mueller N, Rojas-Rueda D, Cole-Hunter T, de Nazelle A, Dons E, Gerike R, Götschi T, Panis LI, Kahlmeier S, Nieuwenhuijsen M (2015) Health impact assessment of active transportation: a systematic review. Prev Med 76:103–114

Mueller N, Rojas-Rueda D, Basagana X, Cirach M, Cole-Hunter T, Dadvand P, Donaire-Gonzales D, Foraster M, Gascon M, Martinez D, Tonne C, Triguero-Mas M, Valentin A, Nieuwenhuijsen M (2017) Urban and transport planning related exposures and mortality: a health impact assessment for cities. Environ Health Perspect 125:89–96

Ogilvie D, Bull F, Powell J, Cooper AR, Brand C, Mutrie N, Preston J, Rutter H (2011) An applied ecological framework for evaluating infrastructure to promote walking and cycling: the iConnect study. Am J Public Health 101(3):473–481

Ogilvie D, Bull F, Cooper AR, Rutter H, Adams E, Brand C, Ghali K, Jones T, Mutrie N, Powell J, Preston J, Sahlqvist S, Song Y (2012) Evaluating the travel, physical activity and carbon impacts of a "natural experiment" in the provision of new walking and cycling infrastructure: methods for the core module of the iConnect study. BMJ Open 2(1):1–13

Panis LI, de Geus B, Vandenbulcke G, Willems H, Degraeuwe B, Bleux N, Mishra V, Thomas I, Meeusen R (2010) Exposure to particulate matter in traffic: a comparison of cyclists and car passengers. Atmos Environ 44(19):2263–2270

Parkin J, Koorey G (2009) Network planning and infrastructure design. Cycl Sustain 131–160

Pikora T, Giles-Corti B, Bull F, Jamrozik K, Donovan R (2003) Developing a framework for assessment of the environmental determinants of walking and cycling. Soc Sci Med 56:1693–1703

Prins R, Pierik F, Etman E, Sterkenburg R, Kamphuis S, Van Lenthe F (2014) How many walking and cycling trips made by elderly are beyond commonly used buffer sizes: results from a GPS study. Health Place 27:127–133

Pucher J, Buehler R, Seinen M (2011) Bicycling renaissance in North America? An update and re-appraisal of cycling trends and policies. Transp Res Part A 45:451–475

Sallis J, Cervero R, Ascher W, Henderson K, Kraft M, Kerr J (2006) An ecological approach to creating active living communities. Annu Rev Public Health 27:297–322

Sallis JF, Owen N, Fisher EB (2008) Ecological models of health behavior. Health behavior and health education: theory, research, and practice, 4th edn. Jossey-Bass 465–486

Schneider RJ (2013) Theory of routine mode choice decisions: an operational framework to increase sustainable transportation. Transp Policy 25:128–137

Singleton PA, Clifton KJ (2015) The theory of travel decision-making: a conceptual framework of active travel behavior. Transportation research board 94th annual meeting. http://pdxscholar.library.pdx.edu/trec_seminar/84/. 21 Dec 2017

Stead D, Marshall S (2001) The relationships between urban form and travel patterns: an international review and evaluation. Eur J Trans Infrastruct Res 1:113–141

Su J, Jarrett M, Mcconnel R, Berhane K, Dunton G, Shankardass K, Reynolds K, Chang R, Wolch J (2013) Factors influencing whether children walk to school. Health Place 22:153–161

Tainio M, de Nazelle A, Götschi T, Kahlmeier S, Rojas-Rueda D, Niuwenhuijsen M, de Sá TH, Kelly Kelly P, Woodcock J (2016) Can air pollution negate the health benefits of cycling and walking? Prev Med 87:233–236

Van Acker V, Van Wee B, Witlox F (2010) When transport geography meets social psychology: toward a conceptual model of travel behaviour. Transp Rev 30(2):219–240

Van Cauwenberg J, Van Holle V, De Bourdeaudhuij I, Clarys P, Nasar J, Salmon J, Maes L, Goubert L, Van de Weghe N, Deforche B (2014) Physical environmental factors that invite older adults to walk for transportation. J Environ Psychol 38:94–103

Van Goeverden C, De Boer E (2013) School travel behaviour in the Netherlands and Flanders. Transp Policy 26:73–84

von Below A (2016) Verkehrssicherheit von Radfahrern: Analyse sicherheitsrelevanter Motive, Einstellungen und Verhaltensweisen. Bundesanstalt für Straßenwesen

Warburton DE, Nicol CW, Bredin SS (2006) Health benefits of physical activity: the evidence. Can Med Assoc J 174(6):801–809

Winters M, Davidson G, Kao D, Teschke K (2011) Motivators and deterrents of bicycling: comparing influences on decisions to ride. Transportation 38:153–168

Wittwer R (2014) Zwangsmobilität und Verkehrsmittelorientierung junger Erwachsener: Eine Typologisierung, Technische Universität Dresden, Institute of Transport Planning And Road Traffic, 16
Wittwer R, Hubrich S (2016) What happens beneath the surface? Evidence and insights into changes in urban travel behaviour in Germany. Transp Res Procedia 14:4304–4313
World Health Organisation (WHO) (2009) Global health risks: mortality and burden of disease attributable to selected major risks. http://www.who.int/healthinfo/global_burden_disease/GlobalHealthRisks_report_full.pdf. 17 Dec 2017
World Health Organisation (WHO) (2011) Action plan for implementation of the European strategy for the prevention and control of noncommunicable diseases 2012–2016. http://www.euro.who.int/en/what-we-do/health-topics/noncommunicable-diseases/cancer/publications/2011/eurrc6112-action-plan-for-implementation-of-the-european-strategy-for-the-prevention-and-control-of-noncommunicable-diseases-20122016. 17 Dec 2017
World Health Organisation (WHO) (2017) Prevention and control of noncommunicable diseases in the European Region: a progress report. 2014 <http://www.euro.who.int/en/health-topics/noncommunicable-diseases/cancer/publications/2013/prevention-and-control-of-noncommunicable-diseases-in-the-european-region-a-progress-report. 17 Dec 2017

A Data Driven, Segmentation Approach to Real World Travel Behaviour Change, Using Incentives and Gamification

Hannah Bowden and Gabriel Hellen

Abstract This paper reviews the data from a 6-month transport project in Bologna, Italy, which aimed to reduce car journeys and increase active travel through gamification and rewards. 667 participants who had registered via a smartphone app, called BetterPoints, and identified themselves as 'everyday car users' were included in the sample. Behavioural categories based on engagement and the frequency and maintenance of tracked sustainable/active trips were proposed as a way to understand the data and to tailor future interventions. 47% of the 667 everyday car users showed some form of maintained engagement and behavioural change throughout the project. It was concluded that gamification and incentives can motivate travel behaviour change, but that more work is to be done to understand the links between user categories and intervention components in order to dynamically adapt and optimise the programme.

Keywords Behaviour change · Voluntary travel behaviour change · Motivation Incentives · Gamification · Segmentation · Persuasive technology

1 Introduction

The evolution of voluntary travel behaviour change programmes (VTBC) from purely campaigns-based social marketing, to mobile phone-based persuasive technologies is proceeding apace within the academic community (Meloni and Teulada 2015). However, academic pilots are currently limited to small sample sizes and have yet to demonstrate scalability to real-world, population level interventions. Intervention design is not the only thing that is lagging behind academia in real world interventions; evaluation methodology is often reliant on retrospective self-report measures, which, whilst still useful, need to be combined with data driven evidence to fully capture the dynamic nature of travel behaviour (Gerike and Lee-Gosselin 2015). This

H. Bowden (✉) · G. Hellen
Durham University, Durham, UK
e-mail: Hannah.bowden.16@ucl.ac.uk

B. Müller and G. Meyer (eds.), *Towards User-Centric Transport in Europe*, Lecture Notes in Mobility, https://doi.org/10.1007/978-3-319-99756-8_12

chapter explores a data driven approach to segmenting participants based on engagement and to understand change by combining this tracked data with in-context self-report surveys. Data is obtained via a smartphone app called BetterPoints. The app is offered as part of a complex behaviour change intervention which employs gamification and incentivisation in a real world setting.

Gamification, as the name suggests, employs game-like approaches in real life contexts to elicit behaviour change, with the objective of improving health or the environment. The application of gamification to transport behaviour change, in order to reduce congestion, improve air quality and encourage greater physical activity, has gained increased interest in recent years. Examples of how gamification has been applied in practice include ‘Beat the Street’, which rewards schools for competing against each other in a 6-week challenge. Participants receive a card which contains radio-frequency identification (RFID) technology and they then tap them against the sensors called ‘Beat Boxes’ located on lampposts across the area. Players receive points for each box they tap and can create or join teams which can receive prizes for tapping the most boxes. Coombes and Jones (2016) showed that weekly active travel increased at the intervention school by 10% in an intervention compared to a control school, where active travel decreased by 7% per child (Coombes and Jones 2016).

The SUPERHUB (Wells et al. 2014) project uses a Points Accumulation Gamification model (PAG-M) to underpin gamified aspects of travel. SUPERHUB offers personalised travel plans upon which users are rewarded for selecting routes and modes of travel that improve upon previous scores for saving money, decreasing CO2 emissions and/or burning calories. This is based on the assumption that sustainable and active travel, whilst being better for the environment, also promotes better health and is more cost effective. Wells et al. point out that whilst gamification has become a popular technique that there is still little knowledge about the effective design of such systems. Highlighting the challenges associated with effective design of rewards based approaches to gamified behaviour change, specifically with regards to the cost, sustainability and scalability of such systems. Other research supports findings that user challenges within persuasive mobile applications are well-received by users but further personalisation is required (Jylhä et al. 2013).

Anable’s (2005) segmentation approach draws on the psychological theory of planned behaviour (TPB) to identify psychographic groups using cluster analysis, pointing out that social-demographic factors have little bearing on the travel profiles of the segments. Forbes et al. (2014) have proposed an algorithm-based approach to tailoring motivational messages to Anabel’s traveler segments. This suggestion holds promise but to be useful in real world travel interventions, it begs the question of whether user categorization based on dynamic, in-context behaviour may deliver better results through personalization and gamification than psychographic profiles.

The smart phone app and technology platform used to engage and reward participants in this study are created by BetterPoints Ltd, an evidence-led health, sustainability and social behaviour change technology company that uses a range of behaviour change techniques, including gamification, underpinned by a points based system. BetterPoints’ first large scale deployment in Bologna, Italy, was delivered in partnership with the local mobility authority, SRM. The 'Bella Mossa' programme, translated as the 'Good Move', ran for 6 months between 1st April and 30th September 2017. The project took place in the metropolitan area of Bologna, a region 3700 km^2 in size, where around 1 million people live and move daily.

This chapter reviews the data collected by participants and proposes a targeted, behavioural category based approach to assessing the data and evaluating the evidence of travel behaviour change in an everyday car users sample. It is hypothesised that behavioural categories relating to engagement; the level and maintenance of tracked positive behaviour, may assist in understanding how the intervention may be tailored for these groups to make it more effective in the future. It reviews the evidence of gamification and incentives in motivating travel behaviour change, discusses the behavioural categorisation approach taken and how to link segmentation to gamification and rewards, to reduce app churn, increase engagement and maintain behaviour change.

2 Methods

Participants were recruited to the Bella Mossa programme through a promotional campaign on radio, newspapers and TV which encouraged people to download the BetterPoints app and register with their name, email, gender and year of birth. All citizens (over the age of 14) living and travelling across Bologna could sign up and manually track their sustainable/active journeys using the free BetterPoints app (on Android and iOS devices). Participants could select from 5 modes of travel; walk, cycle, bus, train and car sharing in order to earn points for tracking journeys which could then be exchanged for real world rewards. Messaging on the users' app timeline/notifications and gamification techniques such as challenges were also employed. In the case of challenges, points earned by participants from different organisations counted towards that organisations' place in a leaderboard which could be viewed via the app and programme web portal.

Participants had several opportunities to earn BetterPoints (BP) throughout the project, primarily through the basic action of tracking a journey, earning BP, and then redeeming vouchers. This "basic loop" was available throughout the 6 months of the programme. BP could also be earned by achieving specific goals, such as the first sustainable trip of the day (25 BP) or cycling/walking at least 150 min per week (1000 BP). Special events were held for the public (e.g. the Flower Festival in Crevalcore), and partner events were held for business associates (e.g. The Extraordinary Blue Night at the Villagio della Salute Più indoor spa centre). Participants could also earn BP if they travelled to the venue sustainably.

Participants of the Bella Mossa programme could convert their BPs at any time into monetary discounts or vouchers (e.g. 5% off/€5 off at a specified supermarket.) The strong and wide public-private partnership between SRM and 85 businesses (ranging from large distributors and major retailers to smaller local stores) provided variety in the types of discounts and rewards offered, meaning every user had something to strive towards—examples include: a €5 voucher for shopping, a free beer, a 2 × 1 ticket for the cinema, a discount for the hairdresser or a free entrance to the spa etc. The inclusion of low cost as well as high value rewards allowed for quick feedback on participants' positive behaviour. The integration of barcode scanning technology into the BetterPoints app allowed vouchers to be redeemed quickly and easily from the participants' smartphone, making rewards very accessible.

Travel behaviour was measured through GPS tracking in participants' mobile phones and in-app surveys and context linked questions. Participants were able to select their active or sustainable transport mode; walking, cycling, bus, train or car share. Tracked journeys were verified by sophisticated algorithms, OpenStreetMap and local bus company routes data. A validation system checked the waypoints against routes for the journey type (tram journeys are on tram lines, bus journeys on bus routes etc…), algorithms checked for speed and acceleration.

There were two methods of collecting self-report data: a selectable and incentivised survey activity and in-context questions. The former selected from the activities menu and the latter appearing after a participant had completed an activity. The in-context questions had to be answered for the points to be received for that activity and appeared in the app when the participant clicked the 'Complete' button upon finishing their activity. The survey questions covered existing travel behaviour such as 'How often do you use your car alone?' at the start of the programme and 'Did you reduce the use of your car/motorcycle because of the BetterPoints app?' at the end. In-context, activity-linked questions were more specific, such as 'Did this journey replace a car journey?'

The behaviour change analysis was based on the tracked GPS data from participants' smartphones. A baseline survey that users could select from the activities menu in the app was used to identify a sample of 'everyday car users'. This sample size was 667 people. Only users who downloaded the BetterPoints app and registered their account for at least 20 weeks of the overall 6-month programme were included in this sample.

The number of sustainable/active journeys these 'everyday car users' made during the beginning (weeks 1–4 from programme inception), middle (weeks 9–12) and end (weeks 17–20) periods of the programme were counted.

The 'everyday car users' were given behavioural categories based on the regularity and maintenance of their sustainable (bus, train, car share) or active (walk, cycle) travel activities, as follows:

- *Maintainers*: recorded at least 3 activities per week for 16 of the 20 weeks.
- *Fluctuators*: recorded at least 1 activity in weeks 1–4, 1 in weeks 9–12, and at least 1 activity in weeks 17–20.
- *Early stoppers*: recorded at least 3 activities in the weeks 1–4 but none after that.
- *Late stoppers*: recorded at least 3 activities in the weeks 1–4 and 9–12, but none after that that.
- *Non-starters*: registered for the project but then didn't go on to track any journeys.
- *Outliers*: didn't fall into any of the above categories.

The smartphone-tracked journey data for the 'everyday car user' group was manually reviewed for patterns in behaviour. Based on observation of data in earlier BetterPoints travel programmes, the database was queried to reveal patterns relating to broad groupings of Maintainers, Fluctuators and Non-Starters behaviour. However, the resulting data set revealed that there were groups of people who had changed their behaviour but then not maintained either the behaviour itself, or the recording of that data at different times. These were then categorized as Late and Early stoppers.

The self-report responses to both the end-of-programme survey question and in-context questions for the 'everyday car users' sample were extracted for each of the behavioural categories identified from the GPS tracked data. This was done in order to ascertain the extent to which the tracked data represented real behaviour change, and not just app use.

3 Results

Of the 667 participants who identified themselves as everyday car users, 596 (89%) demonstrated some level of behaviour change by tracking sustainable or active travel behaviour. 71 (11%) of the 667 self-identified everyday car users registered for the app but did not go on to track any sustainable/active journeys (Table 1).

12% of the 667 everyday car users were classified as Maintainers. They continued their active travel throughout the project, tracking a minimum of 3 activities per week for 16 of the 20 weeks.

47% of the 667 everyday car users showed some form of visible long-term behavioural change throughout the project (Maintainers, Fluctuators and Late Stoppers).

Of the 111 Outliers, 8 people did at least 1 activity in weeks 1–4 and at least 1 activity in weeks 17–20.

Of the 111 Outliers, 19 people did 10 or more activities in weeks 1–4. 3 of these 19 Outliers were still active in weeks 17–20 but had not recorded activity during weeks 9–12 of the programme.

Table 1 Participant behaviour change categories

Category	Count
Maintainer	79
Fluctuator	121
Early stopper	173
Late stopper	112
Non-starter	71
Outlier	111
Total	667

The remaining 84 participants in the Outliers group were recording journeys during weeks 9–12, but only 1 or 2 activities (Table 1).

Participants were asked questions during and after the programme to determine if car use actually decreased. At an interval of at most 3 days, and always after recording an activity, participants were asked 'Did this activity replace a car journey?'

The in-context question "Did this journey replace a car journey" was asked 6657 times and participants answered 'Yes' 81% of the time (N = 5395) (Table 2).

In a survey at the end of the Bella Mossa Programme, participants were asked once 'Did you reduce the use of your car/motorcycle because of the BetterPoints app?' (Table 3).

146 (79%) of the 185 'everyday car users' who responded to the end-of-programme survey said that, as a result of the BetterPoints app, they reduced the use of their car. This represents nearly 22% of the total sample of 667 'everyday car users'.

Table 2 Responses to the in-context question 'Did this journey replace a car journey?'

Category	Count	Number of times the question "Did this journey replace a car journey" asked	Yes—this replaced a car journey	No—This didn't replace a car journey
Maintainer	79	2384	1857	527
Fluctuator	121	2101	1736	365
Early stopper	173	441	350	91
Late stopper	112	1402	1180	222
Non-starter	71	0	0	0
Outlier	111	329	272	57
Total	667	6657	5395	1262

Table 3 Responses to the in-app question: 'Did you reduce the use of your car/motorcycle because of the BetterPoints app?'

Category	Count	Questions asked	Yes	No
Maintainer	79	62	52	10
Fluctuator	121	81	63	18
Early stopper	173	4	3	1
Late stopper	112	24	19	5
Non-starter	71	0	0	0
Outlier	111	14	9	5
Total	667	185	146	39

4 Discussion

The case for using gamification and incentives in voluntary travel behaviour change programmes (VTBC) seems largely closed. The analysis of data from the Bella Mossa programme supports the notion that these techniques appear to shift travel behaviour towards more sustainable modes. 89% of 'everyday car users' answered 'Yes' 81% of the time when asked, in-context, if their sustainable travel had replaced a car journey. 79% of respondents also answered 'Yes', when asked if the BetterPoints app had motivated this mode shift. However, both these questions may be considered to be leading and so it is recommended that future in-context surveys of this kind are more open-ended. Despite this limitation, the use of gamification and incentives in a smart phone app appears to offer a more scalable solution to traditional social marketing campaigns.

The challenge remains that if we are to implement large scale, real world VTBCs such as Bella Mossa, how can we target it in a way that is cost effective and elicits real mode shift as well as maintaining and increasing sustainable and active travel? Are static psychosocial segmentation approaches (Anable 2005) adequate or do we need more dynamic categories based on engagement behaviour to inform the tailoring of the intervention? A data driven approach using tracked GPS data from the BetterPoints app offered 6 potential categories; *maintainers, fluctuators, early stoppers, late stoppers, non-starters* and *outliers*, that could be mapped to 'short term loops' of increased incentives, messaging and links to real world events. For example, the promise of tangible prizes, like a Bella Mossa hat, may not be important to a 'maintainer' who already uses the app frequently, however it may be vital in providing a mid-term goal for an 'early stopper' to strive towards.

Understanding the motivations or barriers that cause 'non-starters' to register on the app but not go on to track any activity may help to identify levels and types of rewards that will provide the initial push to get started. A better understanding of the relationship between engagement and the target behaviour to change is critical to the future design of digital VTBC programmes. The present analysis reviewed the data at the end of the 6 month programme to determine the categories of everyday car users. In order to *dynamically* map rewards to user segments during the programme, new approaches will need to be devised. These may include the development of algorithms for clustering data, assigning categories and delivering rewards such as the approach proposed by Forbes et al. (2014).

Maintaining behaviour change is known to be difficult and this was also observed during the Bella Mossa programme. 285 (43%) of the 667 "everyday car users" were classified as either early or late stoppers, meaning that whilst they showed promising signs of maintained sustainable/active travel, they ultimately gave up tracking this behaviour during the project. It is unknown if this means the sustainable or active travel behaviour itself continued and the disengagement was simply app churn. Churn relates to how many users will re-launch an app after downloading it. This is a persistent problem with apps across many markets, whereby an average of 63% of people don't revisit an app after the first month (O'Connell 2017). If we are to consider the 'Non-starter' and 'Early Stopper' groups to represent app churn, then at approximately 37% of the overall group, this would be better than the industry average. If we add 'Late Stoppers' to this picture the percentage rises to 53%. If those who did not continue to use the app from the outlier group were also considered in this picture, it is likely to be looking similar to the industry average churn rate. Further work may explore the relationship between app churn and behaviour change by following up with churned users with regards to why they stopped using the app and whether the target behaviour had changed and/or new behaviour maintained.

It seems less likely that churn was a result of dissatisfaction with app functionality or the way in which people were asked to record journeys as only 2% of all participants (not just everyday car users) who completed the end of programme survey said that they would not participate again because recording journeys was 'too tiresome'. 84% of people across the whole programme said that they would participate again 'without a doubt'. Despite this positive stance on app functionality, it is recommended that future interventions use more targeted, short-term rewards to maintain engagement. Automatic tracking, rather than manually starting and stopping an activity, may also help to reduce app churn.

As well as the need for better understanding of what constitutes useful user segmentation, it is also important that going forward we define the elements of transport behaviour change interventions in a shared language. Gamification and Incentives or Rewards relate to particular Behaviour Change Techniques. The Bella Mossa research reviewed in this chapter suggests that such a complex real world behavioural challenge requires an acknowledgement of the importance of engagement behaviour, linked with a range of behaviour change techniques drawn not only from gamification and incentives based interventions but extending into social marketing and persuasive mobile apps. In order to test the most effective mix of challenges or games, rewards and messaging—the 'Reward mix' for particular user segments we will need consistent nomenclature to refer to the intervention components (Bird et al. 2013). The Behavioural Change Techniques Taxonomy (Michie et al. 2015) may offer a solution and work to 'code' future VTBC digital programmes such as Bella Mossa, may assist in this mapping.

Motivation is an important factor when trying to induce positive transport behaviour change, but, as we are reminded of by Michie et al.'s Com-B Theory (Michie et al. 2014) the context in which the behaviour takes place should not be forgotten. The holiday season in Bologna (late July/early August) caused significant changes to transport behaviour. It is possible that people were more likely to resort back to old behaviour, such as solo car use and not tracking journeys, because of the changes to routine. To counteract this possibility rewards and event related gamification were increased immediately after the holiday period to encourage people to re-engage with the app. User activity data tracked in the BetterPoints dashboard showed a 'spike' in activity during this period in weeks 17–18, possibly due to the introduction of a competition with prizes, ranging from a Bella Mossa hat to a water bottle, based on how often the app was used. The popularity of this kind of gamification to maintain interest is further supported by the finding that 22% of survey respondents (across the whole participant population) said that they would like to see more 'special occasions' such as these short term challenges and events.

Looking forward, short-term competitions and special events that add variety and interest to the basic loop of track a journey for a set number of points, should be more frequent. The use of behavioural categories such as those proposed in this chapter could be used as a way of more effectively targeting specific sub-groups within the wider everyday car user population with these short-term loops. The ability to identify people based on their behavioural response to a digital intervention and incentives, and dynamically adapt the 'Reward mix', to optimize behaviour change, is considered to be the key to future success in VTBC programmes such as Bella Mossa.

Acknowledgements Marco Amadori and Giuseppe Liguori of SRM. via A. Calzoni, 1/3-40128 Bologna. Joe Oldak, Head of Engineering, Marietta Le, Engagement Manager, Chris Bristow, COO and Rachel Maile, Marketing Officer, all of BetterPoints Ltd.

References

Anable J (2005) 'Complacent car addicts' or 'aspiring environmentalists'? Identifying travel behaviour segments using attitude theory. Transp Policy 12(1):65–78. https://doi.org/10.1016/j.tranpol.2004.11.004

Bird EL, Baker G, Mutrie N, Ogilvie D, Sahlqvist S, Powell J (2013) Behavior change techniques used to promote walking and cycling: a systematic review. Health Psychol 32(8):829–838. https://doi.org/10.1037/a0032078

Coombes E, Jones A (2016) Gamification of active travel to school: a pilot evaluation of the beat the street physical activity intervention. Health & Place 39:62–69. https://doi.org/10.1016/j.healthplace.2016.03.001

Forbes PJ, Gabrielli S, Maimone R, Masthoff J, Wells S, Jylhä A (2014) Towards using segmentation-based techniques to personalize mobility behavior interventions. ICST Trans Ambient Syst 1(4). https://doi.org/10.4108/amsys.1.4.e4

Gerike R, Lee-Gosselin M (2015) Workshop synthesis: improving methods to collect data on dynamic behavior and processes. Transp Res Proc 11:32–42. https://doi.org/10.1016/j.trpro.2015.12.004

Jylhä A, Nurmi P, Sirén M, Hemminki S, Jacucci G (2013) Matkahupi: a persuasive mobile application for sustainable mobility. Paper presented at the Proceedings of the 2013 ACM conference on Pervasive and ubiquitous computing adjunct publication

Meloni I, Teulada BSD (2015) I-Pet individual persuasive eco-travel technology: a tool for VTBC program implementation. Transp Res Proc 11:422–433. https://doi.org/10.1016/j.trpro.2015.12.035

Michie S, Wood CE, Johnston M, Abraham C, Francis JJ, Hardeman W (2015) Behaviour change techniques: the development and evaluation of a taxonomic method for reporting and describing behaviour change interventions (a suite of five studies involving consensus methods, randomised controlled trials and analysis of qualitative data). Health Technol Assess (Winchester, England), 19(99):1–188. https://doi.org/10.3310/hta19990

Michie S, Atkins L, West R (2014) The behaviour change wheel: a guide to designing interventions. Silverback Publishing

O'Connell C (2017) [New Data] Industry App Benchmarks for H1 2017. Retrieved from http://info.localytics.com/blog/vertical-app-benchmarks-first-half-of-2017

Wells S, Kotkanen H, Schlafli M, Gabrielli S, Masthoff J, Jylhä A, Forbes P (2014) Towards an applied gamification model for tracking, managing, & encouraging sustainable travel behaviours. EAI Endorsed Trans Ambient Syst

Part IV
User-Centric, Sustainable And Secure Freight Services

The Applicability of Blockchain Technology in the Mobility and Logistics Domain

Wout Hofman and Christopher Brewster

Abstract Blockchain technology in the last 3–4 years has attracted a great deal of attention due to its potential for the disruption various sectors of the economy including the financial sector, and more recently the mobility and logistics domain. The technology, which was developed as the foundation for the cryptocurrency Bitcoin, has developed much further with creation of Ethereum and the concept of "smart contracts." This chapter discusses the position of blockchain technology in a hyperconnected environment, and analyses a prototypical logistics scenario where the technology could be used. We describe the characteristics of blockchain technology and provide examples of how it is being applied for mobility and logistics. The chapter concludes by considering particular opportunities for mobility and logistics based on the current characteristics of the technology, and also identified a number of challenges faced by the technology.

Keywords Blockchain technology · Mobility · Logistics

1 Introduction

Hyperconnection or universal connectivity is mentioned as one of the most important aspects of the Physical Internet (Montreuil et al. 2013) that can be applicable to both passenger and freight transport. It encompasses 'super-fast connectivity, always on, the move, roaming seamless from network to network, where we go—anywhere, anytime, with any device' (Biggs et al. 2012). Examples of the implementation of hyperconnection can be found in city logistics (Crainic and Montreuil 2016). A hyperconnected world not only comprises individuals with embedded sensors in their smart devices, but includes all types of devices (e.g. vessels, trucks, containers,

W. Hofman (✉) · C. Brewster
TNO, Anna van Buerenplein 1, 2595 Hague, The Netherlands
e-mail: wout.hofman@tno.nl

C. Brewster
e-mail: Christopher.brewster@tno.nl

B. Müller and G. Meyer (eds.), *Towards User-Centric Transport in Europe*,
Lecture Notes in Mobility, https://doi.org/10.1007/978-3-319-99756-8_13

and trains). These devices can be considered as assets used for value delivery, either on an on-demand basis for one or more customers or according to timetables, thereby offering "Mobility and Logistics As A Service" (MAAS and LAAS). Different sensors and supporting communication technology would be used for the identification and tracking of assets, but also as a basis to predict behaviour of these assets like an Estimated Time of Arrival (ETA) at a next destination (e.g. station, port, terminal). Several research papers have identified supply chains and logistics as the main areas for implementing the closely related concept of the "Internet of Things" (Atzori et al. 2010; Gubbi et al. 2013). These developments will probably lead to intelligent objects (Whitmore et al. 2015) or what is otherwise known as ubiquitous computing (Weiser 1991). Cars using the NVIDIA chipset can be treated as hyperconnected computing platforms, thus implementing ubiquitous computing. Already Automatic Identification System (AIS) with Global Positioning System (GPS) is being used for vessels and barges, trucks have on-board tracking units collecting data through CAN bus acting as sensors, and cars and trains have identification mechanisms. The introduction of LoRa technology (www.lora-alliance.org) and 5G (Boccardi et al. 2014) for communication extends the battery life of sensors and thus their utility for machine-to-machine interaction. The combination of ubiquitous computing and long battery life makes possible the decentralisation of decision making to, for instance, transport means (i.e. autonomous vehicles, trains, vessels, and barges) and cargo, where individual boxes can find their way through a logistics network like individuals.

Hyperconnection is mostly described in terms of businesses collaborating in chains (Schonberger et al. 2009) as in the Hyperconnected City Logistics (Crainic and Montreuil 2016) and this is supported by hardware and communication technology providing computational capabilities and level one interoperability (Wang et al. 2009). In the existing literature, neither the information that any two stakeholders have to share, nor their interaction choreography has been explicitly described. Data integration is required to achieve state awareness (McFarlane et al. 2016), also known as situational awareness (Endsley 1995). Conceptual interoperability (Wang et al. 2009), which has not been implemented by supply and logistics stakeholders so far (The Digital Transport and Logistics Forum (DTLF) 2017), is a necessary requirement to support MAAS and LAAS.

It is in this context that we consider Blockchain Technology (BCT) which has received an immense amount of attention and investment in the last 2–3 years. This chapter explores the capabilities of BCT as a means to contribute to situational awareness and decision support by autonomous agents. First of all, the state of the art of blockchain technology is presented, secondly a logistics scenario utilizing a blockchain is described as a basis to identify challenges. Finally, the chapter will explore future potential capabilities of blockchain technology in the area of mobility and logistics.

2 An Overview of Blockchain Technology

BCT uses a combination of technologies that have a considerable history in computer science and in commercial applications. These component technologies include public/private key cryptography (Rivest et al. 1978), cryptographic hash functions (Preneel 1994), database technologies especially distributed databases, consensus algorithms (Vukolic 2015) and decentralised processing. The fundamental purpose is to achieve database consistency and integrity in a context of a distributed decentralised database, where the database nodes are either controlled ("permissioned") or uncontrolled ("unpermissioned"), the prime example of the latter being Bitcoin.

BCT arose out of technology developed in the creation of Bitcoin (Nakamoto 2008). Bitcoin, as conceived by Satoshi Nakamoto, was an attempt to create a "cryptocurrency" outside the control of government, a currency that would operate purely on the Internet (Grinberg 2011). Bitcoin was built on a number of key elements:

- A distributed file called a "blockchain" spread over all computers participating in the system.
- Proof of work—in order to write on the "blockchain" each node needed to complete a complex mathematical procedure (a process which eventually came to be called "mining") in order to have the "right" to write on the blockchain.
- Digital signatures—in order to know which person (using an identity expressed as a number) performed an operation each operation is signed using public-private keys.
- Chained hashes—this technology is widely used in version control and allows each documented to be "hashed" into a digital "summary". A sequence of such hashes are used to construct the blocks in the blockchain.
- Byzantine consensus—the Bitcoin protocol claims to have solved the problem of "byzantine consensus" which prevents "double spend" of Bitcoins.

This enables the creation of a distributed database (a "ledger") which can be used to record transactions of Bitcoins from one person (represented by their public key) to another. This database is immutable and ensures the impossibility of conflicting transactions.

Various people realised that the underlying technology of Bitcoin may have far greater interest. The Bitcoin software provides an "unpermissioned" ledger for the recording of financial transactions but equally that ledger could be used to record non-financial transactions just like any ordinary database can. As the Ethereum White paper states "alternative applications of blockchain technology include using on-blockchain digital assets to represent custom currencies and financial instruments ("colored coins"), the ownership of an underlying physical device ("smart property"), non-fungible assets such as domain names ("Namecoin"), as well as more complex applications involving having digital assets being directly controlled by a piece of code implementing arbitrary rules ("smart contracts") or even blockchain-based "decentralized autonomous organizations" (DAOs) (Buterin 2013). This realisation, which the has been ascribed to multiple authors, has led to a flowering of efforts to use initially the Bitcoin Blockchain for various non-cryptocurrency purposes, and then the creation of alternative platforms or systems (such as Ethereum and Hyperledger cf. below). Together with the realisation that blockchain technology could have a variety of other applications, there arose a number of startups seeking to find opportunities to exploit this technology. The start-ups have grown in number in areas ranging from finance to insurance, from food and agriculture to also logistics.

The key principles of BCT can be outlined as follows:

- **Blocks in the blockchain**: Each block in a blockchain contains (a) an ordered set of records or transactions, and (b) a hash of the previous block in its header (starting from an initial block called the "genesis" block). This means its hash depends on the hash of its parent and so on in turn. This is key to blockchain security and guarantee of permanence since any change in the data of one block would affect all other blocks that follow. Such a change would require a new consensus process (typically involving "proof of work" although not necessarily). A chain of such blocks forms a blockchain.
- **A peer-to-peer network**: A blockchain depends a network of peers or "nodes" who usually provide the computing power to achieve consensus for example by "mining" if consensus is achieved by "proof of work".
- **A distributed immediately replicated file**: Each blockchain is replicated across all "nodes" or computers in the peer to peer network of that blockchain. The presence or absence of a particular node (e.g. being offline) does not affect the operation of the blockchain as a whole, and this ensures guaranteed "uptime".
- **Consensus algorithm**: In order for a new set of transactions to be written to a block, the block must be validated by a consensus algorithm. There are various such algorithms, the most common one being "proof of work" where a node must solve a cryptographic puzzle thus entitling it to validate the new block (and in blockchains based on cryptocurrencies to earn a "coin"). The major issue with "proof of work" is that it does not scale well in terms of the number of transactions. Other consensus algorithms include Byzantine fault-tolerant replication (Vukolic 2015) and "proof of stake" (currently being developed actively within the Ethereum project).
- **Cryptographic signatures**: All transactions in a blockchain are cryptographically signed with public key cryptography to prove identity, authenticity and enforce read/write access rights.
- **Permissioned versus unpermissioned blockchains or ledgers**: As discussed in Walport (Walport 2016) blockchains (or as some people call them "distributed ledgers") can be unpermissioned or permissioned. An unpermissioned blockchain has no single owner fulfilling the ideal that there is no central control. The best example is Bitcoin but the core Ethereum blockchain is also unpermissioned. A permissioned blockchain has a set of owners who control read/write/mining rights and thus operate the consensus algorithm. The Hyperledger Fabric works like this.
- **Smart contracts**: Taking the distributed database concept one step further, Buterin (Buterin 2013) proposed that a blockchain should be a virtual machine, a distributed computer that could run simple programs, so called "smart contracts". This

raises the prospect of writing autonomous pieces of software which run independently of human intervention so called "distributed autonomous organizations".

Next, the blockchain technology stacks Ethereum, Hyperledger, and BigChainDB are presented. There is a variety of technology stacks, like Quorum, Ripple, and IOTA, that could also be discussed. There are thus a great many other BCT platforms or technology stacks, a survey requires an additional section. Ethereum and Hyperledger are however one of the first technology stacks and BigChainDB takes an alternative approach.

Ethereum (https://www.ethereum.org/): One of the most influential is the Ethereum Platform, an initiative of Vitalik Buterin (Buterin 2013) and Gavin Wood (Wood 2015), which was funded by approximately $20 M of bitcoin (Gerring 2016).The vision for Ethereum (Buterin 2013; Wood 2015) was to create a blockchain based distributed virtual machine which would allow "smart contracts" to run as "distributed autonomous" entities. This vision was a significant step in extending the vision as to what BCT was for and how it could be used. A "smart contract" for Ethereum was a small piece of code that would be run "on" the blockchain and crucially would function entirely independently without any possibility of censorship, downtime, fraud or third party interference. This enabled the vision of "distributed autonomous organizations" which would be entities entirely specified in the smart contract code which could run without human interference, and because the blockchain has guaranteed up-time without any possibility of stopping. The creation of the Ethereum Virtual machine as a Turing complete virtual machine was a core innovation, capable of running any program given enough resources ("gas") to run. Ethereum uses a cryptocurrency "ETH" which is publicly traded on cryptocurrency exchanges, and an internal "metering unit" called "GAS". Gas provides a means to provide transaction charges (including running smart contracts) and also allocate incentives for running the Ethereum VM. Ethereum currently uses a "proof of work" mechanism for consensus but as noted above has plans to switch to a "proof of stake" methodology.

Ethereum has had considerable mainstream success in being adopted by companies such as Microsoft and (initially) IBM to provide the underlying system for their own BCT offerings. A large proportion of blockchain start-ups and services are based on the Ethereum platform.

Hyperledger (https://www.hyperledger.org/): The Hyperledger project was founded by the Linux Foundation with the intention of developing cross-industry collaboration in the area of BCT and with a focus on supporting business transactions. This has meant a chief focus on permissioned blockchains. Many major technology companies and financial institutions were among founder members, although the most important and visible is IBM. Hyperledger is designed to be highly modular with the ability to plug in different alternative components for the same basic functionality. IBM has contributed "Hyperledger Fabric" which is the most used blockchain technology stack after Ethereum. Following the modular design, Hyperledger Fabric (https://www.hyperledger.org/projects/fabric) allows components, such as consensus and membership services, to be plug-and-play. It allows for "smart contracts" called "chaincode". The basic consensus mechanism is PBFT[SW1] [LG2] and there is no cryptocurrency because the design philosophy is for permissioned blockchain setups for specific business sectors. Apart from modularity and chaincode, key features of Hyperledger Fabric include:

- **Identity**: A membership identity service that manages user IDs and authenticates all participants on the network.
- **Privacy**: Private channels which are restricted messaging paths that can be used to provide transaction privacy and confidentiality for specific subsets of network members.
- **Efficiency**: Hyperledger Fabric uses a division of labour to assign different roles to different nodes claiming a consequent far greater efficiency of execution.

Partly due to the backing of IBM, Hyperledger has received widespread support and is being used in many different projects, including its use by Walmart in the pork supply chain (del Castillo 2016) and in a major logistics project with Maersk (White 2018). Because of the identity management features and the ability to program chaincode so as to allow access to certain kinds of data to the smart contract participants, we chose to use this technology for the Proof of Concept as reported above.

BigChainDB: In contrast to Ethereum, Hyperledger or other blockchain systems, BigChainDB (https://www.bigchaindb.com/) does not build a full stack of Blockchain technologies, but rather offers an overlay onto existing database technologies to "blockchain-ify" them. "BigchainDB is designed to merge the best of database and blockchain worlds: scale and querying from the database side, and decentralization, immutability, and assets from the blockchain side" (McConaghy 2017). BigChainDB starts with an initial open source database (MongoDB) and then added blockchain characteristics including decentralized control, immutability, and creation and movement of digital assets (McConaghy et al. 2016). The main objective has been to overcome the widely recognised scaling problem that most blockchain projects suffer from. BigChainDB claims to be able to achieve over 1 M transactions per second with this approach. The project sees itself as providing a technological component in a more conventional technology stack. This means in part that BigChainDB explicitly excludes having a virtual machine or other mechanism for running "smart contracts". In their approach, such a functionality would be provided by Ethereum or some other similar technology. The BigChainDB stressed three characteristics as being important:

- Decentralized control i.e. where no single entity controls the network.
- Immutability i.e. where data once written cannot be changed or tampered with
- Transfer of digital assets, i.e. the ability to create an asset and transfer this without central control.

A BigChainDB instance consists of a number of nodes all of which contain parts (but not all) of the complete database. Decentralised control is achieved by this DNS-like federation of nodes which have voting rights in the validation of blocks. Voting operates on a layer above the actual database and in order to achieve speed each block of transactions is written before being validated by a quorum of nodes. Nodes vote to validate a transaction and at validation time "chainify" the block as each block provides a hash id of the previous block. Immutability is achieved through a combination of shard replication, disallowing reversions, database backups and cryptographic signing of all transactions (McConaghy et al. 2016).

3 Challenges for Logistics Inspired by a Business Scenario

Using a business scenario, we consider here the applicability and challenges of BCT and its added value for mobility and logistics. The example focuses on the sharing of the container status amongst autonomously operating enterprises during transshipment via a port. Both the physical and administrative status is considered. Firstly, the current situation is introduced and secondly its possible implementation using BCT.

3.1 The Current Situation

At arrival of a vessel in a port like the port of Rotterdam, various organizations are involved to ensure the container is transported to its final destination on time. Different modalities can be applied like rail, road, and inland waterways. However, transport of a container to its final destination depends on payment of sea transport charges (commercial release), physical availability of the container in the port (physical availability), and release of customs without any further inspection of its content (customs release). This status information is shared amongst the various organizations by messages according to a customer-service provider relation. Figure 1 shows an example of the value chain for transshipment of a container discharged from a vessel and transported by truck to a final destination. A shipping line operating the vessel has a contract with a stevedore for loading and discharging containers on the vessel. A shipping line informs a 'notify' of arrival of its containers in a port of discharge, e.g. a forwarder acting on behalf of a consignee responsible for commercial—and customs release in this example (the transport document lists at least one organization acting as 'notify').

Both a stevedore and a carrier have to receive the current status of a container to pick up the container and transport it to the final destination. However, they are not involved in the commercial and the customs release. The process is normally as follows:

- commercial release is generated by a bank to a forwarder and shipping line involved
- customs release is generated by customs to the forwarder
- a stevedore generates the discharge status to its customer (physical availability), a shipping line

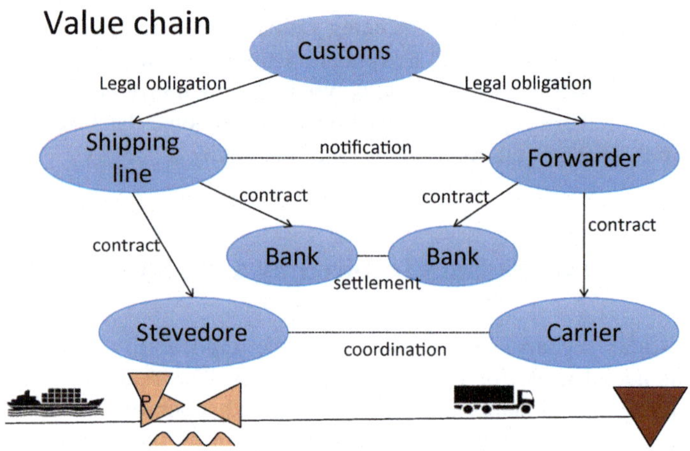

Fig. 1 Value chain for container transshipment in a port

- a shipping line makes commercial release available to a stevedore
- a forwarder send the commercial and customs release to a carrier
- a stevedore still has to receive the customs release and the carrier must know the discharge status to be able to pickup a container. A carrier also must be known by a stevedore to pick-up a particular container.

This demonstrates that a lot of coordination is required between various organizations. Currently, Electronic Data Interchange (EDI) messaging is used to support this collaboration. It causes delays in physical handling due to errors (the wrong carrier received status information), lack of status information (a stevedore is not informed of the customs release), and delays in sharing the status (a stevedore currently submits a discharge list to a shipping line after the vessel has left the port). Delays in the physical processes leading to additional container storage at a terminal are currently caused by delays in information sharing and should be planned based on customer requirements. A (port) community system can address these issues by storing the container status, but it requires trust in the system and clearly specified Identification, Authentication, and Authorization (IAA) mechanisms (Johnson 2010).

3.2 Implementing the Business Scenario with Blockchain Technology

To illustrate the potential added value of the BCT in this domain, we have developed a proof-of-concept application implementing the aforementioned business scenario for sharing status information with Hyperledger Fabric. By sharing real time status data and permissions via a trusted blockchain environment, on-carriage processes

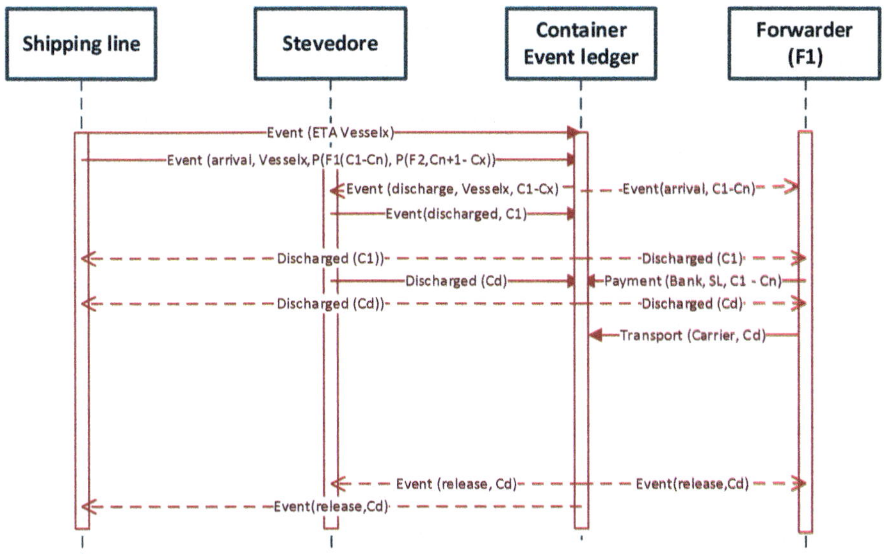

Fig. 2 Sequencing of operations for a container event ledger (selected roles)

can be planned differently, decreasing storage time at a terminal. Each arrow in the sequence diagram (Fig. 2) depicts an event with a function and permissions P of containers C to a role (forwarder = F). The functions of the event reflect milestones in the processes (Hofman 2017).

The first event in the example is the ETA of a vessel, followed by an arrival event. The dotted lines indicate that this information is available to customs and the stevedore, but forwarder F1 only has access to containers C1-Cn based on his permission P. A carrier has to request permission (P-req) on behalf of forwarder F1. By adding the transport order of the forwarder to the event ledger, the carrier could automatically receive the release event and the stevedore would be aware of the carrier picking up container Cd. Smart Contracts provide functionality to the participants, behaving according to rules agreed in a community and participation in a smart contract controls accessibility. Generation of events by trusted sensors (IoT, Internet of Things) could provide validation.

Each event in the sequence diagram of Fig. 2 represents an operation in the smart contract. For the sake of simplicity, we have chosen to develop one smart contract supporting the business scenario. The smart contract includes all relevant data[1] and operations (invoke and query) on these data. All stakeholders have to be registered and assigned roles within this smart-contract. When they are registered, they can query or invoke transactions associated to their roles. Invoke-transactions validate the data entered against the data structure of the smart contract, and the proper

[1] Hyperledger Fabric smart contracts are written in GO Language and each contains its own attribute-value data store.

role assigned to a particular stakeholder. For example, if a shipping line notifies a forwarder, the notifier has to have that role, and participate as a stakeholder in that specific smart contract on the blockchain. If a stakeholder notifying is not registered, an error occurs and the transaction is not added to the blockchain. If this stakeholder is registered as a carrier and not forwarder, the transaction is also not added.

For the proof-of-concept we developed a NodeJS web application that connects to the blockchain. This application exposes a traditional JSON API and has methods for enrolling users on the blockchain, deploying the smart-contract to the blockchain and interacting with our specific smart-contract. For demonstration purposes, we have developed a front-end application with the Angular framework.

3.3 Characteristics of Mobility and Logistics

As the business scenario illustrates, logistics has a number of characteristics which can also be applied in the mobility domain and appear to make the sector potentially a good target for the application of BCT. Characteristics include the following:

- **Heterogeneity and large number of (SME) actors**—there are many different kinds and types of organizations offering different business services from small family run enterprises with 1–2 trucks, to global operating IT driven enterprises and platforms with all types of assets (vessels, planes, trucks, cars, and containers) at their disposal, and those organizing logistics operations and handling all types of formalities. Furthermore there are various supervisory bodies (e.g. port authorities, customs and excise, police, infrastructure managers) who govern processes from various perspectives (safety, security, trafficking, etc.) at a national scale in collaboration with authorities of other countries.
- **Variability of demand**—there is limited predictability of customer demand both in terms of quantity and types of objects and individuals to be moved. It could be improved by sharing or detecting patterns in existing flows, which requires availability and access to data of these flows.
- **Time-sensitivity**—Rapidity of delivery is becoming a key differentiator for suppliers especially online retailers and hence the speed of the logistics component of the complex services provided becomes essential for high customer satisfaction. In E-commerce shipments, transport costs are independent of transport duration, but in all other flows of goods and passengers duration will come with a cost.
- **Lack of ICT integration**—in spite of the availability of sensor technology, current levels of IT integration are severely limited especially those types of integration which allow small-scale providers to be integrated with the large long distance providers.
- **Commercial/economic sensitivity**—for various reasons tied to the business models currently used, there is considerable lack of trust for electronically sharing data.
- **Liability**—sharing specific data might lead to liability in case of accidents, delays or incidents.

- **Commercial and regulatory pressures**—there are considerable pressures to reduce waste (empty transport) and to make the sector more efficient. Regulations forcing this are the major fear of Logistics Service Providers.
- **Local versus global information**—local, national goods movements need not always be available globally, e.g. waste transport or retail supply from national warehouses is local.

These are general characteristics of logistics that boil down to issues like (relative) low IT maturity and skills, lack of trust because business models are built upon lack of transparency, lack of uniformity in data semantics and process alignment, etc. Lack of trust prevents organizations in sharing (high quality) data. It requires an infrastructure for controlled data access and sharing that addresses these issues and provides easy to implement and easy to use applications to SMEs. Most often, centralized, competing IT systems, each with their governance model, are implemented to address all the aforementioned characteristics.

4 BCT for Mobility and Logistics

4.1 Examples in Mobility and Logistics

There are already many projects and BCT applications under development for the mobility and logistics domain. Most of them are still in the experimental phase. For instance, the Rotterdam Port Authority and the Municipality of Rotterdam have raised the Dutch BlockLab for logistics, experimenting with BCT for business scenarios. IBM and Maersk have raised a joint company to develop and implement the so-called Global Trade Digitization (GTD) blockchain providing visibility of container trips and share references to administrative documents (White 2018). Shipchain provides another competing visibility solution for shipments. Another application developed by Marine Transport International is to share the Verified Gross Mass of cargo loaded on a vessel, which is mandatory for sharing. Furthermore, there is a Blockchain in Transport Alliance whose members include various large platform providers like Uber4Freight and Descartes.

Similar developments can be observed in the mobility. For instance, the Blockchain Mobility Consortium (https://blockchain-mobility.org/) develops a distributed solution empowering vehicle owners in monetizing their assets and reducing insurance costs. DOVU is a similar example (https://dovu.io/). Also companies like Toyota are investing in BCT for mobility services for their (autonomous) cars. Other examples of applying BCT address issues like counterfeiting and theft. Examples include the Halal Chain raised to prevent counterfeiting by integrating production, inspection, insurance and logistics, and retail and consumers. Similar BCT applications are being developed for other products like coffee (Coffee 2017). There is also a generic solution focusing on for instance counterfeiting and stolen goods called BlockVerify (http://www.blockverify.io/).

4.2 Relevant Functionality and Opportunities

BCT can serve five basic functions relevant to mobility and logistics, namely:

- **Non-repudiation**. A BCT application can provide an immutable ledger providing proof that data was entered in the ledger at a certain time. It can be used in case of disputes between a customer and service provider or an enterprise and an authority. As such, data quality has to be optimal, the application needs to contain a complete log and audit trail. Non-repudiation can help address certain aspects of counterfeiting and theft which is of interest to consumers or authorities.
  Non-repudiation can be used to solve conflicts like related to service delivery, infrastructure utilization, etc. For instance, it can be used to proof a vehicle was driving at a particular road when a traffic ticket was issued. Another is the proof that goods have been delivered or have been handed over to the responsibility of a carrier.
- **Resilience and robustness**. A BCT application will run on several nodes that all contain the data or part of the data, depending on the type of technology used. There is, for instance, BCT that duplicates data to at most eight nodes of the network. These nodes can operate in different (cloud) environments, where the cloud operators would not have access to the data stored by the nodes. The network of nodes is resilient to failure of one node, such as a hardware failure or cyber-attack.
- **(Near) real-time data sharing**. Data entered in a BCT application is available to all participating users, based on the latency of the BCT utilized. Therefore, data is near real-time available to all users, depending on their access rights.
  There are several applications of this functionality. For instance, infrastructure managers have a real-time view of infrastructure utilization, release data of cargo is real-time available to a terminal and carrier (see the example), and customers of carriers or Logistics Service Providers are able to improve their planning and reduce express deliveries.
- **Billing and payment**. The log and audit trail can contain all details of using a particular service, e.g. for mobility and logistics. These could trigger billing and payment, for instance with one payment card (or and app on a smart device) that is used for MAAS. Of course, different service providers will have different pricing strategies linked to their service like first class travelling outside rush hours, which implies that a customer has to select a particular product and link it to its account.
  This type of function can be used for many purposes. Examples are in payment of services like energy consumption of a vehicle, energy production of an electric vehicle, infrastructure utilization, an delivery of goods according (standardized) payment conditions triggered by for instance Proof of Delivery. Especially in logistics, this functionality can also be applied for financing supply chains. Transparency of deliveries to a customer can be used to finance other activities, which can also be applied by vehicle manufacturers.
- **Decision support**. Entering status information like the position of a vehicle and its ETA at a location will improve decision making. In case a person knows a train will

be too late or not arrive at all due to an incident, that person may decide to choose another transport mode available. The same applies for the logistics example in the beginning of this section: if stakeholders have instantaneously access to status information, they can improve their decision, optimize planning, and reduce costs. There are various applications of this type feasible like predictive Maintenance, Repair, and Operation (MRO) of vehicles and infrastructure, (synchromodal) planning based on predictions of infrastructure availability and available transport capacity, optimization of behavior by selecting alternative routes (Montreuil et al. 2013), etc.

As the examples of applying the functionality show, there is a variety of options available to apply BCT I mobility and logistics. All options are on the various collaboration interfaces between stakeholders along three dimensions. These three dimensions are: (1) capacity utilization, (2) decision support in terms of planning and rerouting, and (3) payment, like indicated before. The following interfaces are foreseen:

- **Infrastructure user and—manager**: use of an infrastructure by any transport means, e.g. cars, trucks, barges, and trains. This functionality can also make use of data of all infrastructure users in a specific area at a given time interval. Storage of continuous data streams of IoT in a BCT application is not recommended, since the transaction rate of BCT is still too low.
- **Asset owner and—user**: use of an asset for its particular purpose like a car or truck. The user, i.e. the driver, can also be the asset owner. In this context, decision support relates to for instance MRO of the asset.
- **Customer and Logistics Service Provider (LSP)**: a providers offers capacity to customers for transport of freight and/or passengers, think for instance of public transport like train or metro and passenger transport by air.
- **Hubs and transport legs**: process synchronization of hubs like terminals, airports, and railway stations with respect to transport means that call upon these hubs and (potentially) have to pay for services provided by those hubs.
- **Governing and inspection authorities**: various authorities have specific tasks from a societal perspective as laid down in laws and regulations, e.g. (food) safety, security, and taxation. These authorities can use the data stored in private sector BCT applications to support their tasks, e.g. by risk assessment from a customs perspective or by monitoring passengers and freight movement from a security perspective assessing necessary licence information.

A blockchain application will automatically provide data integrity and—provenance. Data integrity is provided since data stored on a blockchain is immutable. Since each stakeholder enters its particular data to the blockchain, it will always be clear who the owner of a blockchain is (data provenance).

Non-repudiation and decision support are also relevant to authorities governing person-, goods-, and related money flows. Persons travelling internationally can store their details on a blockchain (or make them available using a blockchain) thus providing information to destination countries for risk assessment. The same is applicable to goods flows from a safety and security perspective and other risks assessed by for instance customs authorities. Tax authorities may be able to utilize blockchain data for VAT purposes. Finally, billing and payment could be used by auditors to automatically validate the administration of a company. These types of functionalities could dramatically reduce administrative costs for enterprises.

4.3 Issues and Challenges

There are still many challenges with respect to trust and economic sensitivity. Trust is (partly) addressed by permissioned blockchains, where there are procedures for joining a blockchain instance. The use of BCT as Identity Provider and Certification Authority could facilitate trust. The Sovrin Foundation (https://sovrin.org/) is such an example, but there others using Distributed Identity (DID) scheme developed by the World Wide Web Consortium (W3C).

Economic sensitivity implies that complete transparency is inappropriate current service providers and customers. Especially in logistics, lack of transparency is the basis for many business models of Logistics Service Providers (LSPs). Applying BCT would imply that significant amounts of data from all stakeholders would become transparently accessible, which is a large barrier. In a similar manner, this would apply to mobility. Thus, mechanisms have to be developed for controlled data access amongst stakeholders.

Governance is another significant issue in the use of BCT in mobility and logistics (van Deventer et al. 2017). The core issues are who has final decision making power in a networked system scenario, to change the system, to update or otherwise alter processes, and is this even allowed due to immutability of a blockchain. The procedures to install and operate a node have to be governed, especially in the context of operating nodes provided by a commercial provider like Amazon, IBM or Microsoft. To ensure proper operation of a BCT application, the code (e.g. Smart contracts or chaincode) has to be developed in a controlled manner by a team of developers. These developers should be independent of the business, because they control the functionality of the solution. Finally, the developers need to have establish transparent procedures to update the functionality according changing user requirements. These procedures should also include version management of the data structures stored on a blockchain application to adhere to immutability of blockchain data.

Most of the current BCT applications have very limited throughput and storage capacity, storing only small amounts of data, i.e. hashes which provide a link to actual data of an object like a car or container. This implies that a separate conventional database stores all data related to a particular object. This linked data is not stored on the blockchain. It would be worthwhile to explore whether or not this data can also be stored on the blockchain, thus providing non-repudiation similar to a cadastre for property.

Finally, another issue is data quality, which has three aspects. The basic functions that we mention depend on the willingness of users to enter data on a blockchain application, and this affects the completeness of the data on the blockchain. For logistics, the willingness to put data on the blockchain depends on transparency and the added value of a particular blockchain application to an organization. Correctness of data is the second aspect relevant to blockchain technology. All validation rules of data written to a blockchain are validated by a smart contract or chaincode. It makes these pieces of software code complex, combining process rules with data validation rules. There are also blockchain technologies like BigChainDB that are able to run a (JSON) validator as an extra module used before storing data in a blockchain. These mechanisms have to assure that whatever data is written to a blockchain is correct when retrieved. A final aspect of data quality is consistency of data. The data stored in a blockchain needs to reflect the physical world. The aspect of trust is relevant, but in logistics other mechanisms have been developed in the past to address this issue. For instance, there are at least three stakeholders involved in transport of cargo: a shipper, a carrier, and a consignee. If each of them enters data, the consistency can be validated.

5 Conclusions

This chapter has explored the potential of blockchain technology for mobility and logistics. We presented a business scenario for logistics and provided a brief overview of current developments in BCT including a number of examples where the technology is being used for mobility or logistics. As many authors have suggested, BCT could be significantly disruptive to mobility and logistics. Large car manufacturers like Toyota have already invested in this technology. This chapter also identifies potential applications for mobility and logistics, which can dramatically reduce administrative costs for enterprises, improve risk assessment by authorities like customs and food inspection authorities, and contribute to preventing counterfeiting and have proper VAT payment of services and products. BCT for mobility and logistics is still in its experimental phase. There are many different experiments, but there is not yet a large scale blockchain application operational to address the functionality identified before. We have also noted a number of limitations concerning governance, scalability and throughput which need to be overcome for the technology to demonstrate its full potential. Given current interest and widespread investment in the technology, we expect to see significant impact in medium term.

References

Atzori L, Iera A, Morabito G (2010) The Internet of Things: a survey. Comput Netw 54:2787–2805

Biggs P, Johnson T, Lozanova Y, Sundberg N (2012) Emerging issues for a hyperconnected world. The global information technology report, pp 47–56

Boccardi F, Heath RW Jr, Lozano A, Marzetta TL, Popovski P (2014) Five disruptive technology directions for 5G. IEEE Commun Mag 52(2):74–80

Buterin V (2013) Ethereum white paper—a next-generation smart contract and decentralized application platform [Online]. Available: https://github.com/ethereum/wiki/wiki/White-Paper

Coffee M (2017) World's first blockchain coffee project—Moyee Coffee Ireland, Medium [Online]. Available: https://medium.com/@MoyeeCoffeeIRL/worlds-first-blockchain-coffee-project-cd04fff9e510. [Accessed 16 Feb 2018]

Crainic TG, Montreuil B (2016) Physical internet enabled hyperconnected city logistics. In: Transport research procedia—the 9th international conference on city logistics

del Castillo M (2016) Walmart blockchain pilot aims to make China's port market safer, Coindesk [Online]. Available: http://www.coindesk.com/walmart-blockchain-pilot-china-pork-market. [Accessed 21 June 2017]

Endsley M (1995) Toward a theory of situation awareness in dynamic systems. Hum Factors 1:32–64

Gerring T (2016) Cut and try: building a dream—Ethereum blog, Ethereum blog [Online]. Available: https://blog.ethereum.org/2016/02/09/cut-and-try-building-a-dream. [Accessed 11 May 2017]

Grinberg R (2011) Bitcoin: an innovative alternative digital currency. papers.ssrn.com, 9 Dec 2011 [Online]. Available: https://papers.ssrn.com/sol3/papers.cfm?abastract_id=1817857. [Accessed 10 May 2017]

Gubbi J, Buyya R, Marusic S, Palaniswami M (2013) Internet of Things (IoT): a vision, architectural elements, and future directions. Future Gener Comput Syst 29(7):1645–1660

Hofman W (2017) Improving supply chain processes by subscription to milestones. In: 12th ITS European Congress, Strasbourg, France

Johnson B (2010) Information security basics. Information Secur Assoc J 8:28–34

McConaghy T (2017) BigChainDB 2017 roadmap, The BigChainDB Blog [Online]. Available: https://blog.bigchaindb.com/bigchaindb-2017-roadmap-d2e7123f9874. [Accessed 1 May 2017]

McConaghy T, Marques R, Muller A, De Jonghe D, McCullen G (2016) BigChainDB: a scaleable blockchain database [Online]. Available: https://www.bigchain.com/whitepaper

McFarlane D, Giannikas V, Lu W (2016) Intelligent logistics: involving the customer. Comput Ind 81:105–115

Montreuil B, Meller RD, Ballot E (2013) Physical internet foundations. In: Service orientation in holonic and multi agent manufacturing robots, Heidelberg, Springer-Verlag, pp 151–166

Nakamoto (2008) Bitcoin: a peer-to-peer electronic cash system [Online]. Available: https://bitcoin.org/bitcoin.pdf

Preneel B (1994) Cryptographic has functions. Eur Trans Telecomm 5(4):431–448

Rivest R, Shamir A, Adleman L (1978) A method for obtaining digital signatures and public-key cryptosystems. Commun ACM 21(2):120–126

Schonberger A, Wilms C, Wirtz G (2009) A requirements analysis of business-to-business integration. Fakultat Wirschaftsinformatik und angewandte Informatik Otto-Friedrich-Universitat, Bamberg

The Digital Transport and Logistics Forum (DTLF) (2017) An outline for a generic concept for an innovative approach to interoperability in supply and logistics chains. Brussels

van Deventer O, Brewster C, Everts M (2017) Governance and business models of blockchain technologies, TNO [Online]. Available: http://publications.tno.nl/publication/34625060/VN3AE9/deventer-2017-Whitepaper-Blockchain-images-spreads.pdf

Vukolic M (2015) The quest for scalable blockchain fabric: proof-of-work vs. BFT replication. In: Open problems in network security, Springer, pp 112–125

Walport M (2016) Distributed ledger technology: beyond blockchain [Online]. Available: https://www.gov.uk/government/publications/distributed-ledger-technology-blackett-review. [Accessed 2018]

Wang W, Tolk A, Wang W (2009) The levels of conceptual interoperability model: applying systems engineering principles to M&S. In: Spring simulation multiconference

Weiser M (1991) The computer for the 21st century. Sci Am 3:94–104

White M (2018) Digitizing global trade with maersk and IBM—blokchain unleashed, Jan 2018 [Online]. Available: https://www.ibm.com/blogs/blockchain/2018/01/digitizing-global-trade-maersk-ibm. [Accessed 16 Feb 2018]

Whitmore A, Agarwal A, Xu LD (2015) The Internet of Things—survey of topics and trends. Inf Syst Frontiers 17:261–274

Wood G (2015) Ethereum: a secure decentralised generalised transaction ledger—Homestead revision [Online]. Available: http://gavwood.com/paper.pdf

The Physical Internet from Shippers Perspective

Carolina Ciprés and M. Teresa de la Cruz

Abstract The Physical Internet will change the way that goods are handled, stored, packaged and transported across the supply chain. It mimics the Digital Internet, as freight in the Physical Internet would travel seamlessly as data is exchanged in Internet. Physical Internet has become a key element to achieve ALICE vision (Alliance for Logistics Innovation through Collaboration), representing shippers and logistics service providers. In order to achieve this vision, research challenges in five different areas should be tackled: sustainable, safe and secure supply chains; corridors, hubs and synchromodality; information systems for interconnected logistics; global supply network coordination and collaboration; and urban logistics. A survey launched by ALICE gathered the shippers' perspective on the realization of the Physical Internet, including as key factors the transition required (business and governance models, regulation) as well as the barriers/triggers for its implementation. Future steps will focus on gaining endorsement on this vision.

Keywords Physical internet · Supply chain · Transport · Logistics · Freight Mobility

1 Introduction

One of the first outcomes of Mobility4EU project[1] was to identify trends and drivers that will impact transport and mobility in Europe in 2030. Moreover, it has identified user demands that call for solutions. In the area of freight transport, a series of promising novel and innovative transport solutions were derived. The Physical

[1] http://www.mobility4eu.eu.

C. Ciprés (✉) · M. T. de la Cruz
Fundación Zaragoza Logistics Center, Calle Bari 55, Nayade 5, Saragossa, Spain
e-mail: ccipres@zlc.edu.es

M. T. de la Cruz
e-mail: mdelacruz@zlc.edu.es

B. Müller and G. Meyer (eds.), *Towards User-Centric Transport in Europe*, Lecture Notes in Mobility, https://doi.org/10.1007/978-3-319-99756-8_14

Internet can be found among those solutions, fulfilling user needs related to freight transport. Those needs are, among others, efficient transport flows & networks, real-time travel info & services, interoperable and seamless journeys, data security and/or privacy & transparency, protecting climate, environment and health, and empower new players & innovations.

Logistics and supply chain management is currently full of inefficiencies: even though the figures vary among countries, one in every three trucks carrying goods across Europe is travelling empty,[2] the average load factor is 50%,[3] manufacturing and storage facilities are underused resources and the use of multimodal transport systems still faces many challenges, not being a reality for freight yet. Inefficiencies not only have an economical cost, but also an environmental impact, increasing greenhouse gases emissions while governments aim at its reduction. Moreover, labour conditions of logistics workers can still be substantially improved.

The Physical Internet initiative tackles the above mentioned limitations, changing the way that goods are handled, stored, packaged and transported across the supply chain.[4]

This chapter will provide the vision of Physical Internet from the shippers' perspective. Section 2 will explain the basis of the Physical Internet concept; Sect. 3 will provide insights on each one of ALICE ETP roadmaps and how they help to fulfil the vision of The Physical Internet; Sect. 4 will detail the shippers' position on the Physical Internet based on the outcomes of a survey launched by ALICE; and Sect. 5 will state future steps towards gaining shippers' endorsement to the Physical Internet paradigm.

2 The Physical Internet Concept

This section provides an overview of the origin and basis of the Physical Internet.

The Physical Internet Concept was outlined in 2011 through the Physical Internet Manifesto,[5] and has been further developed since then, led by the Professors Benoit Montreuill and Eric Ballot. The idea behind this new paradigm is to mimic the way that the Digital Internet works. Thus, goods would travel through synchromodal hubs across an "open global logistics system", encapsulated in special designed containers to be modular, standard and smart-so called π containers (Fig. 1). Therefore, a Logistics web which is efficient, sustainable, adaptable and resilient would be deployed. Physical Internet should be inclusive, open and for the benefit of all stakeholders.

Handling and storage systems and facilities would be conceived to manipulate only π containers, being cheaper, easier, faster and more reliable than they currently are. The hubs are the nodes of the network where the containers are being stored or

---

[2]EC Press Releases, http://europa.eu/rapid/press-release_MEMO-88-81_en.htm?locale=en.

[3]Load factors for freight transport, European Environment Agency, 2010.

[4]See footnote 3.

[5]The Physical Internet Manifesto, www.physicalinternetinitiative.org.

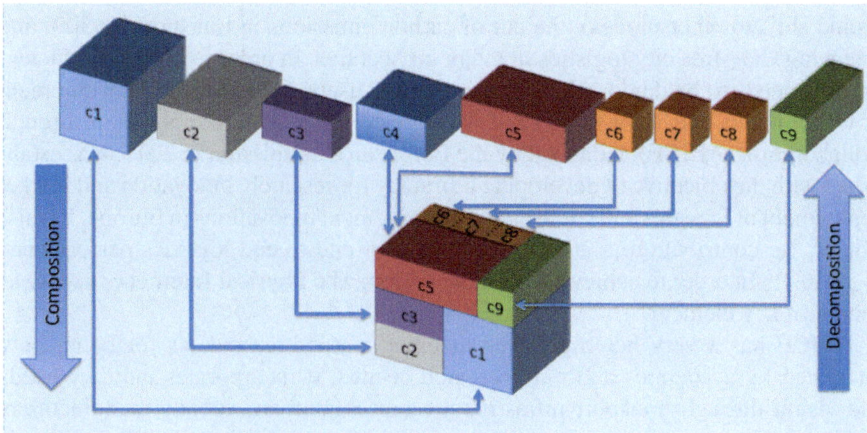

Fig. 1 Modularity of π containers (Montreuil et al. 2010)

dispatched according to the optimized flows, and are also where the interaction with the entities out of the Physical Internet takes place (Montreuil 2011).

In order to be realized, this new and advanced logistics system will need the support of information and communication technologies and other developments such as Internet of Things, big data analytics, 5G mobile networks, additive manufacturing, 3D-printing, and robotics.

Even though The Physical Internet Concept may seem very futuristic, the most innovative shippers and retailers are starting to test this model. New companies are blooming inspired in this concept, such as MonarchFx,[6] Carrycut[7] or CRC-Services.[8] MonarchFx is a brands and retailers alliance operating in the US and concentrated on e-commerce logistics. In a collaborative way, members use the alliance's shared routes and network when some of the specific orders cannot be fulfilled in the expected timeframe by their current network. The company has also a special focus on intelligent technology and automation. Carrycut is a digital marketplace for freight services, in the form of an app that brings together retailers, transporters and private users. Finally, CRC-Services (which stands for Collaborative Routing Centres) aims to increase the logistics performance for small and medium size deliveries creating an open network of regional hubs to connect logistics flows (multi-manufacturer and multi-retailer) achieving truck loading rates above 90%. In Europe, The European Commission envisions a series of measures in order to build a competitive transport system by 2050.[9] Some of the key goals include the optimization of the performance of multimodal logistics chains (more than the 50% of road freight over 300 km

[6]www.monarchfxalliance.com.

[7]https://www.carrycut.com.

[8]www.crc-services.com.

[9]Roadmap to a single European Transport Area-Towards a competitive and resource efficient transport system 2011.

should shift to other modes), the cut of carbon emissions in transport by 30% and achieving CO_2-free city logistics in major city centres. In order to do so, an efficient strategy needs to be deployed, bringing together users, operators, and development of technologies and infrastructure. The Alliance for Logistics Innovation through Collaboration (ALICE), launched by the European Commission in 2013, was established with the objective of developing a strategy for research, innovation and market deployment of logistics and supply chain management innovations in Europe. Its mission is: "to contribute to a 30% improvement of end to end logistics performance by 2030".[10] In order to achieve ALICE's mission, The Physical Internet concept has become a key element.

ALICE has a very heterogeneous member's structure, and far for being only integrated by academics and pure research centres, it incorporates industry needs and vision through transport infrastructure representatives, vehicle manufacturers and handling companies, logistics service providers and shippers. It is the shipper's perspective that will be taken into account in the rest of this Chapter.

According to ALICE, to become a reality, this new model needs to evolve from the current situation in which supply chains are fully-owned (vertical integration) to a full collaborative network. In order to do so, a series of milestones have been defined within the coming decades so as to ensure that The Physical Internet concept is implemented for freight transport and logistics in 2050.

ALICE research activities are articulated around five Working Groups. Those groups match the five specific areas that need further research and innovation to achieve its vision (Fig. 2). Those specific areas are:

1. Sustainable, Safe and Secure Supply Chains
2. Corridors, Hubs and Synchromodality
3. Information Systems for Interconnected Logistics
4. Global Supply Networks Coordination and Collaboration
5. Urban Logistics.

Insights on each one of the five roadmaps and how they help fulfil the vision of The Physical Internet with a special focus on shippers will be provided in the next section.

3 Roadmaps Towards the Physical Internet

Each one of the five ALICE Working Groups, which bring together industry, academia and public agencies, published a roadmap to reach the vision of The Physical Internet. These roadmaps not only identify key research areas towards Physical Internet, but also specific projects that contribute to its implementation. This section reviews the aim and key milestones for each one of the five ALICE roadmaps and

[10] ALICE ETP Mission & Vision www.etp-logistics.eu/?page_id=114.

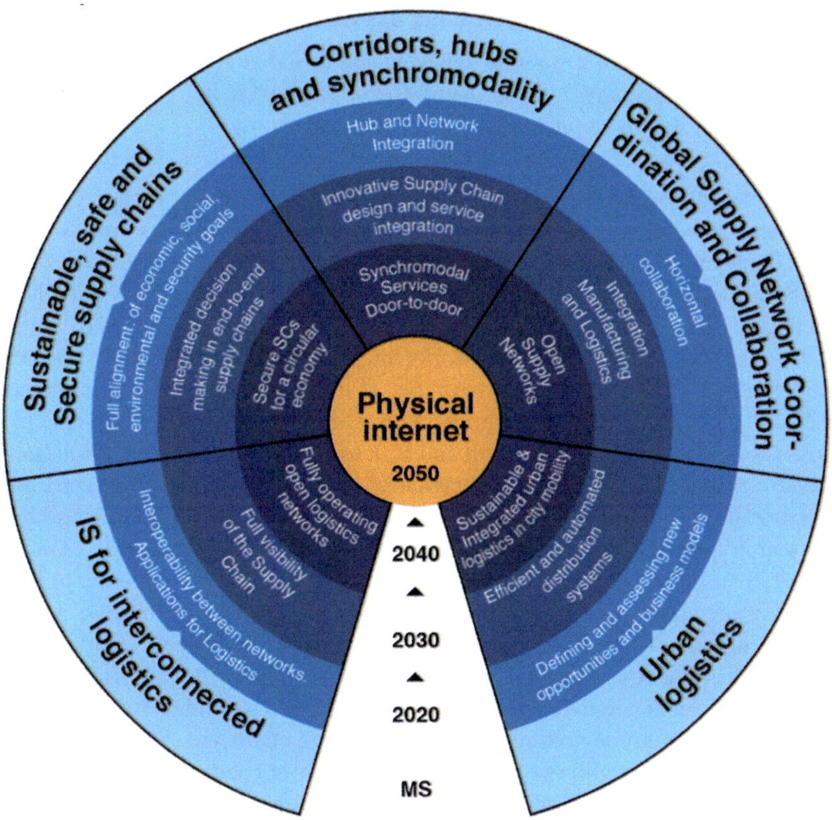

Fig. 2 Milestones towards the Physical Internet as a function of ALICE's roadmaps (ALICE ETP ROADMAP S www.etp-logistics.eu/?page_id=24)

illustrates them through related European projects and their contribution to the Physical Internet vision. The five roadmaps are the following:

1. Sustainable, Safe and Secure Supply Chains
2. Corridors, Hubs and Synchromodality
3. Information Systems for Interconnected Logistics
4. Global Supply Networks Coordination and Collaboration
5. Urban Logistics.

3.1 Sustainable, Safe and Secure Supply Chains

The first of ALICE roadmaps covers two areas; supply chain sustainability, and security.

Key milestones before the Physical Internet realization in 2050 are:

2020—Full alignment of economic, environmental, social and security goals
2030—Integrated decision making in end-to-end supply chain
2040—Safe and secure supply chains for circular economy.

As it is stated in the roadmap, in the design of sustainable logistic systems the role of shippers and manufacturers is crucial, as they decide what to transport and how. For instance, a wider and holistic product point of view can address this issue. This can be done by improving sustainability, taking into account possibilities such as modular design, late postponement closer to the customer, alternative and environment friendly packaging or sustainable sourcing strategies. The design and operation of the production and distribution networks (the flow perspective) is also a key factor, since it strikes on both costs and environmental impact. Lastly, the process perspective can also contribute to sustainability, with all the links in the chain connected in a closed-loop. Likewise, information and ICT tools for real-time data exploitation, can also improve the efficiency, effectiveness and sustainability of the processes. However, all those options face challenges in order to become real and tangible.

Research projects try to overcome those difficulties, for instance, ManSYS project[11] focused on using additive manufacturing, or 3D printing in order to reduce material usage and waste, providing a set of e-supply chain tools for mass adoption of this technique. The project developed decision-support software for metal 3D printing, achieving cost savings of around 30% in production and improving responsiveness to customer needs. The development of new and intelligent load carriers and standardized load unit devices represents also a huge challenge. The project Modulushca,[12] focused on the flows of fast-moving consumer goods, addressed the development and operation in real conditions of ISO-modular logistics units with digital interconnectivity. Building upon the work done before in other projects, Modulushca carried out two pilots involving companies such as Procter & Gamble, Chep, Jan de Rijk Logistics and Poste Italiane. The first pilot showed the benefits of handling and transporting the ISO-modular units within one company, and a second pilot evaluated the impact of these logistics units in cross docking and transhipment processes. Now, the project CLUSTERS2.0[13] aims at developing new modular loading units together with innovative handling and transhipment technology, to accelerate handling processes within clusters for road and intermodal modes. Similarly, the project Less ThanaWagonLoad[14] will develop new rail transport solutions for single pallets. These kinds of initiatives are tiles on the path to the Physical Internet realization.

Supply chain security tries to boost legitimate trade and logistics whilst safeguarding supply chain safety and security. According to ALICE's roadmap, there are three main innovation drivers on this area: enabling IT technologies, proactive and responsive supply chain concepts and coordinated border management. The latter refers to

[11] www.mansys.info/.

[12] www.modulushca.eu/.

[13] cordis.europa.eu/project/rcn/209715_en.html.

[14] cordis.europa.eu/project/rcn/209714_es.html.

a coordinated approach by border control agencies (domestic and international) that pursuit to increase efficiencies over managing trade and travel flows, while maintaining a balance with compliance requirements.

Challenges such as the development or use of low intrusive security technologies or end-to-end supply chain risks visibility are addressed by different projects. SAFE-POST[15] aimed to raise the level of postal security by integrating innovative screening solutions suitable for an uninterrupted flow of parcels and letters. Also, a Postal Security Platform enabling interoperability between postal actors and increasing the security and visibility in the postal chain was developed, CASSANDRA[16] focused on the supply chain visibility needs of business and government in international flow of containerized cargo, thus improving business operations and cross-border security inspections. The challenge of enhancing supply chain resilience in order to increase the robustness of supply chain operations is addressed by CORE[17] project, developing and demonstrating a series of resilience tools and instruments. All these projects contribute to the creation of a robust physical network of networks, the concept where the Physical Internet rests upon.

3.2 Corridors, Hubs and Synchromodality

The roadmap developed by ALICE's working group on corridors, hubs and synchromodality aims to reach the ambition of "EU wide co-modal transport services within a well synchronized, smart and seamless network, supported by corridors and hubs, providing optimal support to supply chains." Synchromodality refers to the synchronization of intermodal services between modes and with shippers.

Key milestones before the Physical Internet realization in 2050 are:

2020—Hub and network integration
2030—Innovative supply chain design and synchromodal service integration
2040—Synchromodal services door to door.

There are two main areas of innovation in this roadmap: the integration of transport services and supply chain, and the integration of transport services and infrastructures (for example, TEN-T network).

In order to achieve the former, there is a need of understanding the demand for the synchromodal freight transport system, the alignment between supply chain and transport system, and also to explore the new roles for the hubs in the entire supply chain. Similarly, the integration of transport services and infrastructures depends on deploying an integrative freight network strategy, a transport chain designed and fully operational for synchromodality, and deploying ICT as integrating technology.

---

[15] cordis.europa.eu/project/rcn/102916_en.html.

[16] cassandra-project.eu/.

[17] coreproject.eu/.

The above mentioned CLUSTERS 2.0 will develop an open network of hyper connected logistics clusters and hubs towards Physical Internet while keeping neutral the environmental and local impacts (congestion, noise, land use and local pollution levels). The initial network will comprise Zaragoza (PLAZA), Duisburg (Duisport), Lille (Dourges), Bologna-Trieste (Interporto/port of Trieste), Brussels (BruCargo), London (Heathrow), Pireaus (PCT) and Trelleborg (Port).

Many projects have addressed the role of ICT as integrating technology. MIELE[18] designed the architecture and developed a pre-deployment pilot interoperable ICT platform as interface to ICT systems (i.e. single windows, port community systems) in Italy, Portugal, Spain, Cyprus and Germany. NEXTRUST[19] develops interconnected trusted collaborative networks along the entire supply chain, horizontally and vertically built to integrate shippers, LSPs and intermodal operators.

3.3 Information Systems for Interconnected Logistics

The roadmap developed by the Information Systems for Interconnected Logistics Working Group aims to define the R&I pathway to achieve real-time (re)configurable supply chains in (global) supply chain networks, with available and affordable ICT solutions for all types of companies and participants. This is a key requirement in order to enable the concept of Physical Internet.

Key milestones before the Physical Internet realization in 2050 are:

2020—Interoperability between networks and IT applications for logistics
2030—Full visibility throughout the supply chain
2040—Fully functional and operating open logistics networks

To achieve this ambition, not only innovations in the area of ICT are needed (Internet of Things, Smart-devices, data analytics, Big Data, new support and planning systems) but also in the areas of new collaborative business models and data governance.

SELIS[20] and AEOLIX[21] projects will address many of these challenges by developing ICT common communication and navigation platforms for pan-European logistics applications. Smart-Rail project[22] also tackles data privacy and security and collection, distribution, management, and analysis of data in rail freight transport environment. The project identifies the need for a data platform, where all necessary information would be gathered and it also explores the implementation of blockchain technologies in rail freight transport.

---

[18] ec.europa.eu/inea/en/ten-t/ten-t-projects/projects-by-country/multi-country/2010-eu-21105-s.

[19] nextrust-project.eu/.

[20] selisproject.eu/.

[21] aeolix.eu/.

[22] smartrail-project.eu/.

3.4 Global Supply Network Coordination and Collaboration

This roadmap addresses coordination and collaboration among stakeholders in global supply networks. Supply Network Coordination refers to vertical synergies along supply chain (different modes and players), whereas supply network collaboration refers to horizontal synergies across supply chains (maximizing resource utilization and minimizing overall costs). Those are steps needed to transition from the current individually managed supply chains to open networks in the Physical Internet model, thus increasing efficiency and sustainability.

Key milestones before the Physical Internet realization in 2050 are:

2020—Horizontal Collaboration
2030—Integration Manufacturing Logistics
2040—Open Supply networks.

In order to achieve supply chain network collaboration, the R&I activities identified by ALICE's Working group are related to the strategic collaborative (multi-stakeholder and multi-criteria) logistics network design, the tactical planning and execution of collaborative networks involving new tools to maximize the use of resources across different stakeholders and systems, to build resilience capabilities and risk management of the collaborative networks which are expected to be at least as secure as the traditional supply chains, and to develop and implement new profitable business models for collaborative services.

To reach a coordinated supply chain network, efforts are needed to implement coordinated planning along the entire chain. In order to do so, it will be essential to overcome the current barriers to data sharing among stakeholders and systems. An effort needs to be done in order to integrate flexible manufacturing into open supply chains. This could be an opportunity to re-shore manufacturing to Europe, by means of building "manufacturing villages" sharing non-unique resources or agile and modular manufacturing.

All the above must be supported on drivers and enablers to facilitate the adoption of coordinative and collaborative approaches among stakeholders, creating awareness, favouring a management mind-set, disseminating good practices and exploring the implementation of policies to incentivize their adoption.

A good example of this was CO3 project[23], that developed and validated new value chains and business strategies between major shippers, resulting in logistics cost reductions of 10–20%, with carbon footprint reductions of 20–30%. An interesting outcome of this project was the definition of the role of a neutral trustee, not involved in the operational area, which would maximize the total synergy gains of the network while keeping its impartiality. The project showed that the most critical issue to make horizontal collaboration successful was the mental shift, and that trust amongst the partners could be achieved through gain sharing tools.

Smart-Rail project aims to develop approaches for demand driven rail service innovations with the focus on supply chain, optimizing the alignment between supply

[23]www.co3-project.eu

chains and rail transport services. Cooperative business models developed in this project show that a neutral coordination body is required in order to improve rail freight services offered to the shippers.

Inspire[24] project is also developing innovative business models, creating flexible manufacturing networks and integrating them into the entire value chain. The ultimate goal is to facilitate more local production in Europe in the medium term.

3.5 Urban Logistics

The fifth roadmap was developed jointly by ERTRAC, the European Road Transport Research Advisory Council, and ALICE. Its objective is to increase the efficiency, sustainability and security of urban logistics by means of identifying its main research priorities.

Key milestones before the Physical Internet realization in 2050 are:

2020—Defining and assessing new opportunities and Business Models
2030—Efficient and automated distribution systems
2040—Sustainable and integrated urban logistics in the city mobility system.

This roadmap identifies research needs aiming to increase the energy efficiency of the whole urban logistics system, improve the urban environment by reducing emissions and noise, improve the reliability of the urban delivery system and increase safety of people and security of goods.

The European Commission's objective to improve the quality of life in urban areas (which was materialized in the pact of Amsterdam with the Urban Agenda for the EU)[25] has led to many European projects funded in this area.

The project U-TURN[26] tackles the urban food logistics through advanced tools and supply chain collaborations, involving four pilots in Germany, UK, Italy and Greece. SUCCESS[27] is focused on construction consolidation centres, and how this concept would address current challenges in the construction supply chains in city urban areas. SUNRISE[28] project explores new ways of supporting development and implementation of neighbourhood-level and urban-district-level transport innovations, including freight transport. All those mentioned innovations contribute to the development of the Physical Internet concept.

Table 1 summarizes the main results/expected results of the above mentioned projects linked to the Physical Internet.

[24] inspire-eu-project.eu/.

[25] Urban Agenda for the EU, 2016.

[26] u-turn-project.eu/.

[27] success-urbanlogistics.eu/.

[28] civitas-sunrise.eu/.

Table 1 European projects impacting on the Physical Internet development

Acronym	Name	Main results/expected results linked to the PI
ManSYS	Manufacturing decision and supply chain management system for additive manufacturing	Development of a decision support software for metal 3D printing, increased responsiveness to customer needs, transport reduction leading to less emissions
Modulushca	Modular Logistics Units in Shared Co-modal Networks	Framework on how Physical Internet can enable an interconnected FMCG logistics system. Development of modular boxes in the FMCG sector. Sensoring and communication approach for modular logistics units, algorithms for digital interconnectivity between different IT systems. Recommendations for the standardisation of ISO-modular containers
CLUSTERS 2.0	Open network of hyper connected logistics clusters towards Physical Internet	To develop prototypes of New Modular Load Units. To develop governance models introducing the role of a neutral agent that will form the basis for new collaborative business models. To establish a Pan-European community approach of shippers bundling their regular transportation demand with other shippers and to favour intermodal alternatives
LessThana WagonLoad	Development of 'Less than Wagon Load' transport solutions in the Antwerp Chemical cluster	To develop a smart specialized logistics cluster for the chemical industry in the Port of Antwerp in order to shift transport volumes from road to rail freight by means of new rail transport solutions for single pallets and new added value rail freight services for the industry within the Antwerp chemical cluster
SAFEPOST	Reuse and development of Security Knowledge assets for International Postal supply chains	Promote the seamless flow of parcels and letters by: identification of the main security threats in postal supply chains. Description of security measures to maintain or increase security in postal supply chains. Development of a platform enabling interoperability between postal actors and increasing the security in the postal chain
CASSANDRA	Common assessment and analysis of risk in global supply chains	Development of the data pipeline concept to enable secure data sharing. Implementation of dashboards for supporting businesses and customs for risk management and supply chain visibility for international trade lanes
CORE	Consistently Optimised Resilient Secure Global Supply-Chains	Increase effectiveness of security & trade compliance, without increasing the transaction costs for business. To increase co-operative security risk management. Integrate compliance and trade facilitation concepts with supply chain visibility and optimisation by business communities

(continued)

Table 1 (continued)

Acronym	Name	Main results/expected results linked to the PI
Smart-Rail	Smart Supply Chain Oriented Rail Freight Services	Contribution to a mental-shift of the rail sector towards a client oriented and supply chain focus. Development of working business models for cooperation of different stakeholders. Development of a methodology and architecture for exchange of data/information between stakeholders, required for the optimisation process
Miele	Mediterranean Interoperatibility E-services for Logistics and Environment sustainability	Software tool to enable ICT systems to be interoperable at ports
NEXTRUST	Building sustainable logistics through trusted collaborative networks across the entire supply chain	To develop horizontal and vertical trusted networks bundling freight volumes and shifting them off the road to intermodal rail and waterway
SELIS	Towards a Shared European Logistics Intelligent Information Space	Development of a 'lightweight ICT structure' to enable information sharing for collaborative sustainable logistics for all at strategic and operational levels
AEOLIX	Architecture for EurOpean Logistics Information eXchange	To develop a cloud-based collaborative logistics ecosystem for configuring and managing (logistics-related) information pipelines
CO3	Collaboration Concepts for Comodality	Development and validation of new value chains and business strategies between major shippers allowing them to have logistic cost and emissions reductions
INSPIRE	Towards growth for business by flexible processing in customer-driven value chains	To develop new innovative business models for more flexible and sustainable manufacturing value chains
U-TURN	Rethinking Urban Transportation through advanced tools and supply chain collaboration	To develop innovative collaboration practices for urban freight distribution with a focus on food logistics. To design and implement a collaborative platform for supporting information sharing and the creation of appropriate logistics sharing partnerships
SUCCESS	Sustainable Urban Consolidation CentrES for conStruction	Development of new collaborative business models for construction of Urban Consolidation Centres
SUNRISE	Sustainable Urban Neighbourhoods—Research and Implementation Support in Europe	To develop Neighbourhood Mobility Labs with a focus on urban logistics among other issues

4 Shippers Position Towards the Physical Internet

According to the European Shippers Council,[29] manufacturers, retailers and wholesalers are collectively referred to as shippers. ALICE ETP[30] defines shippers as "manufacturers, retailers and wholesalers, and in general cargo owners who send goods for shipment, by packaging, labelling, and arranging for transit, or who coordinate the transport of goods".

This section will show the shippers position regarding the Physical Internet, taking into account the results of pilot tests performed with different manufacturers and retailers, as well as the outcomes of a survey launched among ALICE members regarding the key areas for implementing Physical Internet, as well as its drivers and barriers.

The Physical Internet will mean a paradigm shift for mobility and logistics. As logistics decisions are made by shippers and not by logistics service providers, shippers' position on Physical Internet can either foster the required developments or hinder its realization, should not be consensus on its benefits.

The Physical Internet metaphor relies on the principle that logistics could function as the Digital Internet, where freight would be exchanged seamlessly just as data is exchanged in Internet. Some shippers are already on board with this innovative concept, such as Boeing, CHEP, P&G, Volvo, HP, Walmart, and J.B. Hunt (Carlson 2016) (Fig. 3).

As stated by the Physical Internet manifesto, the Physical Internet would be an open system founded on interconnectivity, enabled through standardization and protocols, just like the Internet. Continuing with the metaphor, the Physical Internet would move freight through the logistics web utilizing open and shared networks fully interconnected. This would not only result on efficient logistics, but also on an economically, environmentally and socially sustainable system.

Retailers such as Carrefour and Casino have already experienced first-hand the benefits of Physical Internet. Indeed, in 2015 a study simulated the effects of Physical Internet on their distribution networks with their top 100 suppliers (Ballot et al. 2015). Results from this simulation showed potential cost savings of 32%, a 60% reduction of greenhouse gas emissions and 50% of shifted volume from road to rail (Fig. 4).

Prior to that, CO3 project gathered different shippers and demonstrated the impact of horizontal collaboration, a required step in the path for Physical Internet. Indeed, the case between Nestlé and Pepsico[31] earned the CO3 Award for Horizontal Collaboration. This case built a horizontal collaboration community in fresh & chilled retail distribution between two Fast Moving Consumer Goods shippers (Nestlé & PepsiCo), a logistics service provider (STEF) and a neutral trustee (TRI-VIZOR). The distribution synergy between the Belgian fresh networks of Nestlé and PepsiCo was simulated, and the logistics transport synergy of setting up horizontal collaboration was calculated by the neutral trustee in an 'offline' mode. In this example,

[29] europeanshippers.eu/about/.

[30] Available: ALICE Research and Innovation Roadmap on Physical Internet, 2017.

[31] Horizontal Collaboration in Fresh and Chilled Retail Distribution. D4.3, CO3 project.

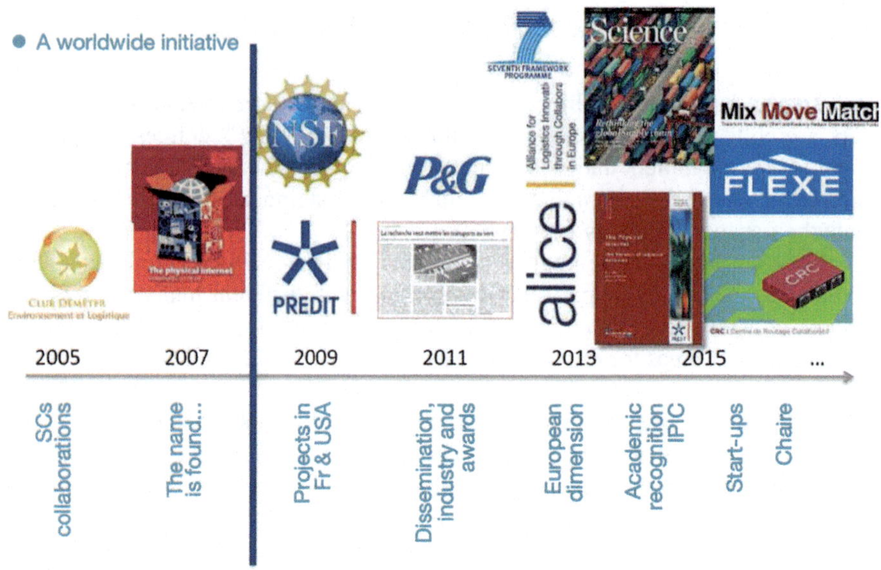

Fig. 3 Physical Internet development timeline (Ballot 2015)

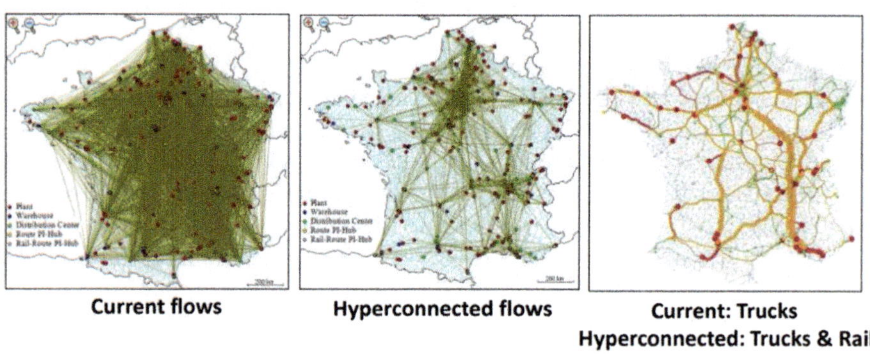

Fig. 4 Results from a simulation experiment with top retailers Carrefour and Casino in France and their 100 top suppliers (Ballot et al. 2015)

shippers were initially very apprehensive due to anti-trust. Horizontal collaboration in this case delivered 10–15% transport cost savings and even more significant CO_2 reductions.

Other example of horizontal collaboration in the framework of CO3 project was the one involving shippers P&G and Tupperware. P&G selected the Belgium-Greece corridor due to its low load vehicles. P&G shipments consisted of detergents being moved between Mechelen (Belgium) and Athens (Greece). These products are quite heavy, which yielded that the 45 feet containers used for the transportation were full in weight but half empty in volume. Tupperware, in turn, was manufacturing

Fig. 5 P&G and Tupperware business case on horizontal collaboration (See footnote 23)

in Belgium and delivering plastic boxes to Thiva, 100 km. away from Athens. This made Tupperware suitable for horizontal collaboration with P&G. The business case involved removing all direct Tupperware truck shipments to Greece, which were being loaded as bulk in the vehicles (Fig. 5 left). Instead, the Tupperware products were shipped to the P&G distribution centre in Belgium. There, P&G detergents were being palletized (Fig. 5 right) and loaded in containers. The horizontal collaboration required that the Tupperware plastic cases were top-loaded on the detergents pallets and transported to Greece. The resulting collaborative supply chain saved 150,000 truck-km, improving from 55 to 85% cube and weight fill, and attaining 17% cost savings.

Initiatives such as MixMoveMatch[32] have increased truck fill rates and reduced emissions and costs for shippers such as 3M or L'OREAL. Indeed, 3 M reported having reduced transport costs by 35% and CO_2 emissions by 50% since the Mix-MoveMatch system was implemented.

In order to move from these isolated studies to a more comprehensive view of shippers on the Physical Internet, ALICE ETP launched a survey among its members. The results of the survey were collected and yielded a roadmap towards achieving the Physical Internet Vision.

The survey[33] focused on different areas towards the implementation of Physical Internet.

- Components & technical developments (including standards) needed to achieve PI implementation
- Transition management: business models, regulations and governance
- Expected impacts of Physical Internet realization
- Barriers, opportunities/triggers and Infrastructural Investments for PI.

[32]mixmovematch.com.

[33]ALICE Research and Innovation Roadmap on the Physical Internet. D3.1 SETRIS project.

This section analyses the main outcomes from the shippers perspective regarding the transition required (business and governance models, regulation) as well as the barriers/triggers for implementation.

The responses provided to the transition section of the survey concluded that business models should be in place to ensure the economic sustainability of the system. Indeed, an open and shared system as the Physical Internet would require a fair cost and value sharing. The allocation of costs along the interconnected network could also become an issue if not clearly defined.

The concept of coopetition was also identified by the respondents as a potential scenario in the framework of the Physical Internet. According to Forbes, "Coopetition" is "a term used to describe unconventional collaboration and cooperation within an otherwise competitive field of players".[34] Indeed, competitive shippers may distribute jointly to a common retailer and thus save transportation costs and reduce their carbon footprint. But for this to happen there is need for a neutral party to solve anti-trust issues.

In general, to better understand the key business models for Physical Internet, the survey outcomes showed that it was necessary to create simulation or successful case studies showing potential benefits that may drive early adopters to start implementing the principles of Physical Internet, in the form of small scale networks, new start-ups, etc.

But for the successful realization of the Physical Internet, not only business but also governance models should be in place. According to the survey results, a system based on the sharing of assets, data, cost information, would require a neutral body to define the governance rules for the Physical Internet.

Governments should also encompass these developments, adapting regulations to fit this new paradigm, and enabling the implementation of new technologies, collaboration with competitors or efficient assets/infrastructure sharing.

In any case, realizing such a paradigm change must overcome different barriers. Respondents considered data sharing as a key barrier that would make systems more vulnerable, and thus cyber security developments should advance to face them.

Cultural change was also identified as a barrier when implementing such a groundbreaking concept. Sharing assets may generate distrust and the feeling that the distribution of gains is not fair. Therefore business models should be robust and trustworthy, and governance models should be transparent. Moreover, this paradigm may generate room for new players but leave a segment of the current players off the market.

Nevertheless, respondents concluded that climate change and the societal and political stress on overcoming its effects will ask for new innovative logistics solutions. And in that environment the Physical Internet will continue building consensus and gaining adopters.

The shared economy will be more and more a reality, even more for the younger generation that will not see the principles of Physical Internet so distant from other

[34]https://www.forbes.com/sites/forbesagencycouncil/2017/04/18/challenging-the-competition-coopetition-and-collaboration-within-a-competitive-marketplace/#11fa98de712c.

initiatives such as crowd sourcing. Moreover, technologies such as blockchain will tackle the fear for distrust and unreliability of data.

As stated in ALICE roadmap, shippers have shown "a strong support on Physical Internet benefits as potential cost, energy consumption and emission savings are clear for them". Shippers would be indeed end users of the PI, and could therefore foster its faster implementation. To ensure shippers support, ALICE is deploying new activities to further engage industry, but also to build consensus with academia and governments on the positive impact of Physical Internet.

5 Future Steps

In order to achieve a paradigm shift towards the Physical Internet vision, ALICE ETP has identified a path to be followed starting from the current situation:

- Fully owned supply chains (as is)
- Horizontal collaboration and vertical coordination
- Physical Internet in a full collaborative network.

Figure 6 shows the realization of the Physical Internet with its implications regarding economic and environmental expectations while solving the current challenges and yielding the future network of networks.

As a first step to gain consensus from industry, SENSE project has been launched, funded by the EC under the umbrella of topic MG.5.4—2017 Potential of the Physical Internet—Coordination and Support Action. SENSE will develop a roadmap to move from the current situation through the three stages stated above.

The aim of the topic was to detail the concept of Physical Internet "*into a strategic and operational vision which has the capability to get industry-wide endorsement of all stakeholders*".[35]

Next steps foreseen to do so are developing a roadmap towards the Physical Internet, monitoring relevant research projects, fostering the link between ALICE (who already identified the concept) and other ETPs representing shippers, as well as creating consensus not only among industry but also between public entities and academia.

SENSE project will focus on these activities but other projects will also work on the field. The project funded under MG.5.4—2017 Potential of the Physical Internet—Research and Innovation Action will model the Physical Internet benefits and validate them in specific real life case studies. As stated already in this chapter, demonstrating Physical Internet benefits and creating new associated business models in small scale cases would be one of the key actions to gain industry consensus.

Other European projects will focus on the future of mobility and logistics and will consider Physical Internet roadmap as input to their future work. These projects will

[35]ec.europa.eu/research/participants/portal/desktop/en/opportunities/h2020/topics/mg-5-4-2017.html.

Fig. 6 Physical Internet vision (SETRIS project, Defining the concept of a truly integrated transport system for sustainable and efficient logistics, 2017)

deal with concepts such as the port of the future,[36] big data in transport,[37] integrating urban nodes in the TEN-T network,[38] transport needs[39] and future supply chains.[40]

They will be intertwined with SENSE project mentioned above, as developments made by these Coordination and Support Actions will input the Physical Internet roadmap. The results of these projects will impact the roadmap towards Physical Internet in different ways. Indeed, they will deal jointly or separately with Physical Internet hubs (smart ports, urban nodes); focus on big data, one of the triggers for Physical Internet; analyse supply chain integration, a key factor to unlock the potential of the Physical Internet. Eventually they will provide a research agenda that will pave the way towards the Physical Internet.

References

Ballot E (2015) What is Physical Internet? Potentials of interconnected logistics services

Ballot É, Montreuil B, Meller R (2015) The Physical Internet: the Network of Logistics Networks

Carlson E (2016) Decoding The Physical Internet

Montreuil B (2011) Towards a physical internet: meeting the global logistics sustainability grand challenge CIRRELT-2011-03

Montreuil B, Meller RD, Ballot E (2010) Towards a physical internet: the impact on logistics facilities and material handling systems design and innovation. Progress in Material Handling Research

[36] http://ec.europa.eu/research/participants/portal/desktop/en/opportunities/h2020/topics/mg-7-3-2017.html.

[37] http://ec.europa.eu/research/participants/portal/desktop/en/opportunities/h2020/topics/mg-8-2-2017.html.

[38] http://ec.europa.eu/research/participants/portal/desktop/en/opportunities/h2020/topics/mg-8-7-2017.html.

[39] http://ec.europa.eu/research/participants/portal/desktop/en/opportunities/h2020/topics/mg-8-7-2017.html.

[40] http://ec.europa.eu/research/participants/portal/desktop/en/opportunities/h2020/topics/nmbp-37-2017.html.

Carbon Footprint Accounting in Freight Transport: Training Needs

Susana Val, Beatriz Royo and CarolinaCiprés

Abstract Carbon footprint accounting is becoming a topic of great interest, with special relevance in the freight transport sector. That is not only due to environmental concerns but also to regulations and marketing issues (i.e. brand image, competition or internal procedures for a better development). This Chapter shows that there is no extended knowledge on how to implement common methodologies in the transport sector. Training is crucial, even when using one of the common standards, as depending on the different stages while implementing the methodology, results may differ. Therefore, the transport sector is facing a lack of training and education on the field, and this Chapter intends to provide some of the current requirements that the market is demanding on carbon footprint accounting.

Keywords Carbon footprint · GHG emissions · Emission accounting · Emission reduction · Training & Education · Freight transport · Logistics · Green efforts

1 Introduction

The reduction of Green House Gas (GHG) emissions according to the adoption of the Paris Agreement on climate change in 2016 can only be achieved with the society's firm engagement, from the individual customers to industry levels. Concerning the latter, wide literature has focused on environmental sustainability in the manufacturing industry, while less attention has been dedicated to the service industry and least of all the logistics industry (Lin and Ho 2011, Davarzani et al. 2016). The adoption of green initiatives by logistics service providers (LSP) to reduce environmental

S. Val (✉) · B. Royo · C. Ciprés
Zaragoza Logistics Center, C/Bari 55, Bloque 5, 50197 Saragossa, Spain
e-mail: sval@zlc.edu.es

B. Royo
e-mail: broyo@zlc.edu.es

C. Ciprés
e-mail: ccipres@zlc.edu.es

B. Müller and G. Meyer (eds.), *Towards User-Centric Transport in Europe*, Lecture Notes in Mobility, https://doi.org/10.1007/978-3-319-99756-8_15

impacts is gaining importance (McKinnon 2005, Liu et al. 2010). Some of the gaps identified in literature include the impact of green initiatives on LSP performance, the evaluation of sustainability performance and the information and communication technologies supporting green initiatives (Centobelli et al. 2017).

Several studies have been conducted for the identification of the main barriers and drivers for transport and logistics communities to report their emissions. On the one hand, the complexity of the process, the lack of mandatory regulations and knowledge hinder this practice. On the other hand, the increasing environmental society's awareness, the investors' concerns and Information Communication Technologies (ICT) might foster this behaviour. Furthermore, training and education would help to overcome some of the identified obstacles. Section 2 presents the current state on GHG emissions accounting in the logistics and transport industry; Sect. 3 explains the analysis conducted for identifying the knowledge gaps and Sect. 4 shows the results; Sect. 5 refers to some related studies; Sect. 6 describes the proposed education programmes' contents and structure features for covering the needs and Sect. 7 states the conclusions.

2 Carbon Accounting: Status Quo

The magnitude of climate change requires developed countries to make changes to lifestyle choices—from the products they consume to where they spend their vacations, to the building in which they live and work. GHG standards will play a vital role in transition, providing transparency and assurances needed for product labelling, purchasing of carbon offset, regulating business emissions and certifying the GHG accounting (Secretariat 2012).

Carbon accounting requires GHG standards to provide reliable and comparable systems to face the following issues:

- Support many types of mandatory and voluntary government programmes
- Incorporation into legislation and regulations such as cap and trade regional as international trade agreements
- Incentives (production subsidies, tax and other business incentives) to support new industries technologies
- Technology Research & Development (R&D) funding
- Support a range of important business functions
- Carbon labelling of products and events for consumer and stakeholder communications avoiding "greenwashing" (Alejos 2013)
- Support new financial products
- Support practitioner's competence and certification regarding quantification, auditing, reporting, labelling and communications.

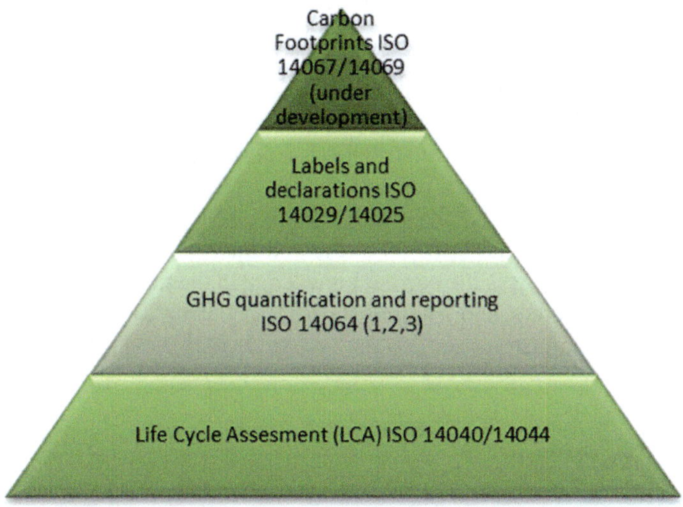

Fig. 1 Environmental ISO regarding LCA. (based on Camarero 2011)

The International Organization for Standardization (ISO) provides the ISO 14000 family with practical tools for companies or organizations to manage their environmental responsibilities.[1]

Figure 1 shows the ISO related to carbon accounting; at the base of the pyramid are ISO 14040-14044, which describe the principles and framework for Life Cycle Assessment (LCA). The second row of the pyramid contains more specific ISO for organizations to implement an auditable GHG inventory, verify and validate GHG projects and define the baseline scenarios for monitoring projects; above are the ISO related to labels and declarations and finally, at the top are the ISO regarding carbon footprint accounting, 14067 for products and 14069 for organizations, both of them still under development. As a consequence of the lack of standardization, GHG programmes proliferated with climate change (McKinnon 2018).

Other programmes such as GHG Protocol—developed by the Worldwide Resources Institute and the World Business Council for Sustainable Development (WRI/WBCSD)—establishes the global standards to measure, manage and report GHG emissions for private and public sector operations. GHG Protocol offers online training for their standards and tools and provides the accounting platform for virtual GHG reporting. These international standards give guidance documents, which provide useful information, but can leave room for the interpretation (Davydenko et al. 2014). Below them, there is a wide variety of freight sector methods and tools that offer specific guidance for the sector (Smart Freight Centre 2016) and make it difficult to provide a comparable framework for companies to demonstrate their green efforts. As they cannot be audited and certified, and there are no mandatory

[1] https://www.iso.org/iso-14001-environmental-management.html.

regulations, companies disregard these practices and in consequence they do not make the effort of reporting the environmental performance.

Nevertheless, the Global Logistics Emissions Council (GLEC) framework[2] provides a harmonized basis from existing methodologies—as consistent as possible—-for the calculation of emissions from freight transport chains across modes and global regions. It proposes an approach for data format, collection, analysis and reporting within the logistics sector.

The following Section describes the research analysis for identifying the main barriers and drivers, assesses the current knowledge in the field and recognises some of the most relevant areas and training formats according to the target audience's role in the logistics industry.

3 Survey Analysis

This Section shows the main barriers, drivers and evaluation of the knowledge level and engagement of stakeholders regarding carbon footprinting. It also provides the specific aspects of carbon emissions accounting which require a consolidated training, as well as the best training formats to teach how to calculate and ultimately reduce GHG emissions.

This survey was a Web questionnaire translated into eight languages (German, Swedish, French, English, Italian, Spanish, Greek and Romanian), built with Survey Monkey[3] and circulated to the transport and logistic community among different European Countries. It was open throughout November 2017.

The questions were assessed using the Likert scale (1: Not at all—5: Very much) and an open-ended option for the alternative "others". The questionnaire was divided into three sections:

- General Information, just to establish the company framework and role of the same
- Status of Knowledge, with the aim of understanding the "status of knowledge" regarding GHG emissions accounting as well as identifying the possible "drivers" and "barriers" to calculate emissions. There were three head-questions with their corresponding alternatives to assess according to a Likert Scale.
- Training and Education needs, to understand what are the most important training and education needs for each of the following working positions:
  - Administrative employees (i.e. customer accounts, buyers, Human Resources (HR) managers, etc.)
  - Operational employees (i.e. drivers, workers, etc.)
  - Middle managers (i.e. transport managers, fleet managers, etc.)
  - Top managers
  - Environmental/Corporate Social Responsibility (CSR) specialists.

[2]http://www.smartfreightcentre.org/glec/what-is-glec.

[3]https://www.surveymonkey.com/.

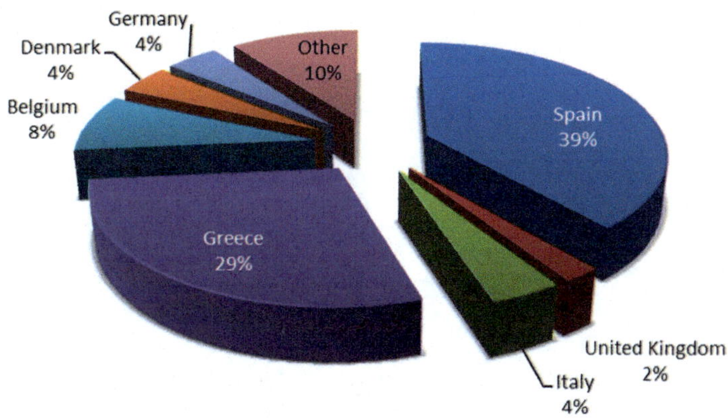

Fig. 2 Web questionnaire: % of participation by country

4 Survey Results

Results are shown divided into the three sections mentioned above.

4.1 Survey Results: Section 1

There were 54 participants coming from different countries. Nevertheless, most respondents were from Spain, Greece and Romania (see Fig. 2). The target audience within these countries was the logistics and transport community, which consisted on manufacturers, logistics service providers, retailers, public administration, research centres, Non-Governmental Organizations (NGOs) and other stakeholders. The participation by company type is shown in Fig. 3.

4.2 Survey Results: Section 2

The first question in the second section of the survey aims at assessing the main drivers for companies to report GHG emissions. As mentioned above, a Likert Scale was used (1: Not at all important—5: Very important) and the responses showed a small difference between those with the above-three score and the rest (Fig. 4). These answers show the average value for each option, and indicate that the most encouraging factor for companies to engage in GHG emissions accounting is helping to identify opportunities for cost and emission reduction, followed by the opportunity for improving the business strategy. Requests from customers and public or mandatory reporting seem to be slightly less important than the rest of the alternatives.

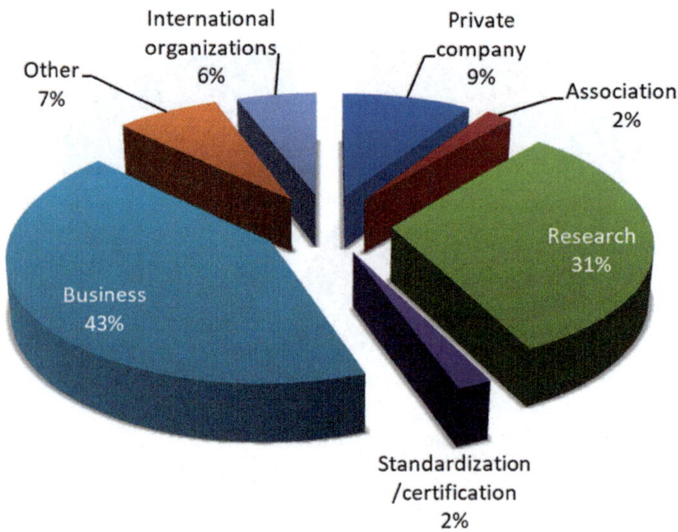

Fig. 3 Web questionnaire: % of participation by company type

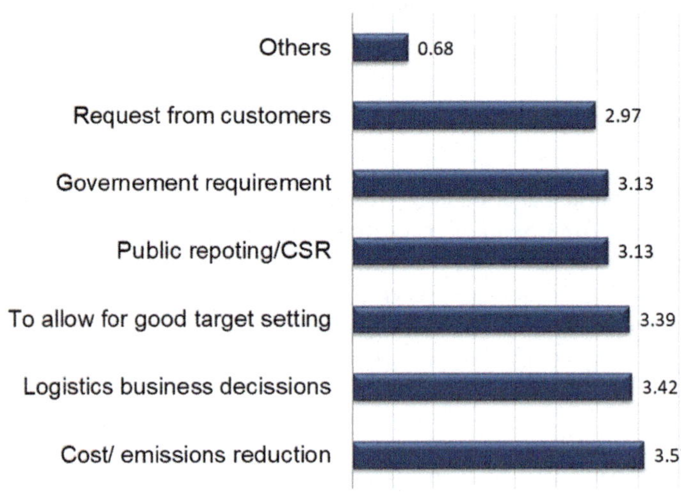

Fig. 4 Emissions accounting drivers

The second question in the second Section of the survey aims at identifying the main barriers for carbon footprint accounting and reporting using the same Likert Scale (1: Not at all important—5: Very important) mentioned above. The following graph, (Fig. 5), shows the average values for the options given. According to the results, it can be easily understood that interviewees lack available resources—as the survey was conducted among a sample of Small and Medium Enterprises (SMEs) and big companies-, they experienced difficulties in calculating such emissions and

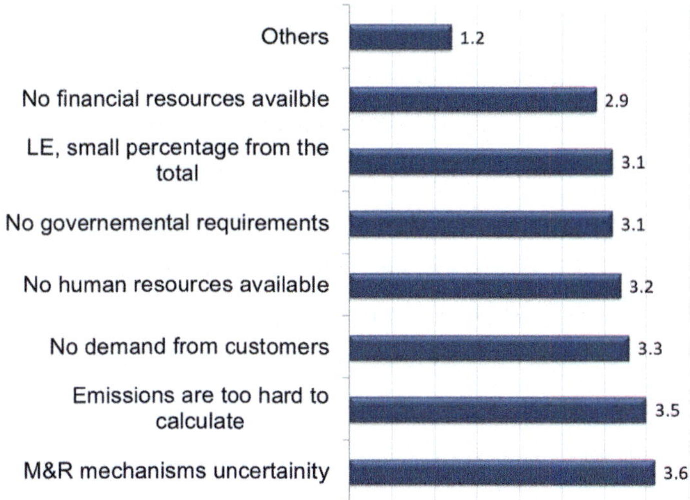

Fig. 5 Barriers for accounting emissions

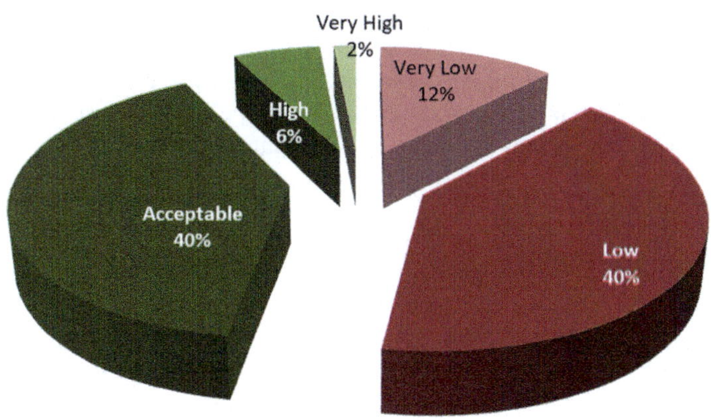

Fig. 6 Survey's participants' knowledge level

they have some uncertainty regarding measurement and how to use mechanisms to report—due to the lack of a standardized methodology and training.[4]

The third question in Sect. 2 asked participants for their specific level of knowledge. It is remarkable that over 50% of the answers report having an insufficient knowledge about measuring, verifying and reporting carbon emissions from transport and logistics activities in their organizations, which leads to strong training needs in this particular field (see Fig. 6).

[4]LE: Logistics Emissions.
M&R: Measurement & Reporting.

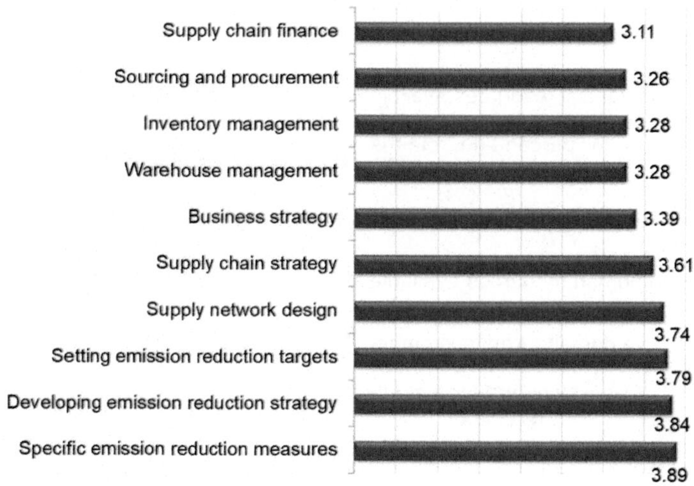

Fig. 7 Best practices assessment

As the majority of freight in Europe is still transported by road, this was the preferred mode to receive training on accounting emissions. It is also due to the fact that this sector is currently very fragmented in Europe, with almost no possibilities to get on board on the recent developments and standards.

4.3 Survey Results: Section 3

The third section asked for assessing the interest in receiving training related to "Best Practices". Figure 7 shows the average values using the Likert Scale (1: Not at all interesting—5: Very interesting). The three options with the highest score "specific emissions reduction measures", "developing emissions reduction strategy" and "setting emissions reduction targets" represent the companies' main reasons for measuring GHG emissions. By adopting these three best practices, companies will be able to "identify opportunities for cost and emissions reduction" and "improve their business strategy".

Furthermore, the survey provides information about the areas for ensuring a correct measurement process according to all the different roles in an organization (Table 1). The darkest colors reflect the preferred option for each role in the organization and red bold text the three ones with the largest score.

Last but not least, the survey results indicate the participants would rather attend Face to Face (F2F) meetings than any other format, regardless the role. Gaming experience obtains a good score for all roles except for operational employee, the eLearning, an option continuously gaining popularity, is only selected for middle

Table 1 Training areas by role

	Definitions and measure for messiosn	Specifics standards or policies in your industry	Best practices for emissions reductions	Tools to calculate emissions in the supply chain (i.e. LCA)	Most appropiate data to use	Tools to calculate emissions in a particular mode or segment	Reporting standards of emissions for clients or company mangemant	Specific regulation regarding esmissions for your industry	General regulatory regarding emissions in your country/ EU	Verfication processes	Certifications
Administration Employees	2.93	3.00	3.38	2.92	2.92	3.17	2.75	2.91	2.92	2.58	2.83
Operational Employees	3.15	2.85	3.00	3.00	3.67	3.45	2.83	2.42	3.09	2.58	2.83
Middle Managers	2.93	3.77	4.25	3.69	3.38	3.33	3.64	4.50	3.33	3.33	3.50
Enviromental/ CSR Specialists	4.23	4.00	3.38	4.00	3.36	3.82	3.75	3.60	4.27	3.50	3.55

managers and environmental/CSR specialists. The least preferred alternative is paper materials (see Table 2).

5 Related Studies

Other surveys have been conducted concerning the GHG emissions accounting. Specially, regarding the second section of the survey analysed in Sect. 4 of this Chapter, there are particular studies showing the lack of incentives and motivation for reporting GHG emissions.

The study developed by CE Delft in 2014 (Delft 2014) gathers similar surveys showing that environmental efficiency is not considered yet as an important factor when purchasing transport in a company or making a buying decision.

Specifically, Byrne and Wolf (Byrne et al. 2013; Wolf and Seuring 2010), concluded that, even though third party logistics (3PLs) and shippers' environmental concerns have increased, traditional performance objectives such as price and timely delivery are still much more relevant.

Table 2 Preferred formats and materials

	Paper materials (manuals, Leaflets)	Online trainins (slides, case studies, videos, live webinarr,etc)	F2F Meeting	Individual/ Small Group coaching	Moderated peer to peer workshops	Gaming experiences (e.g. Webb apps)
Administration Employees	2.79	3.00	3.31	2.83	2.83	3.08
Operational Employees	2.93	2.77	3.43	3.08	2.75	2.92
Middle Managers	3.00	3.38	3.38	3.15	3.31	3.46
Enviromental/ CSR Specialists	3.33	3.75	3.55	3.00	3.45	3.64

Table 3 Outcomes from surveys conducted in 2007 and 2013 (Delft 2014)

	Lammgård 2007 (%)	Lammgård 2013 (%)
Price	51	54
Transport time	23	16
Delivery	18	22
Environmental efficiency	8	8

Lammgård (Lammgård et al. 2013; Lammgård 2007) surveys were very similar to the one previously described and were conducted in Sweden by the same authors. Although they were performed in a different year the outcomes were quite similar, showing that the environmental performance is neglected (see Table 3).

Moreover, another survey sent to Swedish shippers (Borlänge: Trafikverket 2013) obtained the same results giving the highest priority to financial and operational aspects whereas the environmental performance was ranked relatively low.

An additional survey was conducted by Lieb (Lieb and Lieb 2010) with a slight difference to the surveys explained above. The purpose was to provide evidence of the importance of increasing financial incentives in the market. The study performed a survey to the shippers twice with a small modification. The first time they were asked whether they would consider an operator with a better sustainability performance under equal price and quality conditions whereas the second time they were asked the same question with the exception that the more sustainable operator would cost 5% more. The results showed a significant change in the priorities, the first time one-tenth of the shippers would always use this operator, and over half of the shippers would maybe do so. However, if the cost increases not one of the shippers would definitely use the greener company and only 23% would still consider choosing this company. However, the majority of the shippers (77%) would not consider a more sustainable operator at all if it was 5% more expensive than its competitors.

After analysing all the surveys (Delft 2014) it was concluded that the incentives to report GHG emissions are insufficient from a financial point of view (agree 63%; disagree 15%) and from a social point of view (agree 47%; disagree 14%).

However, the trend is changing. More recent studies are focusing on the increased companies' interest in reducing the GHG emissions and controlling the sustainability performance management where calculating and measuring emissions is essential. Concerning the former, a recent study (Rogerson 2016) examined the impact of increasing load factor for reducing GHG emissions, "*a high load factor is beneficial for decreased CO_2 emissions*" where shippers' freight-transport-purchasing process play a crucial role. Concerning the latter, Supply Chain Management (SCM) objective has been to increase profitability. However, the impact of logistics on climate change is increasing the attention. On the one hand, to control pollution and on the other hand, to increase road safety, (McKinnon 2015) aims at focusing on reducing the environmental impact of logistics although inevitably and frequently refers to the economic and social implications. In fact, (Beske and Seuring 2014) recommends companies moving towards reaching Sustainability Performance in Supply Chain Management identifying key aspects such as "orientation toward SCM and sustainability, continuity, collaboration, risk management and proactivity".

6 Education and Training Programs

As a summary, the conclusions from the survey performed reveal that some training modules should cover the audiences' common aspects while more specific ones would be more oriented to environmental specialists/CSR and middle/top managers. Specifically, contents related to the technical information such as appropriate data to use, tools to calculate emissions in a particular transport mode or segment and general regulation could involve most of the roles. Contents about strategical decisions, verification, certification and reporting processes could be more oriented to managers and specialists. With regard to training formats and materials, F2F meetings are suitable for all the roles; more innovative formats of training include online training, which is considered appropriate for managers and specialists.

Taking advantage of the market research performed to other institutions in terms of similar programs related to GHG emissions accounting, it can be concluded that there is a geographical concentration of such universities in English speaking countries, so that there is little access to the target audience coming from very fragmented sectors. However, this analysis can be taken as a first step by means of a preliminary collection of insights regarding needs towards the harmonization of the training programs in emissions accounting.

According to the survey results, the main training idea to fit the companies' needs at all levels is to focus on training course(s) on accounting and emission reduction analysis for road freight carriers to help them in a practical way. Based on the survey, topics to be covered are business strategy; supply chain strategy; verification & certification; best practices; policies and standards; strategies for companies and customer management; available tools for emissions accounting.

Training materials need to be developed in different languages as the road transport consists of a very fragmented sector, focused on SMEs mainly. Based on the survey results among stakeholders, training should have specific content and specific formats. Indeed, preferred formats include classroom training, website and webinars.

In addition, supplementing educational materials are also targeted at stakeholders that work with carriers such as customers (shippers/cargo owners), industry associations and green freight programs.

According to the stakeholders needs, a proper set of materials should include the following aspects:

- Training of trainers course development
- Education of stakeholders, who work with carriers to create demand
- Sectoral tailoring of training course and marketing/outreach
- Embedding in or recognition by green freight programs

7 Conclusions

Having in mind the long process to standardization and the current situation of the market, one of the main conclusions of this Chapter is the difficulty for professionals to provide reliable and accurate carbon footprint accounting in freight transport. Concerning the former, the European Commission (EC) funded several projects and methodologies for harmonizing the GHG emissions in all transport modes and transhipment centres such as Carbon Footprint of Freight Transport (COFRET)[5] and the GLEC framework. In fact, the current European project, Logistics Emissions Accounting & Reduction Network (LEARN)[6] aims at covering some research gaps from the GLEC framework. With regard to the latter, providing accurate GHG emissions values requires learning which data and information exchange interfaces are the most appropriate to use.

Emissions reduction not only implies costs savings, but also improve brand image, attract financial resources and compliance with future mandatory reporting and regulations. Providing that many companies are following the greening path, they will be able to reach strategic objectives requiring a good Sustainable Supply Chain Performance Management (SSCPM). Specifically, Environmental Performance Management (EPM) using representative, comparable and addible data helps decision

[5]http://www.cofret-project.eu/.

[6]http://www.learnproject.net/.

makers to select more efficient alternatives such as products, carriers, lanes or modes of transport. Lack of human resources and knowledge, due partially to the lack of directives, hinder companies from moving towards decarbonisation. It shows the necessity of education and training materials to increase companies' involvement and speed up the process to zero emissions goal. Education is needed due to the lack of awareness about the topic or simply because outsourcing hinders the actual vision; training is needed once the company is concerned about decarbonisation and really wants to fix goals to gain competitive advantage in the short and medium terms and by means of the ecolabel in the long term.

Furthermore, there is a long path to a globally and recognized certification, willing to provide an extended procedure. Training is an asset to provide skills and competences in carbon footprint accounting and reducing emissions.

New education and training programs covering the identified logistics and transport industry's educational requirements will be designed. The curricular is expected to be divided into sequential modules, which could serve as the basis for designing tailor-made training considering the profile of the business professionals, the sector they belong to and the role they play in it, to ensure a better implementation of carbon footprint accounting in real life. The first module will cover carbon footprint accounting foundations and the new harmonization GLEC framework; the second module, certification and verification processes and the third module will develop strategic thinking training.

Not only in situ training and education are considered, but also new formats such as webinars, e-learning platforms and massive open online courses (MOOC).[7] The latter will permit to reach a large number of learners increasing society's greener practices to reduce emissions, which in turn will speed up the process to stop global warming, improve air quality and reduce pollution-related diseases.

References

Alejos C (2013) Greenwashing: ser verde o parecerlo. Cuadernos de la Cátedra "La Caixa" de Responsabilidad Social de la Empresa y Gobierno Corporativo

Beske P, Seuring S (2014) Putting sustainability into a supply chain management. Supply Chain Manage Int J 19(3):322–331

Byrne P, Ryan P, Heavey C (2013) Sustainable Logistics: a Literature Review and Exploratory Study of Irish Based Manufacturing Organizations. Int J Eng Technol Innovation 01(3):200

Camarero R La (2011) Huella de carbono en las actividades logísticas. Guía de calidad medioambiental. Laureano Vegas, Departamento de comunicaci6n Centro Español de Logística

Centobelli P, Cerchione R, Esposito E (2017) Environmental sustainability in the service industry of transportation and logistics service providers: systematic literature review and research directions. Transp Res Part D 454–470

Davarzani H, Fahiminia B, Bell M, Sarkis J (2016) Greening ports and maritime logistics: a review and network analysis. Transp Res Part D Transp Environ 48:473–487

[7] http://mooc.org/.

Davydenko I, Ehrler V, de Ree D, Lewis A, Tavasszy L (2014) Towards a global CO_2 calculation standard for supply chains: suggestions for methodological improvements. Transp Res Part D Transp Environ 32:362–372
Delft CE (2014) F. I. LOT 3: introduction of a standardised carbon footprint methodology. In: En C. D. Delft, Fact-finding studies in supportof the development of an EU strategy for freight transport logistics. European Commission DG for Mobility and Transport
Lammgård C (2007) Environmental perspectives on marketing of freight transports-the intermodal road-rail case. Ph.D. thesis ed., Göteborg University, Göteborg
Lammgård C, Andersson D, Styhre L (2013) Purchasing of transport services: a Survey among major Swedish Shippers. http://www.chalmers.se/en/centres/lead/research/transportinkopspanel en/Documents/LammgC3A5rdetal_NOFOMA.pdf
Lieb K, Lieb R (2010) Environmental sustainability in the third-party logistics (PL) Industry. Int J Phys Distribution Logistic Manage 40:524–533
Lin C, Ho Y (2011) Determinants of green practice adoption for logistics companies in China. J Business Ethics 67–83
Liu X, Grant D, McKinnon A, Feng Y (2010) An empirical examination of the contribution of capabilities to the competitiveness of logistics service providers: a perspective from China. Int J Phys Distrib Logistics Manage 40(10):847–866
McKinnon A (2005) The economic and environmental benefits of increasing maximum truck weight: the British experience. Transp Res Part D Transp Environ 10:77–95
McKinnon AC (2015) Environmental sustainability: a new priority for logistics managers. In: McKinnon AC, Browne M, Piecyk M, Whiteing A (eds) Green logistics: improving the environmental sustainability of logistics. United Kingdom, Kogan Page, London, pp 3–31
McKinnon A (2018) Decarbonizing logistics: distributing goods in a low carbon world. London, Kogan Page
Rogerson S (2016) Thesis for the degree of Doctor Philosophy:environmental concerns when purchasing freight transpor. Department of Technology Management and Economics, Chalmers University of Technology Sweeden
Secretariat IC (2012) GHG Schemes addressing climate change—How ISO standards help. Switzerland. ISO, 2010-12/2000
Smart Freight Centre (2016) GLEC Framework for Logistics Emissions Methodologies, Version 1.0
Trafikverket (2013) Conlogic. Företagens logistikanalyser—åtgärder för bättre resurseffektivitet och mindre miljöpåverkan. Borlänge
Wolf C, Seuring S (2010) Environmental impacts as buying criteria for third party logistical services. Int J Phys Distrib Logistics Manage 40(1–2):84–102

Part V
Personalised and Seamless Services in Passenger Transport

Mobility as a Service—Stakeholders' Challenges and Potential Implications

Juho Kostiainen and Anu Tuominen

Abstract The transport sector is going through major changes. Global challenges and megatrends such as climate change, urbanization, safety and security concerns, and servitization are driving transport system developments towards sustainability and user-centricity. Over the past few years, Mobility as a Service (MaaS) has become a common term for the idea of providing the end user a one-stop shop for discovering, accessing and paying for a wide variety of transport options with varying pricing models. While multimodal information and local ticketing integrations are common, scalable comprehensive solutions are still lacking. This paper identifies key challenges for different stakeholders and their potential implications on the way towards integrated and inclusive sustainable servitization of mobility.

Keywords Mobility as a service · MaaS · Challenges · Implications
Combined mobility services · Integrated mobility services

1 Introduction

Transport is one of the key fields in which sustainability transition is urgently needed to limit climate change and improve resource efficiency. The current car-based transport system is the outcome of an extremely successful, globally adopted 100-year-old tradition, and characterized by fossil oil dependency, high transport costs for households, and adverse environmental, safety and health impacts. The mobility market structures, regulatory regimes, individual daily mobility patterns and the planning and management procedures for mobility create a lock-into the present inefficiencies that is hard to crack. Nevertheless, it is clear that challenges such as climate change,

J. Kostiainen (✉) · A. Tuominen
VTT Technical Research Centre of Finland Ltd., Vuorimiehentie 3, Espoo P.O. Box 1000, FI-02044 VTT, Finland
e-mail: juho.kostiainen@vtt.fi

A. Tuominen
e-mail: anu.tuominen@vtt.fi

B. Müller and G. Meyer (eds.), *Towards User-Centric Transport in Europe*,
Lecture Notes in Mobility, https://doi.org/10.1007/978-3-319-99756-8_16

economic uncertainties, social and demographic change and political turbulence call for new solutions and knowledge for society, communities and individuals to better organize transport and mobility (L'Hostis et al. 2016). Digitalization and servitization are seen as potential means to help tackle these challenges.

The European Strategy for Low-emission Mobility (European Commission 2016) is one of the key documents framing the Commission's initiatives for the coming years in the field of transport. The strategy has three main elements: (1) Increasing the efficiency of the transport system; (2) Speeding up the deployment of low-emission alternative energy; (3) Moving towards zero-emission vehicles. Especially the first element is relevant in the context of digital services, referring to technological and other innovations for improving mobility of people and goods. Digitalization aims to enhance the productivity and effectiveness of the transport system as well as help in promoting more environmentally friendly, customer oriented and safer transport, while creating new business opportunities. Transport-related services leveraging digitalization include sharing services; mobility and transport services; information services; traffic control and traffic management services; mobility management services; charging and ticketing; and supplementary remote services (Tuominen et al. 2016).

Looking beyond transport services and individual digital solutions, the concept of Mobility as a Service (MaaS) envisions a system where mobility operators provide a comprehensive range of mobility services to the users (Heikkilä 2014). Over the past few years, MaaS has become a common term for describing the general idea of providing the end user a single service for discovering, accessing and paying for a wide variety of transport options (MAASiFiE 2016). Other terms referring to more or less the same meaning of combined offering of public and private transport solutions include Combined Mobility Services (UITP 2011) and Integrated Mobility Services (Lund et al. 2017).

While multimodal information services (e.g. trip planners) and local ticketing integrations (e.g. travel cards) are quite common, combining and integrating these with emerging car and bike sharing schemes and on-demand services into all-encompassing and scalable solutions has many challenges.

In this paper, we aim to expand the often technically oriented discussion into the key societal challenges and barriers for achieving a truly comprehensive and integrated vision of MaaS. We will also show how a systematic analysis of socioeconomic challenges can promote the deployment of MaaS concept by clarifying the views, risks and fears of different stakeholders. Our approach is based on the national experiences on the rise of the MaaS concept and practices in Finland as well as knowledge from both national and European research projects in the field.

The paper is structured as follows. Various perspectives on MaaS are reviewed in Sect. 2. Section 3 presents our approach to analyse the emerging societal barriers and challenges. Then, we present the results of our analysis in a form of six challenge categories. Finally, in Sect. 5 we conclude with a discussion on the practical implications of the results.

2 Perspectives on Mobility as a Service, MaaS

2.1 The Scope

The scale and scope of what is understood as Mobility as a Service (MaaS) varies a great deal ranging from simple information aggregation and trip planners all the way to comprehensive one-stop shop applications covering booking and payment as well as pricing models for the whole mobility offering. In between lie different on-demand or sharing services (e.g. taxis, ride-hailing, ride sharing and car sharing), information aggregators that link to the various transport services (e.g. apps such as Urbi (2018) and Transit (2018)), integrated payment schemes (e.g. travel cards for different types of public transport), and collaborative offerings covering certain services (e.g. local public transport with taxis or trains with car sharing).

The MAASiFiE project (2016) defined MaaS as "multimodal and sustainable mobility services addressing customers' transport needs by integrating planning and payment on a one-stop shop principle." Transport Systems Catapult (Transport Systems Catapult 2016) described MaaS as a new concept that offers consumers access to a range of vehicle types and journey experiences by putting the customer first, defined as "using a digital interface to source and manage the provision of a transport related service(s) which meets the mobility requirements of a customer". The MaaS Alliance white paper (2017) similarly describes the processes in a MaaS offering to include registration and user accounts, multimodal journey planning and real time information, booking and payment, and the ticketing and access during the journey.

Already over a decade ago HANNOVERmobil, for example, included car-sharing memberships, discounts for rail services, taxis, rental cars etc. in annual public transport passes, which resulted in half the users abandoning private cars or skipping the purchase of one—albeit with limited market success and take-up (Röhrleef 2017). However, while convenience and seamless user experience through integrating the search, booking and payment of all public and private transport options into one application is the main point in the MaaS definitions, it is not the full picture. Rather, a driving idea and value proposition of MaaS, when taken furthest, is providing mobility as a subscription package comparable to a mobile phone plan that includes calls, SMSs and data for a fixed monthly price (Hietanen 2014). For example, in November 2017, the Whim app (2018) published a monthly package including unlimited rental car use and unlimited taxi rides (within a 5 km radius, and not overlapping with car rental). Results from different forms of mobility package integrations (from travel cards to season passes and capped prices as well as including car sharing and rental services with public transport) have shown before that subscription-based pricing can be more appealing than pay-as-you-go ones (Kamargianni et al. 2016).

One of the main aspirations of MaaS is to provide a competitive alternative to a private vehicle. This means users saving the cost of the often under-utilized asset and instead spending on services that best meet their mobility needs. MaaS does not aim to replace or compete with any existing transport service but rather refers to an integrated way of accessing them. For the user it means a one-stop shop, and for the transport providers a new sales channel. Agreement on and understanding of this is necessary in order for the different transport providers to wish and agree to allow the integration of the services (Kamargianni et al. 2015).

2.2 *The Value Network*

Realizing the MaaS vision requires collaboration from different service layers ranging from the infrastructure and fleets to the transport providers and operators as well as the final integration, and user interfaces and experience. No matter how good the usability and packaging, the usefulness cannot be resolved if the available choice or capacity of underlying transport services is poor. While the wording and grouping varies a bit, the overall structure and elements of a MaaS service types are often described in a very similar manner (e.g. MAASiFiE 2016; Finnish Transport Agency 2015; McKinsey & Company 2017; Holmberg et al. 2016; Jittrapirom et al. 2017) and can be broadly categorized into four service layers as follows.

The first layer—producers and manufacturers—includes vehicle manufacturers, component producers and system integrators. These physical assets are a necessity for transport services but have little to do with the form in which they are utilized (e.g. whether vehicles are sold as private cars, taxis or to be used for car-sharing).

The second layer—transport services—comprises mobility and logistics service providers offering the actual transfer services to the end consumers, such as taxis, public transport, car sharing, bike sharing, and so on. Holmberg et al. (2016) make a distinction between public transport and other transport providers in that, while providing mobility services like the others, public transport and transport planning (e.g. local authority) is also tasked with managing negative externalities and providing often subsidized services based on societal needs rather than users' ability and willingness to pay. That is, taking into account wider economical, societal and environmental perspectives than what purely commercial service providers might consider.

The third layer—supporting services—includes supporting service providers. These are e.g. suppliers of hardware, software and services such as platforms, payment and clearing, and data management. The platform or IT-infrastructure providers are in a central role for service integration and the interoperability and (de facto) standardization of data and application programming interfaces (APIs).

The fourth layer—MaaS operators—represents actors that may offer organizing services that function as brokers, resellers and integrators of the transport and mobility services of others. They may also serve as the one-stop shops sourcing the transport services of other providers and managing the booking, payment and service level agreements and guarantees towards the end user. In addition, the MaaS operators can take different roles. Holmberg et al. (2016) suggest that there could be operators at a local level (integrating local transport services) and providers connecting the end users to the locally aggregated services. Essentially, this means that the local operators could be franchises of a global provider.

Within the above categorization, one stakeholder can have activities in several or even all layers either by themselves or through partnerships. The reason for differentiating the individual and supporting services from the "pure MaaS layer" is that the challenges are different. For example, starting a new car rental or shared taxi company versus setting up deals with those companies to package and sell their services. The MaaS provider needs to find a business model for offering services or packages at a viable price while covering the transport providers' revenues. And when it comes to packaged deals or multimodal trips, this model has to be able to provide a guarantee of a high service level and reliability without managing or having direct control or exclusivity on the actual transport services.

Furthermore, the differentiation is important when considering the market size and business potential for different actors. In 2015, transport in EU-28 accounted for 13% of the total final consumption of households and 2000 euros per head (of course varying greatly between countries, cities and household types), representing the potential revenue to be shared in new ways (European Commission 2017). Even though the integrated or packaged services are the grand vision of MaaS, the different ride-hailing (Marketsandmarkets 2017; Goldman Sachs 2017), on-demand transport (ABI Research 2016) and shared mobility (Grosse-Ophoff et al. 2017) services (i.e. the transport service layer) are expected to be the main business cases and growth areas in the future—whether or not they are interoperable with one another. The MaaS layer on top of them serves as a reseller relying on commission or other forms of revenue.

The shift towards on-demand services is also evident in automakers' investments in their own services and partnerships through which they cover many of the layers in the value network (e.g. Autonews 2015; Insights 2017). With automated driving, further changes in consumer behaviour as well as new cost and revenue structures between stakeholders are expected to arise, increasing the role and market share of infotainment services (PwC 2017), not to mention autonomous fleet operations (Keeney 2017).

2.3 Prerequisites for the Integration of Mobility Services

Apart from the *technical* accessibility, interoperability and interface challenges, the key challenges include *regulation and enablement* by public bodies, *user acceptance* and the *feasibility of the business case for the whole ecosystem* (Lund et al. 2017). These three challenge types comprise, both individually and mutually, the key prerequisites for MaaS to emerge. Due to their varying and potentially conflicting needs and objectives, we have chosen these as the three dimensions for analysis.

Regulation relates here to several issues such as how (and to what extent) governments should regulate entry, requirements and rights of new mobility service entrants, operating permits, transport data and interfaces, public service obligations, taxes and auditing, certifications, and reporting. The public sector should support the realization of new services while avoiding a situation where services compete against public transport in an unsustainable manner. This also applies to the combining of e.g. social care and private taxi rides. Finland provides an example as a forerunner in advancing regulation for digital mobility services. The first stage of the Act on Transport Services (effective in 2018) contains e.g. provisions on opening of information and ticket sales interfaces to third parties (requiring at least single tickets to be available for reselling) as well as provisions on passenger transport and taxi licenses (e.g. removal of fixed number of licenses) (Ministry of Transport and Communications 2017).

For the end user, there are certain characteristics that define the quality of the transport service. If the quality is acceptable compared to the old way of travelling, the new services will gain users. Carse (2011) defined a combination of those characteristics as Transport Quality of Life (TQoL). We claim that Mobility as a Service is most directly related to three of the TQoL characteristics. The first one is functionality, meaning the fluency, predictability, and travel time of the service. The second, information on timetables, delays, etc. And the third is price which can be effected in various ways depending on offering, pricing models and usage.

Feasibility of the business case depends on several issues. First of all, the integrated service must have a customer segment for which it provides clear value over the individual services. The primary customers are unlikely to be those who heavily rely on private vehicles (e.g. families) or those whose needs are well met by public transport (Lund et al. 2017). For transport service providers, integration into MaaS operators' services should provide clear added value in terms of, for example, more customers in return for possibly losing the direct customer relationship and being stacked next to competitors. A MaaS operator relies on a sufficient offering (i.e. transport options and fleets) while feasible pricing may require a significant customer base to be attractive for all parties, creating a chicken-and-egg situation.

An ideal win-win-win situation—and even a necessity—for MaaS services to truly prosper is a case where the above three prerequisites are in balance.

3 Material and Method

To identify and explore the main societal barriers and challenges for deployment of one-stop shop MaaS services, we have used a systematic approach for identification and analysis of the Political, Economic, Social, Technical, Environmental, and Legal (PESTEL) aspects. Further, we have identified potential negative impacts in case the barriers and challenges are not overcome. We also present examples of actions or solutions to address them. Our approach is derived from the risk analysis for new mobility solutions (Chalkia 2017) used in the Mobility4EU project of the European Commission's H2020 programme. The approach has four steps:

1. Identification of the key Political, Economic, Social, Technical, Environmental and Legal challenges for Mobility as a Service, MaaS
2. Characterization of the challenge type and the meaning for public actors, users and business actors
3. Assessment of the potential impact in case the barrier is not overcome
4. Discussion on the implications and potential impacts of the challenges.

We have used data from three sources in the assessment. First, we have interviewed the two MaaS operators with different approaches in Finland, namely MaaS Global (whose model is based on monthly subscriptions) and Tuup (who provide their own on-demand service as well as a one-stop shop type application). Second, we have participated in the preparation of a Finnish growth programme for the transport sector (Ministry of Economic Affairs and Employment, Finland 2018) during fall of 2017 and considered the views of public, private and research organisations participating in and providing input for four workshops on needs of businesses and the roles of stakeholders. Third, we have screened the material collected in the context of five research projects touching the field of mobility services: MAASiFiE (2016), DAC (2018), Mobility4EU (2020), MOBiNET (2018), and VAMOS! (2018).

4 Results

Table 1 shows the key challenges of the Mobility as a Service concept identified (cf. 2.3) within PESTEL categories and the potential negative impacts if not overcome. The challenges are discussed in detail in the following paragraphs.

Political Challenges are largely relevant for public actors such as national, regional and local authorities and private service businesses. Firstly, lack of communication, cooperation, understanding and agreeing on the roles of public and private stakeholders in providing mobility services, ensuring service levels and managing subsidies poses challenges for the service operators. Second challenge is the lack of common agenda and vision and too many individual goals for MaaS among the wide group of stakeholders.

It has been recognized that the "MaaS vision" has not been clearly defined, the interpretations vary (Jittrapirom et al. 2017) and the steps to its realization have been missing (Finnish Transport Agency 2015). As there is no clear and agreed description of what the term comprises and where is the line between the comprehensive user-first vision and individual transport services, the level of needed collaboration and interconnectivity is seen very differently. Due to this, one of the more discussed—or even controversial—aspects are the roles and responsibilities of different public and private organisations even though each consider themselves as proponents and supporters of MaaS.

Potential implications in case the challenges are not overcome are that opportunities to integrate and convey public services by MaaS operators remains unclear and the businesses segmented, which hinders further system developments. Also,

Table 1 Key challenges for different stakeholders in realizing MaaS

Challenge type	Challenge for users	Challenge for public actors	Challenge for businesses	The potential implications in case the barrier is not overcome
Political		The roles of public and private stakeholders in organizing, providing services and managing subsidies in relation to MaaS; Lack of common vision and balance between planning-driven and market-driven systems		Low level of service integration and system (platform) development and inability to create customized offerings and packages; individually offered services not found and used by customers
Economic	A chicken and egg problem in the supply and demand of packaged offerings: no competitive price point and alternative for car ownership. No clear (business) incentives to open services for third parties (incl. competitors) and risking losing direct customer contact			MaaS markets neither develop nor provide the customers with sufficient added value compared to private car or separately offered transport services
Social, behavioural	Lock-ins to private car use; fear of losing flexibility	Reaching the accessibility and service level targets of public services through MaaS; Comprehensive information sharing and awareness raising		Attitudes and travel behaviour do not change; alternatives to private cars do not scale-up; the service level of the few existing services is not adequate
Technical	The capability to use certain devices or applications	Legacy systems	Interoperability of the individual systems is insufficient, lack of common standards	Lack of interoperability and thus coverage of offering does not scale across transport services and regions
Environmental		Unclear environmental impacts		Increase in mileage and in exhaust emissions; Negative impacts on modal shares
Legal		Changes to existing laws are required	Restrictions on services; slow de- and re-regulation process	The supply of or ability to integrate mobility services will not increase due to out-dated legislation

the range, diversity and feasibility of services might not increase and hence services would not attract new users.

Most of the integrated information and payment solutions for transport are currently restricted to public transport (Polis 2017). This, and the fact that public transport is the foundation for mass transportation and for managing congestion and accessibility objectives, puts public transport organizations in a position to expand their already significant service offering by partnering with new actors (Kamargianni et al. 2015). The public transport actor including and offering the services of others, however, poses a challenge for visibility and interoperability for non-partnered services. That is, unless each separate stakeholder (e.g. car and bike sharing companies) joins the partnership. At that point, however, the total offering may become inefficient in managing all the services in relation to the service level objectives (the public actor would offer much more than the desired service level). Local transport planning could procure an efficient system, including subsidy allocations, but it can also limit opportunities for competition and innovation.

The balance between comprehensively and sustainably planned system and an open market for innovations is challenging. As Heikkilä (2014) points out, the risks of an open MaaS environment and appointing operations to the private sector include the issue of serving districts where transport services simply are not profitable, and since transportation is often highly subsidized, there may be little to be gained in the efficiency of operations to begin with. On the other hand, Heikkilä also argues that while the risks and drawbacks need to be carefully considered and examined, the potential benefits of the transformation include new businesses, increased efficiency of operations and a decrease in public funding required for transportation.

Another issue in the roles of transport provides, both public and private, is the direct customer relationship. Service providers want to understand their customers and their behaviour (i.e. obtain user and usage data) and ensure their brand and image is not harmed by a third party providing the end-user interface (e.g. getting the blame for mishandled data or issues with a third party's interface).

Economic Challenges are relevant for the users and the public and private stakeholders for different reasons. Low integration of individual sharing services into multi-modal information, ticketing and sharing systems may cause a chicken and egg problem in the supply and demand of the multi-modal services. In this case, the service offering does not grow and provide a competitive alternative for car ownership.

Potential implications in case the challenges are not overcome are that mobility services do not become commonly used but remain as marginal activity and private cars continue to dominate daily travel. Due to low demand, MaaS businesses do not grow and markets do not develop.

The economic viability of servitization is based on the assumption that users will have more money available for consuming the services as a result of investing less on private vehicles (including maintenance, fuels, taxes, depreciation etc.). That is, the cost and convenience of using the services must be at least at the level the user is used to. Both the mobility needs and expenditure vary greatly depending on the user profile. For example, the need for using a private car—and, therefore, also the cost of providing an effective alternative for it—is vastly different between a sub-urban family with kids and a single person living in an urban area.

Economic impacts of MaaS are very different for the various transport system users. For a person who does not own a car and whose mobility needs are currently covered by public transport (e.g. a travel card with a monthly subscription to local public transport services) the main questions are the changes in current service level and the impacts on price of travelling. The user would get an additional channel for acquiring the transport services and the price difference to the usual should reflect the value of changes in convenience. In order for this to be viable, the MaaS operator likely should need to be able to get the customer's ticket at the normal direct (often subsidized in case of public transport) or even lower cost or receive a sufficient commission for acting as a sales channel. This may require changes in existing subsidy policies, which vary from place to place. In the Helsinki region in Finland the public transport season and senior ticket prices, for example, are heavily subsidized only for those customers living within the serviced municipalities that also pay for the subsidy. Taking customer identity into account requires the technical capabilities for sharing or passing along the customer information. This can be done through either collaboration for perceived common gains or through regulation (see the *Legal Challenges* section). While MaaS operators can provide additional sales channels for public transport tickets, it may not directly reduce the required service levels (e.g. sales points and channels) of the public actor, thus resulting in no incentive as a cost-cutting collaboration.

Then again, the feasibility heavily depends on the selected business model and plan. Ability to resell tickets, for example, might not be a necessity for a MaaS operator should the service be based just on covering the cost of used transport services as a monthly package. Tracking and then covering the costs could potentially be done based on digital receipts sent to a digital wallet or receipt operator (Örn 2017), thus reducing the need for business-to-business contracts.

Social and behavioural challenges of the civil society towards MaaS relate to the lock-ins of the traditional behaviour and attitudes valuing car ownership over mobility services. There are some weak signals of developments such as young urbanites not getting a driver's license (Aretun 2014) and a downward trend in car use among young adults (Kuhnimhof et al. 2012). Still, understanding when and how the changes will take place and transitions happen in mainstream behaviour requires more knowledge.

Potential implications in case the challenges are not overcome are that attitudes and travel behaviour do not change, and car ownership continues to dominate. If the benefits of MaaS are not understandable to the potential customers or it does not reach the end users, integrated and packaged services do not scale-up and the service level of the few existing services is not adequate. If the level of integration of a number of services remains limited, the difference to the status quo is limited—the end user would have multiple MaaS applications with varying offering just as they have multiple mobility applications today.

Technical Challenges from the user's side relate to the capabilities of certain groups to use required devices or applications. The business side, however, is more crucial. The availability of data and APIs for accessing and sourcing third party services is a necessity for new data-driven solutions. Lacking standards, insufficient interoperability of individual systems, data and interfaces are some of the technical challenges faced by both the public actors and businesses.

Potential implications in case the challenges are not overcome are that interoperability between systems remains an issue. This severely hinders aggregation and integration of services, resulting in fewer interoperable services and the offerings of MaaS operators not scaling sufficiently across transport providers and regions.

A broader integration with different types of commercial transport services requires third parties to have access to the information, booking and payment interfaces of both the public and the private service providers. Due to the variety of service types and features, varying protocols and data formats, different regional requirements, varying level of capability and digitalization of the content etc., technical interoperability is a major challenge. It requires a lot of adaptation and conformance to common standards and practices.

Environmental Challenges relate especially to the broader societal challenges such as climate change and local air quality. As the impacts of new services in general on the transport mileage, energy consumption and environment are somewhat contradictory, the environmental burden of MaaS as an overarching scheme can only be even more unclear and heavily reliant on selected approaches and models.

Potential implications in case the challenges are not overcome are increases in transport mileage, energy consumption and in exhaust emissions.

There is currently a lack of research on the different impacts MaaS may have on travel behavior and mode choice of people (Kamargianni et al. 2016). The environmental impacts can be either positive or negative, depending on the MaaS concept design (e.g. incentives, subsidies, regulation). The underlying assumption of the concept—providing people with alternatives for private car use—suggests shift towards more sustainable modes and, thus, reducing negative externalities. At the same time, however, it may also shift some people from public transport to increased use of taxi-type services (Holmberg et al. 2016). The impacts depend on a multitude of factors such as types of service, target customer segments, location and public transport service levels. In some cases ride-hailing services, for instance, have been found to reduce the use of public transport in urban areas while increasing it further from city centres by serving as a complementary first or last mile solution (Babar and Burtch 2017; Cleclow and Gouri 2017; Rayle et al. 2016; Sadowsky and Nelson 2017).

The Polis discussion paper on MaaS (Polis 2017) points out that creation of new markets or promotion of certain technologies is reasonable only if they reduce the environmental impact of transport and increase safety while keeping people moving and supporting economic growth. This highlights the worries on sustainability of prevailing solutions and the fact that market-led transport services would cannibalize public transport in the areas where it is most profitable, thus further increasing the proportion of subsidies needed to serve less profitable areas. The argument for new services, though, is that they can serve the less populated areas where buses running nearly empty are inefficient. For this, however, the key question is how to ensure the provision of the desired level of service by market operators, if not required by the transport planning authorities.

Legal Challenges touch the public sector actors, but also the businesses' opportunities are dependent on the regulatory environment. Changes to existing laws can often be required for mobility services to be fully implemented and operated.

Potential implications in case the challenges are not overcome are that the supply of mobility services will not increase due to difficulties in legal aspects that hinder the growth of service businesses. One of the key reasons may be out-dated or otherwise unsuitable legislation for deployment and up-scaling of services.

When discussing the transport service layer, i.e. the services integrated into a "MaaS service", the regulatory challenges start from how freely new services can be offered or whether authorities should plan and limit for example taxi licenses or operation of bus services. Opening up public transport systems to parallel market operators, for example, needs to consider potential issues and need for regulation so that a service provider could not start a new bus line to run a few minutes ahead of a competitor to gain most of the potential revenues (Heikkilä 2014).

One important question here is how to either force through regulation or to convince through added value the private transport service providers to open up their services for others (including competitors) to re-sell and utilize them. The comprehensive one-stop shop MaaS service calls for more integration than including public transport services into single transport service providers' offerings (as is already the case through simple direct partnerships between public and private transport providers). This is insufficient as it would still leave the end users with a range of applications offering single services or service combinations.

One option for selling tickets at the transport providers' discounted prices intended for specific customers (e.g. the municipality residents of a public transport authority or other loyalty customers) is through sharing or passing along the user identity. The second stage of the new Finnish legislation, Act on Transport Services, aims to enable service providers to act on another's behalf so that a provider of a combined service could incorporate tickets for all modes of transport, including various serial and seasonal products as well as discounts by acting on the customer's wishes and on his behalf (Ministry of Transport and Communications 2017).

5 Discussion

This paper aimed to identify and discuss the key challenges and barriers ahead for demand-responsive and servitized future mobility through MaaS. For the assessment, we used a systematic PESTEL analysis of barriers and challenges. Further, the aim was to show how a systematic analysis of socioeconomic challenges can promote the deployment of MaaS concept by clarifying stakeholder pain points.

The proposed analysis proved to be applicable for identifying and analysing the challenges from the perspectives of users, public sector organisations and businesses. The analysis showed that for these three groups, the challenges of MaaS appear differently as also do the risks and fears related to the new integrated service. The analysis indicated that challenges often relate to more than one PESTEL categories. For example, the chicken and egg problem in the supply and demand of packaged offerings (economic challenge) relate also to accessibility and service level targets for public services (social challenge) and potentially require renewal of some legislation (legal challenge). This indicates that the challenges are systemic and call for systemic policy instruments, meaning e.g. strategic ecosystem building and management. The challenges cannot be tackled individually but rather as interlinked issues. This also relates to political challenges since identification of the common vision and agreeing on the roles of ecosystem stakeholders is of crucial importance for enabling system-level changes in transport sector that are required for MaaS to compete with the present approaches.

The challenges for realizing both a variety of different transport services and integration relate to various socioeconomic issues. Overall, however, the aspects boil down to policy choices and business cases. The key policy issues consist of e.g. sustainability objectives in transport planning and ensuring fluent, safe and cost-efficient daily mobility for end users and thriving cities. The business barriers are related to the operational guidelines and boundaries such as legal obstacles and limitations as well as finding feasible and profitable risk and revenue sharing models among the different stakeholders in the value network. A comprehensive service offering needs a sufficient fleet of underlying transport services to be available at a cost efficient price for each party. While scaling up services and user base is likely to be necessary in order to reach sustainable pricing levels for integrated packages, larger adoption of flexible service offerings also means that the capacity of the offered services needs to meet the demand. That is, if the main selling point for replacing the need for owning or using a private car is based on on-demand services, the fleet capacity, availability and responsiveness needs to be adequate to offer good service quality to the users even at peak hours.

As the experiences with ride-hailing services have shown, the effects for traffic and the use of public transport vary based on context. The planning and specifics of MaaS offerings can play a significant role in shaping the future of mobility by, for example, applying pricing plans that support on-demand services more where public transport is less capable. Systematic and careful participatory planning of MaaS offering can also contribute to reducing the environmental burden of MaaS, a theme which is one of the most urgent research needs in the MaaS context. In addition, the roles of different stakeholders, the ownership or management of customer relationships and the access to and ownership of usage data and statistics are further aspects that need clarifications to help different parties make the most of understanding user needs and behaviour and to apply the knowledge for e.g. traffic planning and user-centric service design.

References

ABI Research (2016) ABI research forecasts global Mobility as a Service revenues to exceed $1 trillion by 2030

Aretun Å (2014) Developments in driver's license holding among young people. Potential explanations, implications and trends Linköping: VTI rapport 824A

Autonews (2015) Automakers spot a profit opportunity in mobility services: Daimler, others test new business models in an evolving market. Automotive News article, March 9, 2015. Available at http://www.autonews.com/article/20150309/OEM/303099959/automakers-spot-a-profit-opportunity-in-mobility-services. Accessed 5 Nov 2017

Babar Y, Burtch G (2017) Examining the impact of ridehailing services on public transit use. September 25, 2017

Carse A (2011) Assessment of transport quality of life as an alternative transport appraisal technique. J Transp Geogr 19(2011):1037–1045

CG Insights (2017) Big auto's startup bets: where they're investing across ride-hailing, AI, and mapping. September 15, 2016. Available at https://www.cbinsights.com/research/auto-corporates-investing-startups/. Accessed 5 Nov 2017

Chalkia et al (2017) D4.1—Report on challenges for implementing future transport scenarios. Mobility4EU-project: action plan for the future mobility in Europe. Horizon 2020—Coordination and Support Action. Draft report, July 2017

Cleclow RR, Gouri SM (2017) Disruptive transportation: the adoption, utilization, and impacts of ride-hailing in the United States. Institute of Transportation Services, UC Davis, Research Report UCD-ITS-RR-17-07

Dweller in Agice Cities project funded by the Academy of Finland, Strategic Research Council (2018) www.agilecities.fi/en/. Accessed 14 May 2018

European Commission (2017) Statistical pocketbook 2017. 2.1 Transport. https://ec.europa.eu/transport/facts-fundings/statistics/pocketbook-2017_en

European Commission (2016) A European strategy for low-emission mobility. COM (2016) 501 final. Brussels, 20.7.2016

Finnish Transport Agency (2015) MaaS services and business opportunities. Research reports of the Finnish Transport Agency 56/2015

Goldman Sachs (2017) Rethinking mobility. Equity Research, May 23, 2017. https://orfe.princeton.edu/~alaink/SmartDrivingCars/PDFs/Rethinking%20Mobility_GoldmanSachsMay2017.pdf. Accessed 14 May 2018

Grosse-Ophoff A, Hausler S, Heineke K, Möller T (2017) How shared mobility will change the automotive industry. McKinsey & Company article, April 2017. Available at http://www.mckinsey.com/industries/automotive-and-assembly/our-insights/how-shared-mobility-will-change-the-automotive-industry. Accessed 5 Nov 2017

Heikkilä S (2014) Mobility as a Service—a proposal for action for the public administration, case Helsinki. Master's thesis, Aalto University School of Engineering

Hietanen S (2014) 'Mobility as a Service'—the new transport model? Euro-transport 12(2):26–28

Holmberg P-E, Collado M, Sarasini S, Williander M (2016) Mobility as a Service-MaaS: Describing the framework. Final report MaaS framework, Viktoria Swedish ICT

Jittrapirom P, Caiati V, Feneri A-M, Ebrahimigharehbaghi S, Alonso-González MJ, Narayan J (2017) Mobility as a Service: a critical review of definitions, assessments of schemes, and key challenges. Urban Plann 2(2):13–25

Kamargianni M, Weibo L, Matyas M, Schäfer A (2016) A critical review of new mobility services for urban transport. Transp Res Procedia 14:3294–3303

Kamargianni M, Matyas M, Li W, Schäfer A (2015) Feasibility Study for "Mobility as a Service" concept in London. Report—UCL Energy Institute and Department for Transport

Keeney T (2017) ARK invest white paper. Mobility-as-a-service: why self-driving cars could change everything. Available at https://ark-invest.com/research/self-driving-cars. Accessed 24 Nov 2017

Kuhnimhof T, Armoogum J, Buehler R, Dargay J, Denstadli J-M, Yamamoto T (2012) Men shape a downward trend in car use among young adults—evidence from six industrialized countries. Transp Rev: Transnational Transdisciplinary J 32(6):761–779

Lund E, Kerttu J, Koglin T (2017) Drivers and barriers for integrated mobility services: a review of research. K2 Working Papers 2017:3. Available at www.k2centrum.se/sites/default/files/driver s_and_barriers_for_integrated_mobility_services_k2_working_paper_2017_3.pdf. Accessed 20 Nov 2017

L'Hostis A (ed), Müller B, Meyer G, Brückner A, Foldesi E, Dablanc L, Blanquart C, Tuominen A, Kostiainen J, Pou C, Urban M, Keseru I, Coosemans T, de la Cruz MT, Val S, Golfetti A, Napoletano L, Skoogberg J, Holley-Moore G, Chalkia E, van der Werf I, Bos F, Grosso S, Stans Y, Langheim J (2016) Societal needs and requirements for future transportation and mobility as well as opportunities and challenges of current solutions. European Comission; Mobility4EU project (2016), Deliverable: D2.1, p 85

MAASiFiE (2016) Mobility as a Service for linking Europe (MAASiFiE) project. http://www.vtt.fi/sites/maasifie/. Accessed 14 May 2018

MOBiNET (2018) European Commission FP7 project. www.mobinet.eu/. Accessed 14 May 2018

MaaS Alliance White Paper (2017) https://maas-alliance.eu/wp-content/uploads/sites/7/2017/09/MaaS-WhitePaper_final_040917-2.pdf

Marketsandmarkets (2017) Ride hailing market by service type. http://www.marketsandmarkets.com/Market-Reports/mobility-on-demand-market-198699113.html

McKinsey & Company (2017) The automotive revolution is speeding up: Perspectives on the emerging personal mobility landscape. McKinsey & Company, October 2017

Ministry of Economic Affairs and Employment, Finland (2018) National growth programme for the transport sector 2018–2022. Available at http://urn.fi/URN:ISBN:978-952-327-317-7. Accessed 14 May 2018

Ministry of Transport and Communications (2017) Act on transport services. www.lvm.fi/lvm-site62-mahti-portlet/download?did=246709. Accessed 1 Nov 2017

Mobility4EU (2018) European Commission's H2020 project. www.mobility4eu.eu/. Accessed 14 May 2018

Örn M (2017) Taltio-hankkeen loppuraportti [Taltio project final report, in Finnish]. 31.10.2017. Available at: https://taltio.net/sites/default/files/taltio-hankkeen_loppuraportti_yleinen_31.10.2017.pdf. Accessed 14 May 2018

Polis (2017) Mobility as a Service: implications for urban and regional transport. Discussion paper offering the perspective of Polis member cities and regions on Mobility as a Service (MaaS). 4 September 2017

PwC (2017) Strategy & Digital Auto Report. PwC Strategy&, September 2017. https://www.strategyand.pwc.com/media/file/2017-Strategyand-Digital-Auto-Report.pdf. Accessed 14 May 2018

Rayle L, Dai D, Chan N, Cervero R, Shaheen S (2016) Just a better taxi? A survey-based comparison of taxis, transit and ridesourcing services in San Francisco. Transport Policy 45:168–178

Röhrleef M (2017) Hanover's 'one stop mobility shop'. Intelligent Transport, vol. 1, Issue 1., In-Depth Focus on Mobility-as-a-Service. Available at https://www.intelligenttransport.com/digital/it-issue1-2017-MaaS-idf/. Accessed 23 Nov 2017

Sadowsky N, Nelson E (2017) The impact of ride-hailing services on public transportation use: a discontinuity regression analysis. Economics Department Working Paper Series. 13

Transit app (2018) https://transitapp.com/. Accessed 14 May 2018

Transport Systems Catapult (2016) Mobility as a Service: exploring the opportunity for Mobility as a Service in the UK. July 2016

Tuominen A, Auvinen H, Aittoniemi E (2016) Esiselvitys liikenteen uusien palveluiden ympäristövaikutuksista ja niiden arvioinnista. Liikenneviraston tutkimuksia ja selvityksiä: 28/2016. Liikennevirasto, Helsinki, 36s [Preliminary report on the environmental impacts and the impact assessment of new transport services]

UITP (2011) Becoming a real mobility provider. Combined Mobility: public transport in synergy with other modes like car-sharing, taxi and cycling. UITP position paper, April 2011

Urbi app (2018) https://www.urbi.co/en/. Accessed 14 May 2018
VAMOS!—Value Added Mobility Services, project funded by the Finnish Funding Agency for Innovation (2018) www.vamosapi.com/. Accessed 14 May 2018
Whim app (2018) https://whimapp.com/. Accessed 14 May 2018

Assessment of Passenger Requirements Along the Door-to-Door Travel Chain

Ulrike Kluge, Annika Paul, Marcia Urban and Hector Ureta

Abstract This chapter discusses the current and future demand side as well as the future supply side of the European (air) transport market, taking an intermodal, user-centric and data driven approach. After introducing future passenger demand profiles, according passenger needs and requirements towards the European transport sector are presented. Having applied passenger-centric assessment areas and respective key performance indicators to evaluate the performance of current as well as future European mobility solutions, several gaps and bottlenecks in need of improvement in order to meet future mobility goals are revealed and discussed.

Keywords Air passengers · Air transport · D2D mobility · Intermodality
Novel mobility concepts · Passenger KPIs · Passenger needs
Passenger profiles · User-centric · 4HD2D

1 Introduction

Mobility in Europe faces many emerging opportunities, such as environmental protection, an increased competitive market, the trend towards connectivity but also multimodality and—consequently—a need for collaborative business models (Chiarini 2017). Thus, the overall mobility sector as well as academics start to broaden their scope beyond looking only at one transport mode but integrating the entire door-to-door (D2D) travel chain, taking railway, buses, cars, air travel but also cutting-edge

U. Kluge (✉) · A. Paul · M. Urban
Bauhaus Luftfahrt, 82024 Taufkirchen, Germany
e-mail: ulrike.kluge@bauhaus-luftfahrt.net

A. Paul
e-mail: annika.paul@bauhaus-luftfahrt.net

M. Urban
e-mail: marcia.urban@bauhaus-luftfahrt.net

H. Ureta
AENA, Ibiza Airport, 07817 San José (Ibiza), Spain
e-mail: hureta@aena.es

B. Müller and G. Meyer (eds.), *Towards User-Centric Transport in Europe*,
Lecture Notes in Mobility, https://doi.org/10.1007/978-3-319-99756-8_17

mobility concepts such as mode sharing, personal air vehicles (PAVs), mobility as a service (MaaS), or other public transport on demand into consideration. Considering the entire travel chain is also essential in order to derive an enhanced understanding of passenger groups, learn more about travellers' profiles, personal requirements and needs towards the European transport system as well as to determine potential bottlenecks and improvement areas for passengers' D2D trips (Paul and Kluge 2016; Kluge et al. 2017). Thus, understanding passenger groups within the context of D2D mobility is crucial for stakeholders in the broader sector of mobility providers, such as those mentioned above. Moreover, the European Union recently demanded an intermodal research approach, more integrated transport solutions and defined multimodality as one of the key challenges for the future of European mobility and connectivity (Chiarini 2017).

Air transport can be a part of a passenger's journey and is often used for leisure purposes, holidays, visiting friends and relatives and connecting faraway places within Europe to be easy reachable. This section mainly focuses on air travel passengers but comprises other modes as well, such as road and rail as well as future mobility concepts. In fact, air transport depends on ground transport as possible access and egress modes to and from the airport and hence, other modes should be considered as well when analysing the current and future mobility landscape, such as in regards to the ambitious four-hours-door-to-door (4HD2D) research goal of the Flighpath2050 document (European Commission 2011). The focus within this chapter will therefore be on the demand side as well as the future supply side of European (air) transport, taking an intermodal, user-centric and data driven approach.

After an introduction in Sect. 1, current and future passenger profiles as well as respective developments are discussed in Sect. 2, including the changing environmental awareness or the increasing usage of information and communications technology (ICT) throughout the entire journey. Section 3 introduces performance assessment categories and respective key performance indicators (KPIs) for evaluating novel mobility concepts from the passengers' perspective, which are presented in Sect. 4. It thus provides an overview of new mobility concepts and solutions addressing air passengers' needs and requirements along the entire D2D travel chain. Section 5 concludes and discusses the findings in regard to potential policy implications as well as research and development actions.[1]

2 Future Passenger Demand Profiles

Everyone travels and can therefore be defined as a passenger, whether one commutes to work every day by public transport or car, visits friends and relatives or travels for holidays. In 2015, 1.2 air trips have been conducted per person per year in Europe (Airbus 2017). However, we all differ in size, shape and form but also

[1]The analyses within this book chapter were conducted within the scope of the EU Coordination and Support Action DATASET2050 project (H2020 "Aviation Research and Innovation Policy," GA640353).

concerning age, background, education levels, attitude as well as personal requirements towards mobility. Looking at the European air transport development within the next 20 years, passenger traffic is forecasted to grow annually by 3.3% (Airbus 2017). Hence, air transport will increase, making it even more relevant to analyse this demand in more detail. At the same time, the mobility sector and policy makers work to improve travel experiences and paving the way for a seamless and efficient journey, such as by introducing the ambitious Flighpath2050 goal of reaching 90% of European destinations within or less than four hours, known as the 4HD2D goal (European Commission 2011). To analyse the current and future European transport system in regards to such goals, the passenger-centric point of view on mobility must be understood fully. For that purpose, it is helpful to group the European passenger market into smaller, more homogeneous user profiles (also called types, groups or segments). Market segmentation is often conducted to derive such profiles, which share similar characteristics and preferences and contribute to the understanding of user needs and requirements. Moreover, those passengers' needs, travel requirements and travel behaviour might also change in the future due to factors such as digitalisation, automation and virtualisation, for instance (Kluge et al. 2018). These developments might be easier to depict using passenger profiles as well. This section thus proposes six possible future passenger demand profiles, focusing on the year 2035 and the EU28 and EFTA countries as a proxy for the European market. All profiles have been developed within the DATASET2050 project and travel by airplane. However, as the whole D2D travel chain is considered, they are also applicable to other modes.

After introducing factors that influence passenger characteristics, the development of the profiles is presented. Selected profiles' characteristics are discussed, like the increasing usage of ICT or the rising environmental awareness. The section closes with an outlook of upcoming work and further applications.

2.1 *Factors Influencing Passenger Characteristics*

To explore how passenger profiles of the future look like, factors that influence passengers' mobility demand need to be discussed. Figure 1 depicts a variety of demographical, geographical, socio-economic, behavioural, and mobility aspects. All of these factors are identified as influencing the demand for mobility in general and for air transport in particular (Paul and Kluge 2016; Kluge et al. 2017). However, some are projected to change in the future and, hence, depending on their progressions, will also drive and change European air transport passenger demand. As a data driven approach was used within the DATASET2050 project, profiles are based on available European data and forecasts only.

Demographical Aspects

First, the size of the population and future demographical developments affect air transport demand (Young et al. 2009). A large population increases general

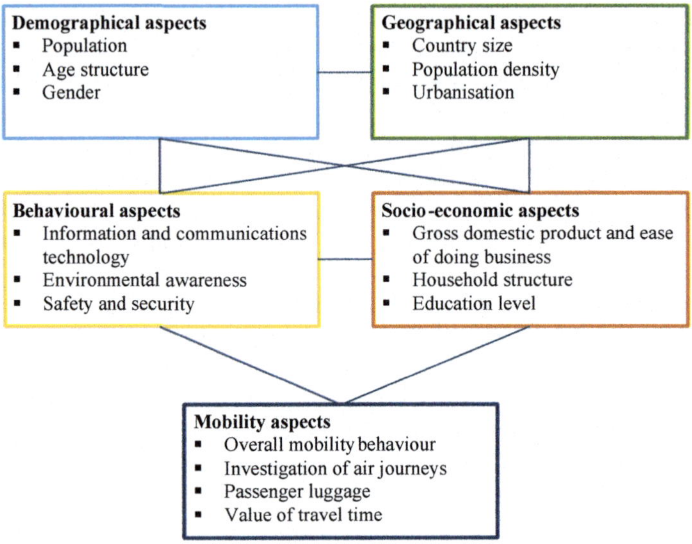

Fig. 1 Factors influencing passenger demand for mobility (Paul and Kluge 2016)

mobility demand and air travel demand. Comparing the six largest countries within the EU28 and EFTA countries region, these make up around 70% of the people from the considered country set (Eurostat Database 2014a) and the majority of mobility and air transport demand is generated within these countries alone. Passengers' age and gender do also influence demand for air travel. Travel behaviour differs by age, like in terms of travel activity, trip duration, disposable income and other factors. For instance, retired passengers have the tendency to conduct trip chaining, combining several destinations in one journey. Public transport, in this context seen as possible access and egress modes, is increasing among passengers at the age of 75+ in urban areas. This could be due to the available infrastructure and services offered by public transport (Alsnih and Hensher 2003). Until 2050, the share of people aged 65 and above will increase from about 20% today to around 30% (Eurostat Population Statistics 2016). In line with the increasing average life expectancy among the considered country set (Eurostat Life Expectancy by Age and Sex 2016) around 1/3 of passengers within Europe will become elderly travellers.

Gender does also influence mobility demand, as seen for instance by the low share of female business travellers (Eurostat Database 2014b) indicating that most passengers travelling for occupational purposes are (still) men. However, this might change in 2035 as the share of females in tertiary education is increasing, which may lead to an increase in working women and hence, an increase in women travelling for business purposes (Kluge and Paul 2017).

Geographical Aspects

Looking at geographical aspects, urbanisation has a strong effect on (air) mobility demand. The degree of urbanisation will increase until 2035 across the considered countries. At that point in time, apart from Liechtenstein, all European countries are predicted to have an urbanisation level of at least 50% (UN 2015). Within urban areas, fast and reliable accessibility to and from airports has to be ensured as well as sufficient infrastructure capacity, traffic and air-traffic routes between these urban areas. Moreover, the distribution of the population across countries (urban vs. rural parts) can help to understand demand for traffic-flows between regions, and hence, provide an indicator about the assessment of the European transport system in regards to the 4HD2D goal. In this respect, one major challenge is to ensure fast and reliable transport options for the "first and last mile", meaning those parts of the trip that ensure a true door-to-door travel experience. This applies especially to final destinations in rural areas.

Socio-economic Aspects

Beside the factors already discussed, one of the most prominent and well-studied driver for mobility and air transport demand is the gross domestic product (GDP). An increase in GDP positively influences air transport demand (Chèze et al. 2011; Dobruszkes et al. 2011; Kopsch 2012; Profillidis and Botzoris 2015). GDP per capita can also be used as a proxy for income per capita (Wadud 2013). For almost all considered countries, GDP per capita values are forecasted to rise constantly, however, current and forecasted numbers still strongly differ from one country to another within the considered country set (Frederick 2014). Generally, it can be expected that a positive GDP growth will drive passenger mobility demand positively in 2035. The business environment is one factor in determining the GDP level, the cooperation with other regions, or the attractiveness in terms of local working conditions. This in turn affects the demand for mobility within and beyond a region. Therefore, this parameter is included by considering the "ease of doing business" index. This index is published by the World Bank and contains almost 200 countries (World Bank 2017). It may be useful for the assessment of future economic development in particular regions, i.e. whether it is attractive for companies to locate subsidiaries or headquarters within particular countries.

The development of European household composition is relevant for the future socio-economic world (OECD 2011) and the distribution of household sizes can provide indicators in regards to the travel party size. For instance, the increasing share of single-households (also called one-person-households) predicted for 2035 (OECD 2011; Euromonitor International 2014) may lead to a higher share of passengers travelling on their own or in pairs. Finally, passengers' education may affect the income levels, family planning and ultimately the household composition as well (OECD 2011). The education level is an indirect factor influencing (air) transport demand, however, also an important variable determining other essential demand drivers as explored above (Kluge and Paul 2017).

Behavioural Aspects

The ICT usage is already an integrated part of the everyday life, at work and in private and omnipresent when travelling (World Economic Forum 2016). In the near future, ICT could also enhance the travel experience, both in terms of comfort and speed, as seen in the example on "smart travel" provided by the World Economic Forum (World Economic Forum 2014). In addition, according to SITA, airports are conducting trials to test how new emerging technology can improve overall travel, such as by using apps supporting along the travel chain with real time information (SITA 2016). In fact, future travellers do wish to use own devices along the travel chain, being in control and planning and preparing their journey on their own (Future Foundation 2015). This digital affinity can also be seen when looking at the new generation of passengers: the Generation Y (Gen Y) and Generation Z (Gen Z) are used to having technology embedded into their day-to-day lives (Civic Science 2016). Hence, ICT usage should be considered as well when analysing the future demand and supply side.

Although consumers are generally eco-friendly and concerned about environmental topics (WBCSD 2008), there is little evidence that this will lead to an actual travel behaviour change, such as the willingness to pay (WTP) for voluntary carbon offset schemes or substitute air travel (Young et al. 2009; Gössling et al. 2009; Hares et al. 2010; Mair 2011). Nevertheless, as within some recent studies evidence emerged showing an increasing willingness to adapt travel behaviour to the pro-environmental mind-set of many passengers (Gössling et al. 2009; Cohen and Higham 2011). Moreover, it is a highly important issue of today's society and hence included as an influential factor on (future) air transport demand.

Finally, perceived safety is one of the top priorities contributing to the overall passenger experience (Gilbert and Wong 2003; Ringle et al. 2011). Digital tools, also using personal data of passengers, will increasingly support travelling today and in the near future. Therefore, data security and privacy of passengers are major requirements for emerging technology supporting travel.

Mobility Aspects

General mobility aspects help to obtain an overall understanding of the current and future European transport market. Demand for different transport modes (rail, road, air etc.), the number of annual trips per capita and per country, the share of private vs. business trips and other general mobility data can provide insights into future mobility patterns and indicate aspects relevant for future demand profiles. For instance, the railway sector is forecasted to increase in passenger kilometres until 2025, making rail even more important as a public transport mode (UNIFE 2016).

2.2 Derivation of Future Demand Profiles

In this section, six future demand profiles for grouping the European passenger market are presented. Within the DATASET2050 project, existing studies of passenger profiles have been researched, collected and clustered around the respective age group, overall trip share, travel purpose, income, and other factors (Fig. 2). Aggregating the results of this meta-analysis, six general passenger profiles have been developed and taken as a basis for future demand profiles.

In the second step, identified passenger characteristics, as elaborated above, have been used to create current passenger demand profiles (base year 2014) (Paul and Kluge 2016). These current demand profiles have been taken as a basis to take account of future developments and forecasts of passenger characteristics. Detailed information of this analysis is available in the DATASET2050 deliverable 3.2 (Kluge et al. 2017) and the next section will elaborate on selected profile characteristics. As a result, six future demand profiles have been developed: (1) Cultural Seeker; (2) Family and Holiday Traveller; (3) Single Traveller; (4) Best Agers (Next Generation); (5) Environmental Traveller and (6) Digital Native Business Traveller. An overview is provided below in Table 1. All profiles conduct at least one air trip, however, the entire D2D travel chain is considered. The most likely access mode choices are provided for each profile, such as public transport, personal vehicle, car sharing or

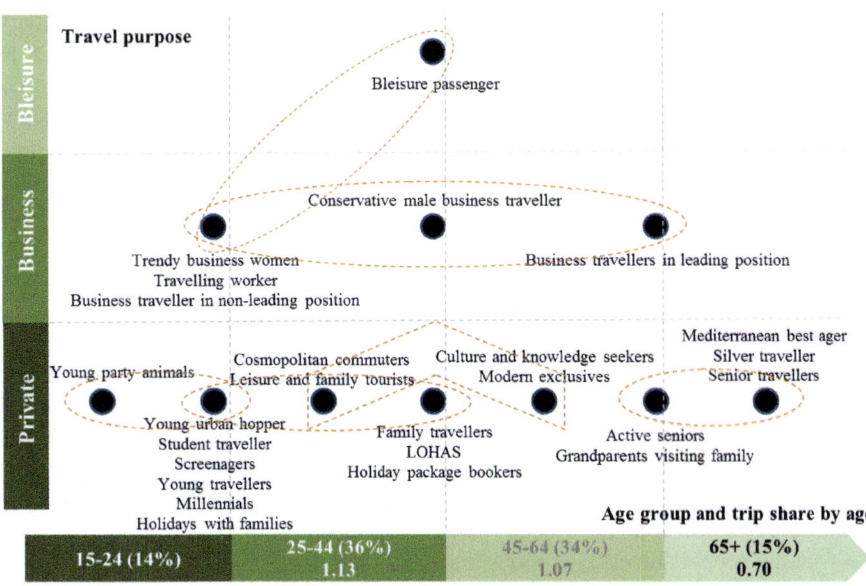

Fig. 2 Clustering of passenger studies according to age group, trip share, travel purpose (Future Foundation 2015; Door to Door Information for Air Passengers project (DORA) 2016; GfK Mobilitätsmonitor 2011; Henley Centre HeadlightVision 2007; Official Airline Guide 2014; Plötner and Schmidt 2014; SITA 2015; Skift 2015)

kiss and fly (meaning a drop off at the airport by friends and relatives). Depending on the context, one person can comprise several profiles. For instance, a person can travel in a work-related context according to characteristics of the "Digital Native Business Traveller". Going on vacation with children, he or she can turn into the "Family and Holiday Traveller" and change travel behaviour accordingly (Kluge et al. 2017).

Within the Mobility4EU project, a high-level picture of societal drivers impacting mobility is discussed. In terms of passengers' needs and requirements towards the transport system, the project confirms these developments as having an influence on passengers (Mobility4EU 2016, 2017). Some of those developments can also be found within the DATASET2050 demand profiles. In particular, the ageing population, the ICT usage and the increasing environmental awareness will be discussed in more detail in the following section.

Due to demographical developments and an increasing life expectancy among the population, Europe faces an increasing ageing population. In 2050, around 1/3 of passengers within Europe will be elderly travellers, who travel differently than their younger counterparts in terms of factors like travel expectations and travel behaviour but also mode choices for access and egress, as already elaborated on in the first part of this chapter. Hence, the "Best Agers (Next Generation)" and the "Single Traveller" out of all demand passenger profiles cover the age range of 65 years and older. 65+ is chosen as an average retirement age, presenting a stage of a new phase in life. The "Single Traveller" does also cover the increasing share of the population living in one-person-households. They travel mostly on their own and have a low travel activity (Kluge et al. 2017). The household composition is also relevant for other socio-economic characteristics (OECD 2011), such as for the low to medium income level of this profile type. New technology is a key influential driver on the future transport sector. Customers will increasingly expect travel information provided in a digital way (also referred to as connected traveller). The growth in ICT usage does also influence the value of travel time positively, as time spend in transport can be used more efficient for working or for leisure purposes (Mobility4EU 2016). The "Best Agers (Next Generation)" belong to a new group of elderly being more confident in the usage of ICT and use technical devices and respective retrieval of information with a medium frequency. Hence, in 2035, people in this age represent a digital savvy passenger group compared to their counterparts today, making this profile type a new generation. The "Digital Native Business Traveller" does also represent a new generation of passenger travelling for business only, using digital tools during travel time mostly to get work done. In comparison to that, the "Environmental Traveller" will try to travel as less as possible, with little luggage and combines private and business trips to "bleisure" journeys (leisure + business = "bleisure"). All eco-friendly access and egress modes are used, such as public transport (Kluge et al. 2017). This pro-environmental travel behaviour is a reaction to the rising public awareness for the protection of the environment (Mobility4EU 2016), however, only a small number of future passengers might belong to the group of "Environmental Traveller".

Table 1 Short description of future passenger profiles for 2035 (Kluge et al. 2017)

	(1) Cultural seeker	(2) Family & Holiday Traveller	(3) Single Traveller	(4) Best Agers (Next Generation)	(5) Environmental Traveller	(6) Digital Native Business Traveller
Main travel purpose	Private	Private	Private	Private	Bleisure	Business
Predominant age group	15–65	30–50 + children below 15	44+	65+	30–44	24–64
Income level	Medium/high	Medium/high	Low/medium	Medium	Medium	Medium/high
Amount for transport expenditure	Medium low	Medium	Low	Medium	Low	Medium/high
Use of technical devices and respective retrieval of information	High frequency	Medium frequency	Medium frequency	Medium frequency	Low/medium frequency	High frequency
Travel activity (trips per capita per year)	0.5–1.5	0.5–1.5	0.25–0.5	0.5	0.5 (as few as possible)	0.5–1.5
Travel party size (in number of people)	1–2	2–3	1	1–2	1–2	1–2
Luggage requirements of passenger	Hand luggage only (short trips) Check-in luggage	Check-in luggage	Hand luggage only (short trips) Check-in luggage	Check-in luggage	Hand luggage only (short trips) Check-in luggage (if necessary)	Hand luggage only (short trips) Check-in luggage
Access mode choice	Public transport Taxi Car Sharing	Public transport Private car (park and travel)	Public transport Kiss & fly	Private car (park and travel) Kiss & fly	Public transport Car sharing Cycling (if possible)	Public transport Taxi Car sharing

This section has presented six possible future demand profiles for the European market. The year in focus is 2035 and the EU28 and EFTA countries are considered and taken as a proxy for the European market. Based on these individual needs and requirements towards the transport system, passenger-centric assessment parameters, measured by respective key performance indicators are presented in the next section. Such KPIs can help to evaluate current and future mobility concepts in regards to user-friendliness and political mobility goals, such as the Flightpath2050 4HD2D goal.

3 Passenger-Centric Assessment Parameters and Key Performance Indicators

In this section, ten assessment parameters, that are extracted from demographical, geographical, socio-economic, behavioural and mobility aspects, are proposed. The parameters cover the demand side only, focus on overall critical passenger-centric components and do match the preferences of the future demand profiles. To measure assessment parameters, 37 respective KPIs are formulated and assigned accordingly (see Fig. 3 for the full list of KPIs) (Paul and Kluge 2017).[2] They served as input for the assessment of novel mobility concepts and solutions in Sect. 4.

As discussed in Sect. 2, Europe is facing a changing age structure towards an aging population and travel behaviour as well as needs and requirements are partly influenced by the life stages of passengers (Alsnih and Hensher 2003). Hence, future mobility solutions should be "Age appropriate & accessible". However, both, elderly travellers as well as travellers with children are essential stakeholders. Families will also make up a fair share of air passengers, such as seen by the example of the "Family and Holiday Traveller" profile (Kluge et al. 2018), and mobility concepts might want to provide "Family friendliness" within transport solutions. Moreover, the European GDP per capita (as proxy for income) will grow constantly until 2050, but does also vary strongly on a country level. Nations such as Croatia or Romania, are already countries with currently low GDP per capita figures and face only a small GDP growth until 2050 (Frederick 2014). At the same time, air travel (especially the emerging LCCs) in Eastern European countries is important to keep migrants and their families mobile, despite low income levels (Kuljanin and Kali 2015). As income is forecasted to differ across Europe, travel options along the travel chain for different budgets should be provided to make mobility affordable and accessible to everyone. "Affordability" of transport modes should be ensured as well.

Beside demographical and economic developments, the degree of urbanisation will increase until 2035 throughout the considered country set, resulting in a high accumulation of passengers in urban areas (UN 2015). Fast, reliable accessibility

[2]Based on the DATASET2050 deliverable 5.3 (Paul and Kluge 2017). For the purpose of this book chapter, two assessment parameters have been removed.

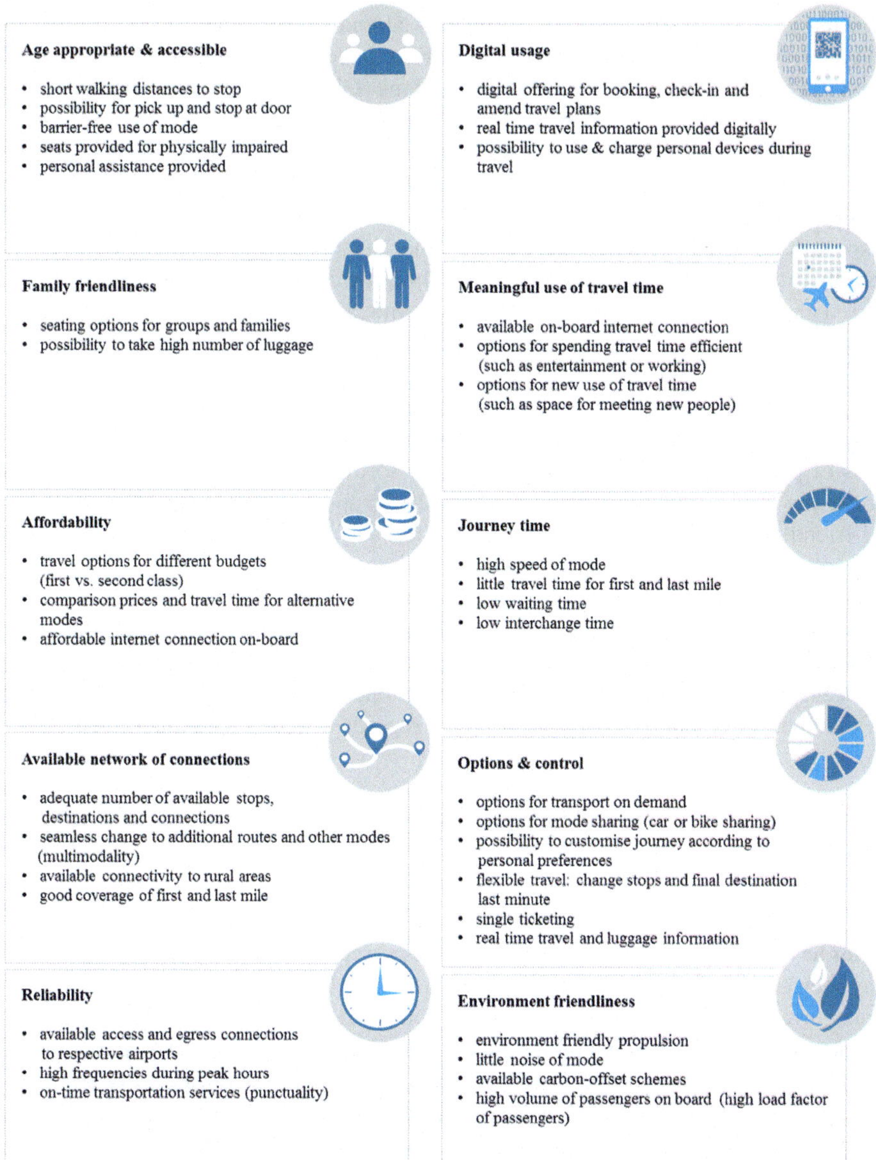

Fig. 3 Passenger-centric assessment parameters with respective KPIs (Paul and Kluge 2017)

to and from respective airports has to be ensured, also between urban agglomerations but also rural areas since lots of trips have private background and holiday destinations are also located in the countryside. As many passengers depend on it, the “Available network of connections” and “Reliability” should also be considered.

The use of ICT within passenger groups is already present and in the future, air travellers wish to complete more duties around their journey using own devices, such as booking, check-in or amend travel plans (Future Foundation 2015). "Digital usage" before departure and while travelling is essential to meet this future need. Moreover, ICT usage does also influence the value of travel time positively, as time spent in transport can be used more efficiently for working or for leisure purposes (Mobility4EU 2016). In this respect, the "Meaningful use of travel time" should also be considered. To reach passengers' final destinations as soon as possible and to fulfil the Flightpath2050 goal of 4H2D2, mobility concepts and solutions should be fast, with little waiting and little interchange time. Simultaneously, as seen in Sect. 2, passengers are increasingly diverse and journeys should be customisable according to personal needs, while still providing access to real time travel information. These requirements are translated into the assessment parameters "Journey time" and "Options & control". Finally, recent studies show first evidence of emerging willingness of passengers to adapt travel behaviour to their pro-environmental mind-sets (Gössling et al. 2009; Cohen and Higham 2011) and for that reason "Environmental friendliness" should be included as well, also matching the preferences, needs and requirements of the "Environmental Traveller".

This section has presented ten passenger-centric assessment parameters, measured by respective KPIs. The proposed list above is extracted from demographical, geographical, socio-economic, behavioural and mobility aspects influencing future air transport passengers. Within the next section, these KPIs have been applied to evaluate mobility concepts and solutions in regards to future passenger needs.

4 Mobility Concepts and Solutions Addressing Air Passenger Needs

After having identified a variety of air passenger requirements, a mobility repository has been acquired within the scope of the EU project DATASET2050, which includes both current as well as planned mobility solutions (Paul and Kluge 2017). Examples of the mobility solutions contained within this database are Uber (2017), DriveNow (2017), Hyperloop (Brücken 2017), or the OysterCard (2017). Mobility solutions have been collected by using a literature review, and by monitoring the current mobility landscape as well as proposed future mobility concepts.

In order to evaluate how well this large variety of solutions addresses future passenger expectations, each mobility solution is assigned to a particular cluster with pre-defined specific criteria. A cluster thus contains a range of mobility solutions that exhibit the same characteristics, as elaborated in more detail in Table 2 (Paul and Kluge 2017). In Sect. 4.2, these clusters are then assessed according to their compliance with the identified user needs discussed in Sect. 3.

Table 2 Definition of mobility clusters and exemplary mobility solutions

Cluster name	Cluster number	Description	Examples
Piloted Flying Mobility	1	A flying vehicle is the main transport mode and the vehicle is non-autonomous, e.g. current helicopter services	• Joby Aviation (2017) • Voom (2017) • Zee Aero (2017)
Autonomous Mobility	2	Mobility solutions operate autonomously and the cluster contains all possible transport modes	• Volocopter (E-Volo and Volocopter 2017) • Lilium and Lilium (2017) • Hyperloop One (Brücken 2017)
On Demand Mobility	3	These solutions are adapted to individual transport demand. There is no fixed schedules. The journey can be monitored in real-time, exchange between provider and user possi-ble. Vehicle is operated by a third person	• Rally bus (2017) • Lyft (2017) • Tuup (Vinka 2017) • Allygator (2017) • Uber (2017) • E-Volo and Volocopter (2017)
Shared Mobility	4	The vehicle not owned by transport users but rented or shared by individual users that require a particular transport service	• car2go (2017) • SnappCar (2017) • Deutsche Bahn (2017) • SocialCar (2017)
Intermodal Mobility	5	Different transport modes are integrated within one mobility solution that is offered to the customer in order to provide door-to-door mobility services	• Busandfly (2017) • AirportLiner (2017) • Mozio (GARA 2017)
Integrated Mobility	6	The mobility solutions offer integrated ticketing across modes and liabilities and responsibilities are shared across ticket providers	• Octopus Card (2017) • Oyster Card (2017) • Clipper (2017) • Mobilitymix (2017)
Smart Mobility	7	A comparison of mobility solutions for specific/ individual transport demand is offered, the user interface is provided via various mobile devices/ online platforms	• CarTrawler (2017) • Free2Move (2017) • 9292 (2017) • Google (2017)
Conventional Mobility	8	Representing traditional transport services, not integrated across modes, fixed schedules, and no on-demand servicess	• Flixbus (2017) • Amtrak (2017) • Thalys (2018) • Network Rail (2017)

4.1 Definition of Mobility Clusters

Each cluster represents a specific form of mobility and mobility solutions within a cluster thus exhibit the same characteristics (Table 2). Considering mobility solutions in such an aggregated way, an overview can be obtained as to which particular developments or forms of mobility can contribute to meeting future passenger demand and requirements. The cluster "Piloted Flying Mobility", for example, contains those mobility concepts that require a pilot operating the vehicle and the only transport mode within this cluster is air. Examples include Joby Aviation or Zee Aero (Joby Aviation 2017; Zee Aero 2017). The cluster "Autonomous Mobility", on the other hand, contains all transport modes and its main characteristic is that the vehicle is operated autonomously, examples here are Volocopter or Hyperloop One (Brücken 2017; E-Volo and Volocopter 2017). Each cluster has specific main characteristics, which are employed to assess future passenger requirements in the following sections. The example mobility solutions are assigned to a cluster whose characteristics best fit the solution's main purpose or features. Having allocated a particular mobility solution to one of the clusters therefore does not necessarily mean that this solution does not potentially exhibit some features of other clusters in the analysis, i.e. some overlap may exist. For the purpose of the analysis within this section, the mobility clusters are considered and not individual mobility solutions.

Analysing the clusters in regard to the number of different modes included yields an initial overview of the level of integration between different transport modes. Figure 4 shows that the clusters "Integrated Mobility" as well as "Smart Mobility" are those which incorporate a high number of different modes. "Autonomous Mobility", "On Demand Mobility" and "Shared Mobility", on the contrary, currently have their focus on a single mode only.

Taking these different clusters allows for a high-level assessment of the passenger requirements elaborated in Sect. 3. The following section therefore focuses on the identification of bottlenecks and gaps concerning passenger requirements.

4.2 Evaluation of Mobility Clusters Applying Passenger Requirements

For each KPI defined in the ten different assessment parameters the individual clusters are ranked on a scale of 0–2, with a cluster obtaining a 0 if it does not meet a particular KPI at all, i.e. this particular passenger requirement is not fulfilled by the cluster. A score of 1 is assigned if the cluster partly meets a KPI, i.e. there are some drawbacks within the cluster regarding this particular requirement, and a score of 2 if the cluster fully meets a KPI. In addition, if a cluster does not focus on a particular aspect, i.e. this KPI is not within the scope of the cluster, it is marked with "non applicable (n/a)".[3]

[3] A detailed outline of the assessment approach can be found in the DATASET deliverable 5.3 (Paul and Kluge 2017). The ranking is based on the information available for each cluster, i.e. the

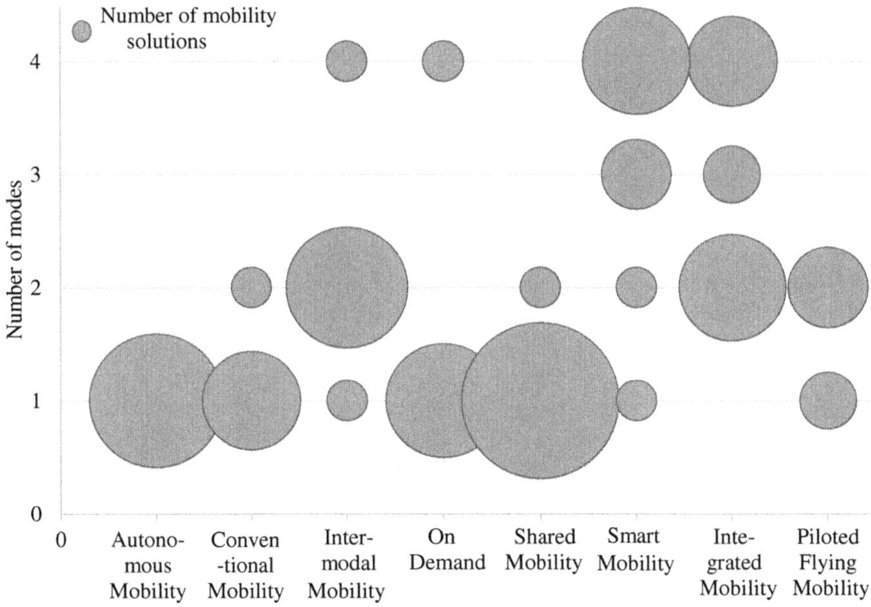

Fig. 4 The number of modes apparent within each mobility cluster (Paul and Kluge 2017)

Following this approach, the ranking is conducted for each cluster for each passenger requirement category and the respective KPI. The ranking is based on the qualitative description and the respective characteristics of each cluster. The example shown here is conducted for all eight clusters across the 37 predefined KPIs. Summing the scores for all KPIs across each cluster provides a high-level comparison of all clusters in regard to their overall performance. With a total of 37 KPIs a maximum score of 74 can be obtained, thus representing 100%. Each cluster's score is depicted in Fig. 5 indicating that none of these meet all passenger requirements at the same time. "On Demand Mobility" ranks highest with an overall score of 60%, followed by "Smart Mobility", as well as "Shared Mobility" with 51% each, and "Intermodal Mobility" with 50%. This result implies that individualised mobility solutions and the comparison of travel options seem to be best capable of meeting diverse passenger needs.

However, although a particular cluster does not rank high in regard to its overall score it might perform well in a particular passenger requirement category such as "Age appropriate & accessible", for example. Therefore, it is worth analysing these different categories individually and compare clusters across these. "Piloted Flying Mobility", for example, has an overall score of about 30% but performs best in the passenger requirement category "Reliability" since it provides passengers with the

---

characteristics described in Table 2, and the requirements outlined in Sect. 3. Within a group of experts of the project consortium, each passenger KPI and its performance within each cluster has been discussed ranked accordingly.

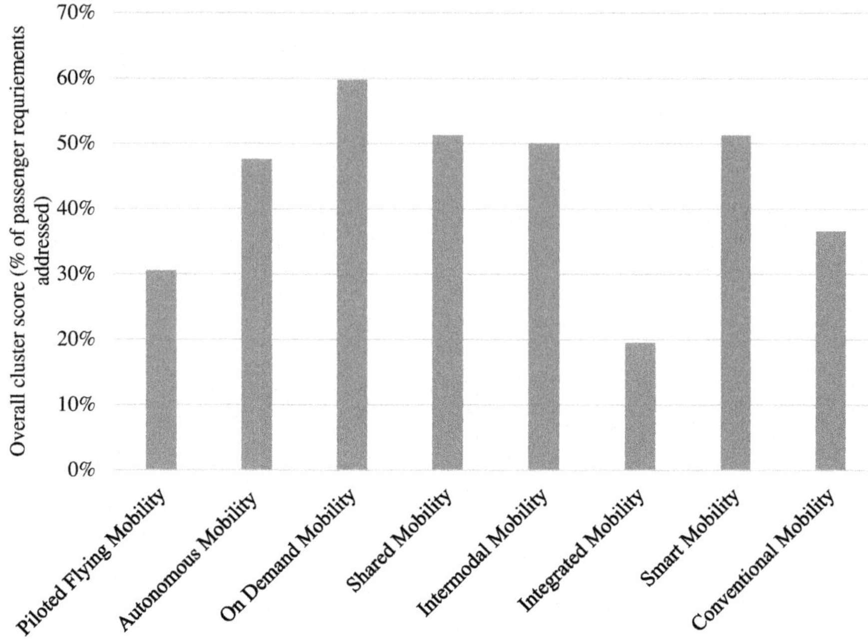

Fig. 5 Overall ranking of mobility clusters in terms of passenger requirements (Paul and Kluge 2017)

ability to access airports in a convenient way without requiring the same infrastructure as road and rail traffic.

Figure 6 shows that six of the eleven overall passenger requirement categories are already addressed quite well, with these obtaining more than 80% of the total score. In the category "Available network connections", the cluster "On Demand Mobility" enables the transport user to adjust schedules to individual departure and arrival times and choose the respective locations accordingly. By this, buffer times passengers incorporate in their journey planning can be eliminated and thus the overall travel time can be reduced significantly, which contributes to moving the European transport system closer to the Flightpath2050 D2D goal. Besides, the cluster "Smart Mobility" enables both a reduction and a more efficient use of overall journey time since transport options can be compared and selected according to specific travel requirements, which is also the reason why this particular cluster performs best in the category "Options & control". In the category "Reliability" a range of clusters meets the user requirement by providing efficient and on-time feeder services to airports as well as the possibility for increased frequencies during peak hours.

Clusters like "On Demand Mobility", "Shared Mobility", "Intermodal Mobility", or "Piloted Flying Mobility" allow for the integration of digital services and products to provide users with different options along the journey. Providing internet accessibility in vehicles would thus allow travellers to use their time in the vehi-

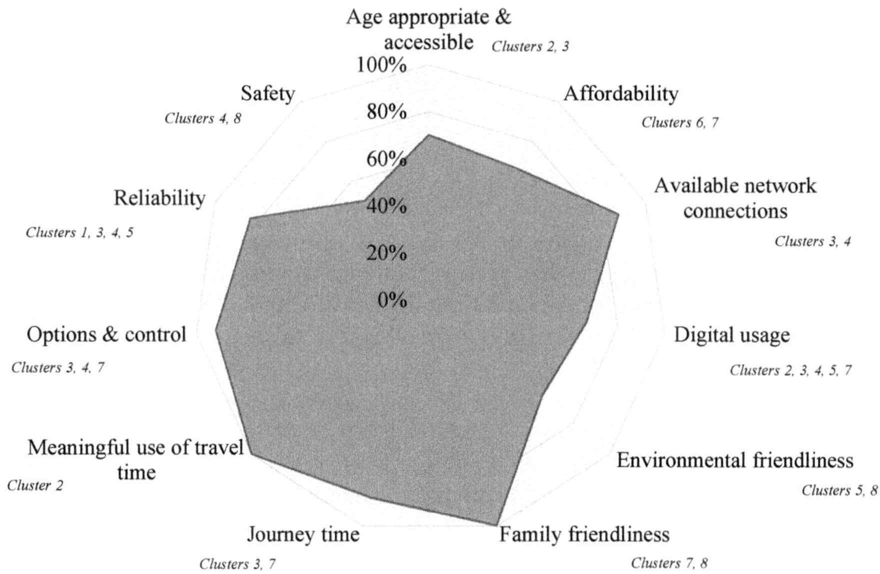

Fig. 6 Identification of gaps concerning passenger requirements (Paul and Kluge 2017)

cle for work, e.g. virtual meetings or telephone conferences. A category currently lacking behind concerning prioritisation by mobility providers is "Environmental friendliness", which includes indicators such as the availability of carbon-offsetting schemes, for example. However, since these aspects will become more important, also driven by politically induced regulation, providers of mobility solutions will be compelled to internalise negative externalities and thus adjust their products and services.

Furthermore, also not related to any passenger KPI, the important aspects of handling passenger data and adhering to privacy regulations as well as maintaining safety standards in regard to technical and operational issues have to be part of the development of new mobility concepts. Gaining an insight into whether passengers' privacy is adequately protected and whether consumer rights are adhered to requires a definition of respective standards providers of mobility solutions have to abide by.

5 Discussion and Identification of High-Level Action Areas

Identifying future passenger requirements and differentiating by passenger groups, which go beyond the distinction between travelling for business and private purposes, provides the basis to detect gaps and bottlenecks along the entire travel chain. In this regard, the derivation of passenger KPIs yields a comprehensive set of assessment parameters that can be applied to test the performance of the European transport system. The evaluation of the latter, represented by currently available as well as planned mobility solutions, thus provides a guideline which areas require improvement and what type of mobility solutions enable enhanced passenger comfort and the potential for travel time reduction. Since the diversity and number of mobility solutions is very large and different solutions have the same underlying concept, such as DriveNow (2017) and car2go (2017) for example, distinct clusters have been defined, which represent a particular form of mobility. By this, passenger requirements can be assessed on a high-level cluster basis, resulting in the detection of gaps and bottlenecks and hence the depiction of future investment requirements.

Based on the evaluation of these clusters, a range of high-level action areas can be identified that enable the enhancement of passenger comfort, a more efficient use of travel time and a reduction of unnecessary buffer or waiting times (Table 3). This list is not exhaustive but provides an initial overview of those aspects that are currently exhibiting the highest improvement potential. For each of the identified gaps action areas and thus improvement potential are proposed, addressing passenger requirements in a better way. In this regard, mobility solutions incorporating barrier-free travel options and enabling passengers to make use of personal assistance services need to be promoted on a larger scale to meet the demand of physically impaired passengers, or those in need of personal exchange during their journey. Relating to this area is the demand for mobility services that truly cover the journey segment to the passenger's door, e.g. an increased diversity and associated business models providing pick-up services from public transport stops to the travellers' home. An important feature here is the potential involvement of the public, e.g. mobility exchange platforms in neighbourhoods to offer on-demand transport services, especially regarding the first and last mile, since some areas are lacking the provision of ubiquitous taxi services (especially in more rural areas).

Other action areas address the demand for the internalisation of negative externalities such as carbon emissions during a journey. Growing environmental awareness among passengers coupled with a potential increase in policy regulations can influence the choice whether to travel and with which mobility options. Pricing and carbon offset schemes may therefore be required to reflect all costs associated with the journey. Furthermore, the use as well as potential reduction of overall travel time are important aspects to be considered in the re-shaping of the European transport system. Concerning potential travel time savings, passengers often plan buffer times in order to account for potential delays or waiting time during transport related processes or the interchange between modes. By supplying real-time information at all times during the journey, passengers can adjust their travel plans accordingly and thus avoid these buffer times. Furthermore, shared liabilities across mobility providers enable an on-demand adjustment of passengers' travel itineraries, thus saving passengers the stress to readjust these by themselves in case of delays or cancellations. Finally, the potential activities and services available for passengers during their travel time have a large impact on how time is valued. Being able to conduct work-related tasks during the journey and thus use this time efficiently may induce travellers to attach a different value to the time they spend within the vehicle. Using travel time efficiently does also depend on other factors, such as available on-board

Table 3 Addressing gaps and bottlenecks in the European transport system

Aggregated gaps and bottlenecks across clusters	High-level action areas
Lack of wide coverage of **barrier-free mobility** solutions and**personal assistance** if required	• Technical/mobility solutions providing easy vehicle access and egress • Availability of provisional personal assistance services
Limited availability of **pick-up services at home** that can be aligned with individual travel requirements	• Mobility solutions covering the first and last mile of a trip, e.g. providing flexible coverage from door to public transport stations
Limited provision of **offsetting schemes** to internalise negative externalities	• Implementation of incentive schemes to integrate negative externalities into the (transport) pricing structure • Transparent information enabling comparison of transport alternatives in terms of negative externalities, e.g. CO_2 emissions
Limited integration of services/products/amenities in the different vehicles, thus enabling **productive travel time**	• Integration of communication technologies and to provide a feasible working environment • Dedicated working or recreation space within vehicles if designed for large groups of people
Unnecessary **buffer times** during exchange between transport modes and mobility solutions	• Solutions providing real-time travel information and according alignment of travel itineraries • Dedicated interchange platforms between modes to reduce walking distances

internet connection, number of stops and possible disruptions, number and character of fellow travellers. To conclude, the high emphasis placed on travel time reduction decreases with increasing services and activities available during the journey.

Having applied a passenger-centric approach to evaluate the performance of the European (air) transport system revealed several areas in need of improvement in order to meet future mobility goals. Although a limited number of mobility solutions already exists that addresses these requirements in some way, more cooperation and coordination across providers is required in order to enhance passenger comfort, reduce overall travel time, and use available transport capacities in a more efficient way.

Acknowledgements We would like to thank the DATASET2050 (GA640353) consortium for their feedback and Viola Gissibl and Fabian Madl for their great work and support for this chapter.

References

9292 (2017) Accessed on 12 Apr 2017. http://9292.nl/en
Airbus (2017) Global market forecast: growing horizons 2017/2036
Airportliner (2017) Accessed on 21 Nov 2017. http://www.airportliner.com/
Allygator (2017) Accessed on 27 Mar 2017. https://www.allygatorshuttle.com/
Alsnih R, Hensher DA (2003) The mobility and accessibility expectations of seniors in an aging population. Transp Res Part A: Policy Pract 37(10):903–916
Amtrak (2017) Accessed on 21 Nov 2017. https://www.amtrak.com/home
Brücken T (2017) So soll der Hyperloop in Dubai aussehen. Accessed on 10 Apr 2017. https://www.wired.de/collection/business/nur-zwoelf-minuten-nach-abu-dhabi-so-soll-der-hyperloop-dubai-aussehen
Busandfly (2017) Accessed on 21 Nov 2017. https://www.yelp.com/biz/busandfly-bamberg
car2go (2017) Accessed on 21 Nov 2017. https://www.car2go.com/DE/de/
CarTrawler (2017) Accessed on 21 Nov 2017. https://www.cartrawler.com/ct/
Chèze B, Chevallier J, Gastineau P (2011) Forecasting world and regional air traffic in the mid-term (2025): an econometric analysis of air traffic determinants using dynamic panel-data models 213266(2025):1–37
Chiarini P (2017) Innovative multimodal transport: the EC's point of view, Presented at DATASET2050 European door-to-door mobility workshop in Madrid. Retrieved from http://innaxis.org/european-door-to-door-mobility-workshop-20th-sept-madrid/
Civic Science (2016) Gen Z is making waves in the tech world. Accessed on 11 Nov 2016. https://civicscience.com/gen-z-is-making-waves-in-the-tech-world/
Clipper, accessed on 21.11.2017, http://www.futureofclipper.com/, 2017
Cohen S, Higham J (2011) Eyes wide shut? UK consumer perceptions on aviation climate impacts and travel decisions to New Zealand. Cur Issues Tourism 14(4):323–335
Deutsche Bahn (2017) Accessed on 04 Apr 2017. https://www.bahn.de
Dobruszkes F, Lennert M, Van Hamme G (2011) An analysis of the determinants of air traffic volume for European metropolitan areas. J Transp Geogr 19:755–762
Door to Door Information for Air Passengers project (DORA) (2016) User groups and mobility profiles, D2.2
Drivenow (2017) Accessed on 21 Nov 2017. https://www.drive-now.com/de/de
Euromonitor International (2014) Datagraphic population and homes: proportion of single person households worldwide. Accessed on 03 Nov 2016. http://www.euromonitor.com/medialibrary/PDF/pdf_singlePersonHH-v1.1.pdf
European Commission (2011) Flightpath 2050 Europe's vision for aviation—Report of the high level group on aviation research, Luxembourg
Eurostat Database (2014a) Population [t_demo_pop]. Retrieved from http://ec.europa.eu/eurostat/de/data/database
Eurostat Database (2014b) Tourism [tour]. Retrieved from http://ec.europa.eu/eurostat/de/data/database
Eurostat Life Expectancy by Age and Sex (2016) Accessed on 10 Jan 2016. http://appsso.eurostat.ec.europa.eu/nui/show.do?dataset=proj_13nalexp&lang=en
Eurostat Population Statistics (2016) Accessed on 02 Nov 2016. http://ec.europa.eu/eurostat/web/population-demography-migration-projections/population-projections-/database
E-Volo, Volocopter (2017) Accessed on 21 Nov 2017. https://www.volocopter.com/en/product/
Flixbus (2017) Accessed on 21 Nov 2017. https://www.flixbus.de/
Frederick S (2014) Pardee Center for International Futures (ifs), GDP per capita (Thousand 2011$)—base case. Accessed on 11 Nov 2016. http://www.ifs.du.edu/ifs/frm_MainMenu.aspx
Free2Move (2017) Accessed on 30 Mar 2017. https://de.free2move.com/about

Future Foundation (2015) Future traveller tribes 2030: understanding tomorrow's traveller, London, Future Foundation. Accessed on 28 June 2016. http://www.amadeus.com/documents/future-traveller-tribes-2030/travel-report-future-traveller-tribes-2030.pdf
GARA (2017) Mozio brings multimodal mobile ticketing pilot to SEPTA airport line. Accessed on 04 Apr 2017. https://www.globalairrail.com/news/entry/mozio-brings-multimodal-mobile-ticketing-pilot-to-septa-airport-line?utm_source=Global+AirRail+Alliance&utm_campaign=4c0adbeada-EMAIL_CAMPAIGN_2017_04_07&utm_medium=email&utm_term=0_9775aa507b-4c0adbeada-47632973
GfK Mobilitätsmonitor (2011) Airport private traveller study Reiseverhalten, Einstellungen und Werte der Privatreisenden am Airport, GfK Mobilitätsmonitor—GfK Roper Consumer Styles. http://www.munich-airport.de/media/download/bereiche/mediacenter/extras/deutsch/Airport_Private_Traveller_Study.pdf
Gilbert D, Wong RKC (2003) Passenger expectations and airline services: a Hong Kong based study. Tour Manag 24(5):519–532
Google, Google Maps (2017) Accessed on 21 Nov 2017. https://www.google.de/maps
Gössling S, Haglund L, Kallgren H, Revahl M, Hultman J (2009) Swedish air travellers and voluntary carbon offsets: towards the co-creation of environmental value? Curr Issues Tourism 12(1):1–19
Hares A, Dickinson J, Wilkes K (2010) Climate change and the air travel decisions of UK tourists. J Transp Geogr 18(3):466–473
Henley Centre HeadlightVision (2007) Future traveller tribes 2020—report for the air travel industry. http://www.amadeus.com/documents/future-traveller-tribes-2030/travel-report-future-traveller-tribes-2020.pdf
Joby Aviation (2017) Accessed on 21 Nov 2017. http://www.jobyaviation.com/LEAPTech/
Kluge U, Paul A (2017) Exploring factors influencing European passenger demand for air transport using a causal loop diagram, Amsterdam, G.A.R.S. In: 14th aviation student research workshop
Kluge U, Paul A, Tanner G, Cook AJ (2017) Future passenger demand profile, D3.2, DATASET2050, Munich
Kluge U, Paul A, Ureta H, Ploetner KO (2018) Profiling future air transport passengers in Europe, Vienna, 7th Transport Research Arena 2018
Kopsch F (2012) A demand model for domestic air travel in Sweden. J Air Transp Manag 20:46–48
Kuljanin J, Kali M (2015) Exploring characteristics of passengers using traditional and low-cost airlines: a case study of Belgrade Airport. J Air Transp Manag 46:12–18
Lilium GmbH, Lilium Jet (2017) Accessed on 21 Nov 2017. https://lilium.com/mission/
Lufthansa (2017) Accessed on 07 Apr 2017. https://www.lufthansa.com/de/en/Lufthansa-Express-Bus
Lyft (2017) Accessed on 30 Mar 2017. www.lyft.com
Mair J (2011) Exploring air travellers' voluntary carbon-offsetting behavior. J Sustain Tourism 19(2):215–230
Mobility4EU (2016) Societal needs and requirements for future transportation and mobility as well as opportunities and challenges of current solutions, D2.1. Retrieved from http://www.mobility4eu.eu/wp-content/uploads/2017/01/M4EU_WP2_D21_v2_21Dec2016_final.pdf
Mobility4EU (2017) Opportunity map for the future of mobility in Europe 2030
Mobilitymix (2017) Accessed on 21 Nov 2017. https://mobilitymixx.nl/home.html
Network rail (2017) Accessed on 21 Nov 2017. https://www.networkrail.co.uk/who-we-are/
Octopus Card (2017) Accessed on 21 Nov 2017. http://www.discoverhongkong.com/de/plan-your-trip/traveller-info/transport/getting-around/octopus-card.jsp
OECD (2011) The future of families to 2030. Retrieved from http://dx.doi.org/10.1787/9789264168367-en
Official Airline Guide (2014) OAG trends report for 2014, what is shaping air travel in 2015?
OysterCard (2017) Accessed on 21 Nov 2017. https://oyster.tfl.gov.uk/oyster/entry.do
Paul A, Kluge U (2016) Current passenger demand profile, D3.1, DATASET2050, Munich
Paul A, Kluge U (2017) Assessing passenger requirements along the D2D air travel Chain, D5.3, DATASET2050, Munich

Plötner KO, Schmidt M (2014) Finale report, FASCINATIONS2050 project
Profillidis VA, Botzoris G (2015) Air passenger transport and economic activity. J Air Transp Manag 49
Rally bus (2017) Accessed on 30 Mar 2017. http://rallybus.net/About
Ringle CM, Sarstedt M, Zimmermann L (2011) Customer satisfaction with commercial airlines: the role of perceived safety and purpose of travel. J Mark Theor Pract 19(4):459–472
SITA (2015) Air transport industry insights: the future is personal, SITA—A 360 degree report. Retrieved from http://www.missionline.it/wp-content/uploads/2015/08/360-report-the-future-is-personal-2015.pdf
SITA (2016) Air transport industry insights. Retrieved from https://www.sita.aero/globalassets/docs/surveys–reports/passenger-it-trends-survey-2016.pdf
Skift (2015) Yearbook for 2015, Megatrends defining travel in 2015, issue: 01, Skift Travel IQ
SnappCar (2017) Accessed on 21 Nov 2017. https://www.snappcar.de
SocialCar (2017) Accessed on 21 Nov 2017. https://www.socialcar.com/en/
Thalys (2018) Accessed on 04 Jan 2018. https://www.thalys.com/de/de/
Uber (2017) Accessed on 30 Mar 2017. www.uber.com
UN (2015) UN world population prospects. Accessed on 10 Nov 2016. https://esa.un.org/unpd/wpp/, 2015
UNIFE (2016) A driver for the EU competitiveness and sustainable mobility worldwide. Retrieved from http://www.unife.org/component/attachments/?task=download&id=110
Vinka (2017) Vinka joins Tuup to launch an on-demand robot bus service in Finland during 2017. Accessed on 27 Mar 2017. http://www.vinka.fi/sohjoa
Voom (2017) Accessed on 21 Nov 2017. https://www.voom.flights/
Wadud Z (2013) Simultaneous modeling of passenger and cargo demand at an airport. Transp Res Rec: J Transp Res Board 2336(1):63–74
WBCSD (2008) Sustainable consumption facts and trends, from a business perspective—the business role focus area, world business council of sustainable development, p 9
World Bank (2017) Doing business measuring business regulations, economy rankings. Retrieved from http://www.doingbusiness.org/rankings
World Economic Forum (2014) Smart travel unlocking economic growth and development through travel facilitation, 17 June 2014
World Economic Forum (2016) Digital media and society implications in a hyperconnected era
Young D, Pilon N, Brom L (2009) Challenges of air transport 2030. Retrieved from https://www.eurocontrol.int/eec/gallery/content/public/document/eec/other_document/2009/003_Challenges_of_air_transport_2030_experts_view.pdf
Zee Aero (2017) Accessed on 21 Nov 2017. http://www.businessinsider.de/larry-page-zee-aero-kitty-hawk-flying-car-startups-registering-new-aircraft-2017-3?r=US&IR=T

Personalised Driver and Traveller Support Systems

Maria Panou, Evangelos Bekiaris and Eleni Chalkia

Abstract Personalised services is a field in rapid development. There are plenty of contexts in transport where personalisation could apply, since not all travellers have the same preferences and needs (due to functional limitations, age, or other reasons). Also, not all drivers drive the same way, even if they belong to the same age cluster. In this article, the personalised HMI in four different mobility-related areas and systems is discussed, i.e. multimodal routing, infomobility services, advanced driver assistance systems (ADAS) and driving training on driving simulators. Relevant developments in various research initiatives are presented as examples, whereas their evaluation results by real users are discussed.

Keywords HMI personalisation · Route guidance · ADAS · Infomobility
Driving training

1 Introduction

Personalisation is the mega trend of today. It has well been realized that the industrial trend "one size fits all" is not anymore acceptable by the clients of today and tomorrow. In transport, this trend led long ago to lean production and even emotional design (Kansei in Japan, Citarasa in Malaysia, etc.). However, in the areas of advanced driver support systems (ADAS), traveller infomobility services and even driver training, personalisation has not been considered. Even adaptability of HMI has been very low, with info provision and warning modes, content and timing being

M. Panou (✉) · E. Bekiaris · E. Chalkia
Centre for Research and Technology Hellas, Hellenic Institute of Transport, Egialias 52, 15121 Athens, Greece
e-mail: mpanou@certh.gr

E. Bekiaris
e-mail: abek@certh.gr

E. Chalkia
e-mail: hchalkia@certh.gr

B. Müller and G. Meyer (eds.), *Towards User-Centric Transport in Europe*,
Lecture Notes in Mobility, https://doi.org/10.1007/978-3-319-99756-8_18

adapted only for specific categories of travellers (i.e. through apps for people with special needs) and not at all for drivers. This chapter focuses on recent efforts to bring personalisation of service to the world of driver/traveller support and training/education systems, trying to answer the following questions:

- What kind of personalisation is needed, should be prioritized and based on which parameters?
- Which personalisation principles/algorithms and thresholds can be used?
- Is personalisation actually needed and/or appreciated by the drivers and travellers?

Personalisation has become a mainstream and highly active area of research in the context of many different domains. From Internet services and e-commerce, music, news recommendation (Linden et al. 2003; Lathia et al. 2013) to mobile location-based services (Quercia et al. 2010), from manufacturing networks (Mourtzis et al. 2014) to digital web maps (Lardinois 2013) and from ehealth and social services (Carr 2010) to tourism (Buhalis and Amaranggana 2015). Personalised solutions increase user acceptance but mainly the effectiveness of the service/system. Specifically for mobility, the role of personalisation is particularly critical, since by providing a service that is customised to each individual traveller and driver needs and preferences, the travellers' satisfaction will increase as the quality of service will be higher and will meet, or even exceed, the expectations of the traveller; leading to a safer, faster and pleasant mobility.

2 State of the Art on Personalised HMI for Mobility

The opportunity to deliver personalised customer service via intelligent routing has been used for years now. Personalised routing that takes into account the dynamics of the personal user profiles has widely been used for commuters, providing the user with the possibilities of programming a tour, i.e. selecting a set of relevant points of interests (PoIs) to be included into the tour, schedule the tour, i.e. arranging the selected PoIs into a sequence based on the cultural, recreational and situational value of each, and determine his/her travel route, i.e. generating a set of trips between the PoIs that the user needs to perform in order to complete the tour (Aksenov et al. 2014). Additional work on personalisation in multi-day GPS and accelerometer data processing, analysing how personalisation in multi-day GPS and accelerometer data can improve the quality of the produced travel diaries has been realised, defining that it is worth collecting annotated data from participants in a multi-day or even multi-week survey. It is reported that one week of smartphone data is not enough to highly personalise trip purpose detection routines (Montini et al. 2014).

Much work on personalisation in routing has also been done in the field of travellers with disabilities, mainly wheelchair users and blind as well as for the older people. Personalised routing models for wheelchair users, focusing on user priorities and sidewalk properties have been researched on a study related to routing for disabled and especially wheelchair users (Kasemsuppakorn and Karimi 2009). In EU projects ASK-IT (Visintainer et al. 2007) and OASIS (Tsakou et al. 2009), routing based on the user's static (needs), dynamic (preferences) and semi-dynamic (based on purpose of trip) profile were developed, using also filtering operations with real-time data. This was realized by the development of specific computing techniques for learning and decision-making, which were based upon multi-agent architecture. OASIS provided a considerably large set of services, including pre-trip services, like Demand Responsive Transport and Special Car Rental services, tailored to the older people user profile. In-vehicle services are related to the users' comfort, safety and driving efficiency and include Emergency Call, Medical Assistance, Vehicle assistance, Navigation, as well as Fatigue Monitoring and the integration of Driving Assistance Systems. All the aforementioned services, aim to support and enhance the personal mobility of older people, taking into account their specific needs.

In addition to the aforementioned, another domain where personalisation occurs rapidly is the infomobility services. The infomobility sector concerns the servicing of mobile users (using a mobile phone, the car's system, etc.) through wireless networks, based on a variety of technologies, like GPS and GIS (routing/mapping systems) and different kinds of distributed content, such as public transport timetables, traffic information, etc. This sector demands users to have access to accurate real-time location-based services that are distributed throughout the web. The personalisation of the infomobility of these services has been addressed from a variety of studies using mainly intelligent agents technologies (Moraïtis et al. 2003; Kauber 2004; Panou et al. 2010). Most projects have coped with this issue, like the Im@gineIT (Spanoudakis et al. 2006), by supporting the following functionalities:

- Support different types of users.
- Receive the user request, as analyzed by the user interface and suggest optimal transportation solutions, tourist events and nearby attractions.
- Adapt the service according to user's habitual patterns.
- Monitor the user's route and automatically provide related events during the journey.
- Support the user while he/she is roaming between regions and countries.
- Support the user while he/she changes travel modes in a multimodal trip.

In the field of personalisation of driver training using simulators, relevant work has been realised in the EU project TRAIN-ALL, where each driver training session uses the driver's training achievements and status as well as individual characteristics, to adapt accordingly the next simulator training sessions (Panou et al. 2010). Additionally, current work on drivers simulator includes information about the user's state of mind (level of attention, fatigue state, stress state) based on video data and biological signals. Facial movements, such as eyes blinking, yawning, head rotations, etc., are detected on video data: they are used in order to evaluate the fatigue and the attention level of the driver. The user's electrocardiogram and galvanic skin response are recorded and analyzed in order to evaluate the stress level of the driver. All the above are taken into account in relevant training personalisation.

In the field of Intelligent Transport Systems (ITS), the technology has allowed the appearance of a number of systems and applications, like GPS navigation, autonomous cruise control, eCall, as well as advanced driver assistance systems (ADAS), like lane departure warning, driver drowsiness detection, or semi-automated driving systems, like platooning, parking assistance, etc. (Schneiderman 2013), that improve comfort, safety and efficiency of transport. The final goal of the vision in the domain of these solutions is to make intelligent vehicles fully automated and achieve accident free traffic scenarios (Nieto et al. 2014).

ADAS are designed to enhance traffic safety by supporting the drivers while driving; however such systems do not take into account the personal driving style and abilities of each driver and, since not all drivers drive the same way, their effectiveness is limited by the manufacturer's, default parameters.

The driver models used at the moment by ADAS are generic, since they represent the way an average driver reacts to a given situation. However, a wide range of tools exist in machine learning which would allow personalising ADAS to various levels. The advantage of personalisation is that the response of the assistance system can be adapted to the specific driving habits of the driver, leading to more efficient interventions (Musicant and Bar-Gera 2011). A number of approaches have been proposed in the literature to realize ADAS personalisation. Two of the more common modelling approaches for ADAS personalisation from the literature are the black boxes (Krajzewicz 2010; Treiber et al. 2000; Salvucci and Gray 2004; Pomerleau 1989; Lefèvre et al. 2014a, b) and stochastic hybrid systems (Shia et al. 2014; Morris et al. 2011; Aoude et al. 2012; Lefèvre et al. 2014c). Additionally, driver models have been used to personalise the control of intelligent vehicles. This strategy has been used in the past using a Piece-Wise AutoRegressive eXogenous model (Lin et al. 2013), artificial neural networks (Pomerleau 1989), Gaussian mixture models (Nishiwaki et al. 2007), and inverse optimal control (Levine and Koltun 2012). The results demonstrated that it is possible to reproduce human driving behavior in a controlled car, by learning from example data. However, these approaches are not directly applicable to driving assistance, since they do not address the problem of safety. For example, there is no guarantee that the lane change controller presented in (Levine and Koltun 2012) will not attempt a dangerous maneuver, or that the cruise control system proposed in (Nishiwaki et al. 2007) will not collide with the preceding vehicle. Nevertheless, understanding individual driver behaviors and development of personalised driver models are critical for active safety control systems (Lefèvre et al. 2015; Butakov and Ioannou 2015; Butakov and Ioannou 2016), vehicle dynamic performance (Fu and Soeffker 2010), and human-centered vehicle control systems (Wang et al. 2015; Wang et al. 2014; Wang et al. 2017), eco-driving systems (Xiang et al. 2015), and automated vehicles (Lefèvre et al. 2016). A learning-based driver model, which can represent human driving control strategies on the highway for personalised driving assistance, where the driver model is combined with a model predictive controller, has been developed towards this goal (Lefèvre et al. 2015). Learning-based driver models have been used for the understanding driver's correction behaviors (DCB), since this is the primary reason for false warnings of ADAS (Wang et al. 2017).

As elaborated above, personalisation of driver/passenger support infomobility services has been recognized as a need and has been approached by a few attempts. The following chapter analyses further some of the most promising research results in this area.

3 Application Areas with Personalised Content

In this chapter, four different applications are presented, where personalised HMI was developed for supporting and enhancing driving and mobility of various users categories, including mobility-impaired and older people. They constitute representative results of each relevant service category and highlight different personalisation scenarios, parameters and customer segments.

3.1 Personalisation of Infomobility Services for Travellers: The Case of Multimodal Routing for Travellers with Restricted Mobility

Multimodal trip planners have been developed in a number of projects, but only a few consider individual needs and preferences. Among these projects are the OASIS and ASK-IT (FP7) projects. In OASIS project a multi-modal trip planner for the older traveller was developed, taking into account the mobility restrictions of the travellers (see Figs. 1 and 2). A sub-menu enables the user to choose the preferred mode of transport in the case of a multimodal trip, as well as based on the accessibility attributes of buses (Spence et al. 2011). For journeys by other modes, the results can be prioritised according to various parameters (as set by the user), e.g. shortest distance by foot, shortest/fastest/cheapest route by car, etc.

After choosing the preferred route plan, it is possible to save the route as a 'favourite'. An important benefit is that the route details become part of the user profile and can be used by other services. This is especially useful for integration, for example, with the pedestrian guidance service developed in the same project, in which different parameters are included, such as calm/safe route. On the map, the sections by bus are shown in orange and the sections on foot (see small sections at the beginning and end of the trip in Fig. 3) in green.

Within ASK-IT project, a personalised route guidance module was also developed, that monitored the persons with reduced mobility (PRM) and provided support and guidance (Kauber et al. 2007). The user can indicate his/her route preference and the type of transport mode he/she would like to use and can also specify his/her disability type (user with walking aids, wheelchair assisted, hearing or visually impaired). The system output provides information on the accessibility level of each proposed route, based on a colour coding (e.g. green colour indicates a fully accessible route, yellow

Fig. 1 Screenshot of OASIS multi-modal trip planner application

colour implies a semi-accessible route and red colour is attributed to an inaccessible route). Screenshots of the application are presented in Fig. 4.

3.2 Personalisation of Infomobility Services for Travellers: The Case of Tourists

Infomobility services present to the user information on Points of Interest (PoIs) on his/her smartphone/tablet. The points of interest for which a traveller might be interested in getting information about their address, the distance from him/her and the route that he/she must follow may vary and can be too many to be presented at limited mobile or pad screen; thus, a proper prioritization is needed, following the user's preferences and/or special requests. An analytic list of such PoIs and their sub-categories has been developed in ASK-IT project, as an initiative to connect touristic and transportation related content with accessibility information for mobility-impaired users. This list forms the ontology, i.e. the formal description of

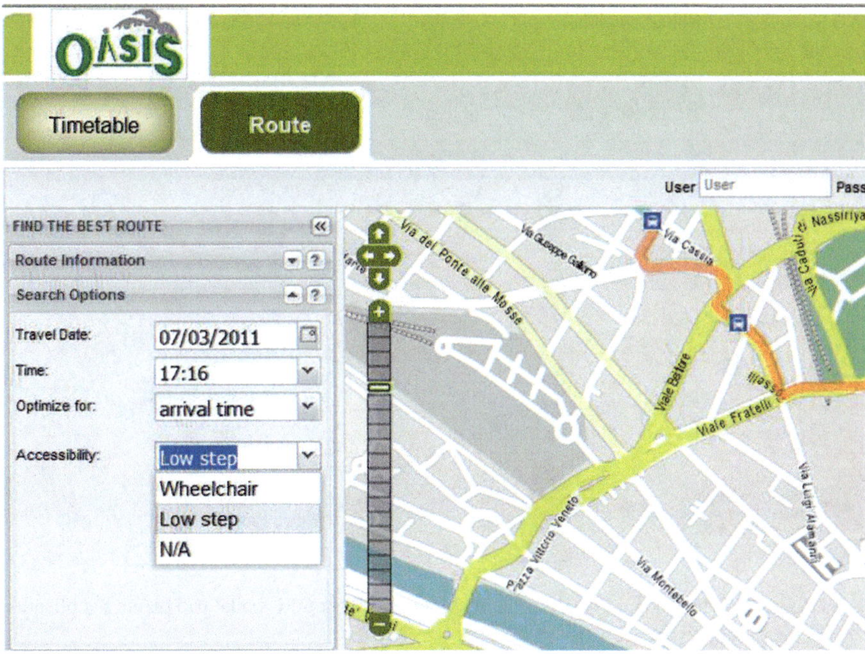

Fig. 2 Accessibility search options sub menu for bus journeys of OASIS multi-modal trip planner application

a field/domain that contains terms and definitions and semantic relationships among them. A part of this list follows below, as example:

- Point of interest: Hotel
- Sub-categories: 1-star, 2-stars, 3-stars, 4-stars
- Point of interest: Museum
- Sub-categories: Modern art, history, war, etc.
- Point of interest: Restaurant
- Sub-categories: Asian, Italian, traditional, etc.

An infomobility personalisation algorithm was developed within a study (Panou 2008). The algorithm is based on the previous selections of specific PoIs by the user (and their sub-categories), taking into account the day of the week, i.e. weekday versus weekend. The logic is that the needs and preferences of the user differ on the weekends and the weekdays. Similarly, the purpose of traveling (e.g. travelling for going to work, for recreational reasons, for emergency reasons, etc.) has been taken into account. The algorithm was applied to the user's mobile device after the system has been used for certain time and the user's preferences are known. Thus, the types and sub-types of PoIs that have been chosen more often are proposed by

Fig. 3 Route description and visualisation on map in OASIS multi-modal trip planner application

the system, rather than all the available PoIs in an area. Until enough data for service personalisation exist, the system presents to the user the PoIs that are closest to him/her.

In the following Figs. 5 and 6, the difference between the non-personalised and the personalised search results is clearly shown, with the later giving much less results, satisfying the user's preferences.

3.3 Personalisation of Driver Assistance Systems: The Case of CAS and LDW

Although there is a plethora of Advanced Driving Assistance Systems (ADAS) in the market today, all are aimed for the average driver, without taking into account individual human factors of each driver, such as the driving style, the driving experience, the driving capabilities, etc. An optimization of the drivers warning systems has been suggested (Panou 2008), through their personalisation, both for the longitudinal and lateral road axes. Specific algorithms were proposed for a Collision Avoidance System (CAS) and a Lane Deviation Warning (LDW) system for a holistic personalisation.

To arrive to a personalised driver support system, such a system needs to know certain parameters regarding the driver attributes, preferences as well as the context of use. These parameters can be static, semi-dynamic or dynamic:

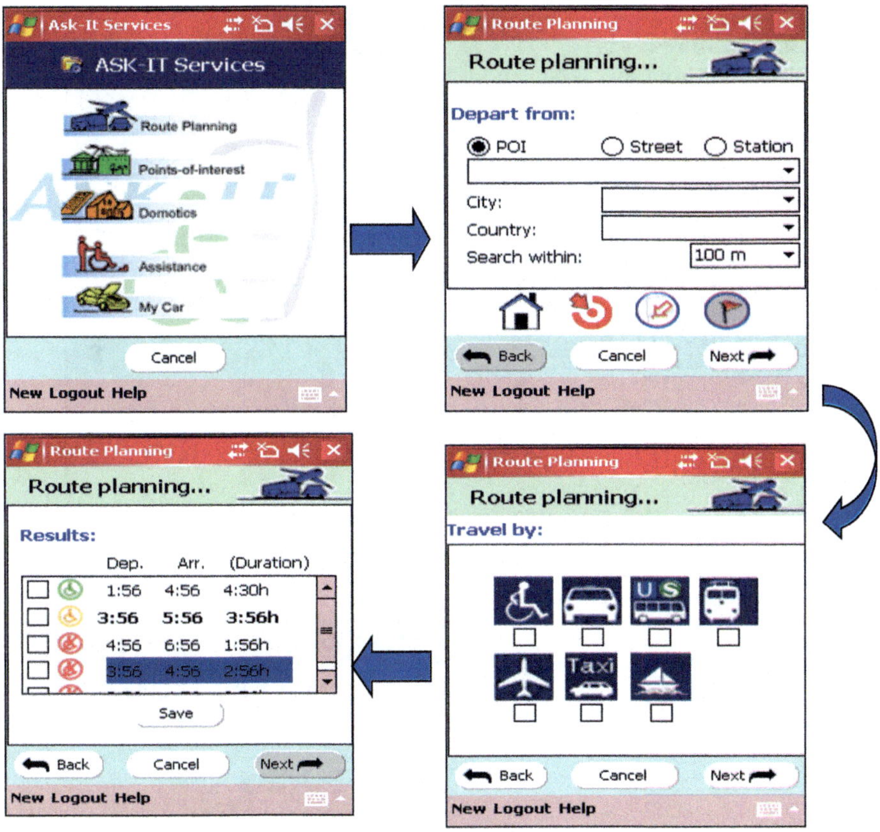

Fig. 4 Route preferences and planning functions of the ASK-IT personalised route guidance

Static parameters are inserted once by the driver, e.g. driving experience, language, any disabilities, etc.

Semi-dynamic parameters need to be defined at the start of a new trip, unless they remain the same with the previous trip. Main parameters in this category are the destination, the reason for travelling (e.g. emergency, commuter to work, tourism, leisure), etc.

Dynamic parameters are calculated automatically by the system. They may be continuously altered and updated when the system is activated. There are 7 dynamic parameters for driving behaviour modeling that the proposed system uses. Six of them are automatically calculated by the car itself or from connected sensors; only the driver's reaction time is self-measured through an algorithm. These are:

- Speed (m/s)
- Time instant of driver's braking (s)

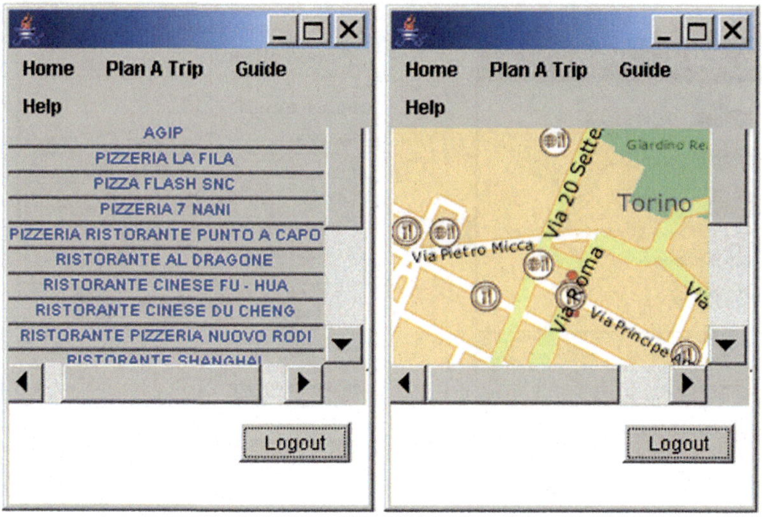

Fig. 5 Search results of PoIs without personalisation and geographical representation on the map

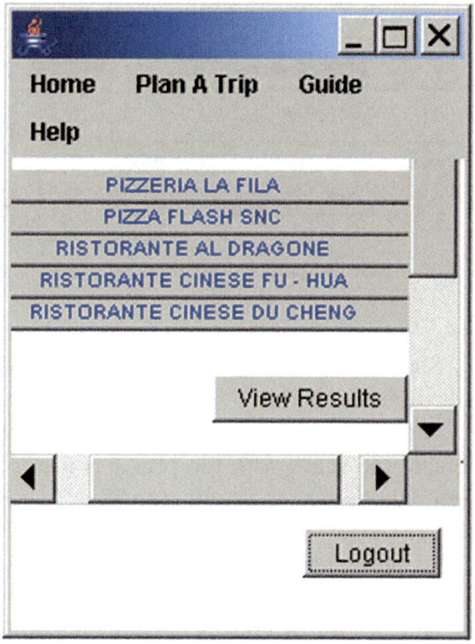

Fig. 6 Search results of PoIs with personalisation algorithm

- Time instant of frontal vehicle braking (s)
- Time GPS (s)
- Time to Collision (TTC)
- Time headway (T_headway)

- Time to Line Crosing (TLC)
- Distance to Line crossing (DLC)
- Reaction time.

For the CAS personalisation, the algorithm is based on the Time to Collision (TTC) or Time Headway measurement, under specific scenarios with car following events (when $TTC \leq 4$ s.). The parameter that is used for the LDW personalisation is the Time to Line Crossing (TLC). At the end, two personalised thresholds are calculated for each driver (one for each system), based on which the warning is provided to the driver at the proper time instant. It should be underlined that for both systems safety margins are put in the personalisation algorithm (i.e. $TTC \geq 2$, $T\_headway \geq 1$).

3.4 *Personalisation in Driver Training: The Case of Adapted Driving Simulator Scenarios*

The simulation technology is grossly characterised by fixed databases with limited possibilities in generating complex traffic scenarios and the simulator driving environment is always smooth and the driver focuses only on his/her own driving and the specific tasks that are requested by the respective program/exercise that he/she follows. However, since real driving conditions are not always smooth, and there are drivers who do mistakes or disobey traffic rules, the drivers must be trained, in order to be ready to compensate such unexpected situations. Moreover, not all drivers drive in the same way, i.e. the driving style differs heavily among drivers, also due to age, experience and even cultural differences. As an example, it is useful to model the typical behaviour of tourists, as drivers, when driving in foreign countries.

Within the TRAIN-ALL project, a module was developed, aiming at capturing the individual drivers behaviour in specific driving simulator scenarios and then using it both for personalised driver training, as well as for constructing a more natural traffic environments at the simulator, with surrounding vehicles performing as would other real drivers (instead of having an artificial and homogeneous, rules-based behaviour). This module has been successfully installed in the car driving simulator at the premises of the Hellenic Institute of Transport. The inclusion in the simulation of other traffic participants, that interact naturally (not always legally though) with the trainee, is possible to "fully immerse" the driving simulator into a realistic traffic environment. By mixing "true" traffic simulation and multi-user driving simulator, a new generation of computer-based training tools has emerged.

Personalised profiles are built with the AI module, which define the personal driving style of each driver. The stored driving profiles of the drivers can be used for many applications in the future, such as the detection of driver's problematic areas that can be used for suggesting appropriate training measures, the development of driver-specific personalised ADAS for a more efficient driver warning (see Panou 2008), statistics and analysis of main factors that contribute to traffic accidents, etc.

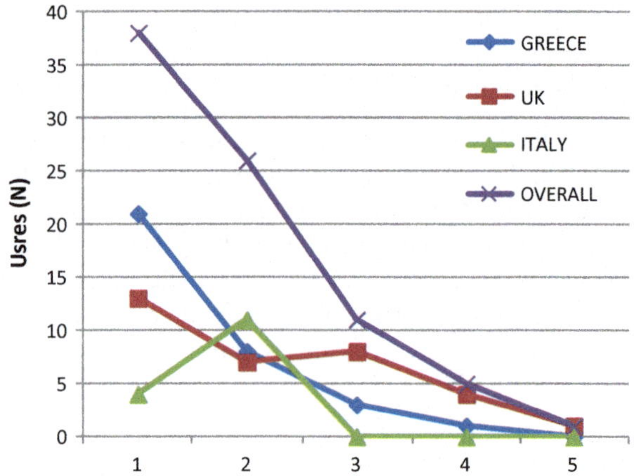

Fig. 7 Usefulness of the OASIS elder-friendly route guidance application (1: very useful to 5: not useful at all)

In order to create a first personalised profile for the driver, three parameters were selected which are based on the most important driving ability parameters of the DRIVABILITY model (Bekiaris et al. 2003), each one corresponding to a different type of driver behavior. These are:

- Longitudinal behaviour: TTC and/or T_headway
- Lateral behaviour: TLC (left and/or right)
- Emergency behaviour: RT.

The mean values of the driving behaviour parameters per driver collected under specific use cases were compared to normative values from the literature.

4 Evaluation Results

In order to prove that personalised systems are really needed/appreciated by the target end users, evaluations took place in all the above-mentioned systems with various clusters and types of users. The hypotheses here is that in all cases, the participants would show a clear preference to the personalised systems, versus the non-personalised ones. Mainly the Van der Laan (Van Der Laan et al. 1997) user acceptance scale indicators were used for most of the evaluations below; here the most relevant and important indicators are presented per application/system.

In OASIS project, the multimodal route guidance application (described in Sect. 3.1) was tested in three pilot sites, namely Greece, Italy and the UK, with 106 older users (aged above 55 years old). It is evident from the graph below (Fig. 7), that increased usefulness was reported in all sites and most (79%) participants were positive and interested about elder-friendly and oriented multimodal route guidance application. No differences among pilot sites were found (Panou et al. 2012).

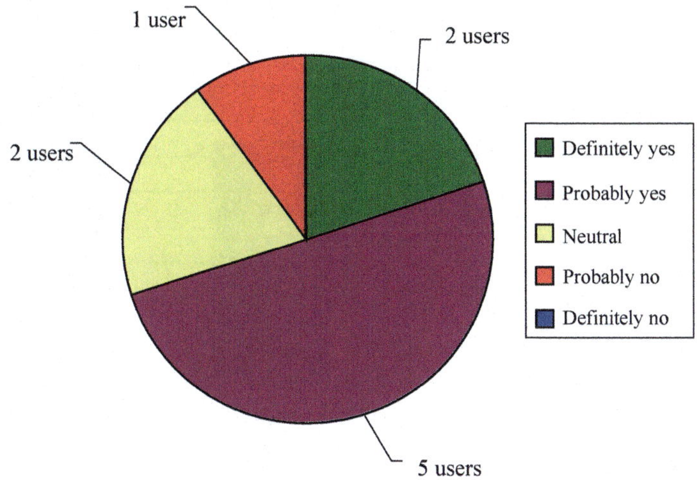

Fig. 8 Users trust on the personalised infomobility system versus the standard one (Panou 2008)

The personalised route guidance module of ASK-IT project (described in Sect. 3.1) was evaluated by 319 mobility-impaired users in 8 European pilot cities and was the most highly valued service developed in the project (among many). Most end users opted for the personalised services. Although, there was a lack of understanding amongst many participants as to how accurate the personalisation service was regarding personal preferences, in general, user acceptance and usability were almost universally high (Edwards 2008).

Ten users tested the personalised infomobilty application described above (Sect. 3.2) (Panou 2008). In general, they appreciated mostly the system effectiveness, usefulness and pleasantness. The following diagram (Fig. 8) presents the users trust to the personalised system, which is positive.

As concerns the ADAS personalised warning algorithms of Sect. 3.3, the systems were tested by 10 drivers under real on-road conditions. For the CAS, the drivers that drive either too close or too far away from the leading vehicle have a higher preference (higher improvement rates of the user's confidence level over the system) for a personalised system than the drivers that drive near to the default value of the warning limit (Fig. 9). This observation was expected. As 4 out of 10 persons of a random sample can be considered to deviate significantly from the 'average driver' regarding their behaviour on the longitudinal road axis, such personalised systems seem to have a big application range.

In terms of the LDW system, the percentage of the improvement of the acceptance level of the personalised versus the standard one was positive for 9 out of the 10 drivers (Fig. 10).

When the participants were questioned about their trust to the system, overall there was a clear preference to the personalised system, while all 10 drivers reported that they expect traffic safety to increase with its use.

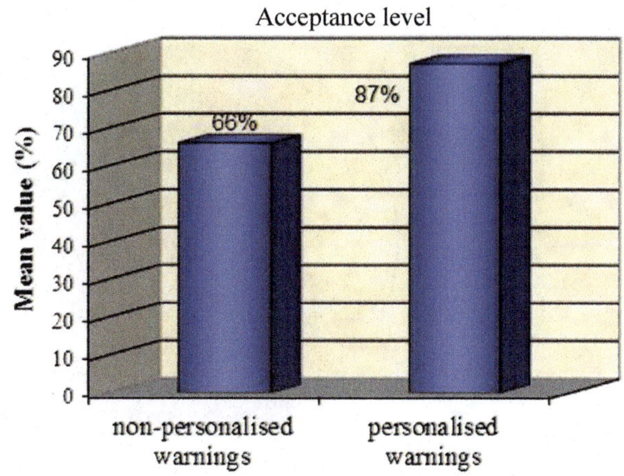

Fig. 9 Mean values of the acceptance levels on the CAS, as indicated by the users, with/without the personalised algorithm

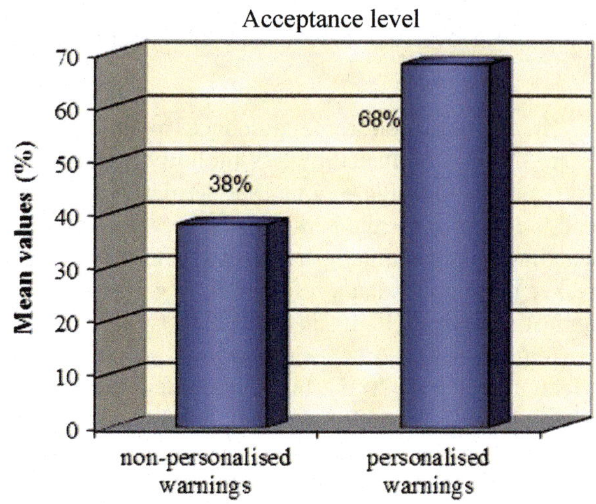

Fig. 10 Mean values of the acceptance levels on the LDW, as indicated by the users, with/without the personalised algorithm

For TRAIN-ALL module (of Sect. 3.4), the scenarios of the Ambient Intelligence (AI) module were rated by 25 novice drivers as more natural (in terms of the driving behaviour of the participating vehicles in the traffic), compared to the standard ones (Fig. 11). In this case, usability is not relevant as the developed system is not about HMI related personalisation, but concerns driving training scenarios. The scenario with the personalised module was considered more natural in comparison to the one without personalisation (Peters 2009).

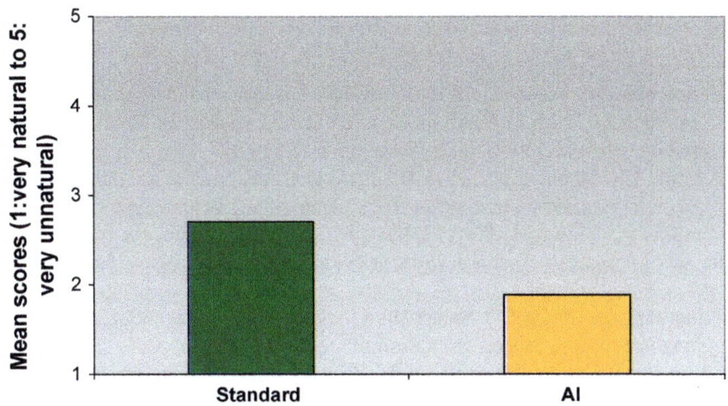

Fig. 11 Mean naturalness ratings of AI module in TRAIN-ALL (standard vs. AI environments)

5 Conclusions and Next Steps

In the application examples presented, all users clearly appreciated and preferred the personalised services/systems, of course with varying acceptance level.

Personalisation is entering all the more infomobility services in the market, especially for trip planning, route guidance and navigations applications. Also, driving simulator developers have started integrating personalised elements in the scenarios for their products; as an example, out of TRAIN-ALL project, 6 driving simulators companies in 4 European countries implemented the personalised elements developed in the project in their training simulation scenarios (Vanhulle 2009).

Personalisation in the area of transport and mobility is a key issue for increasing traffic safety and market uptake/diffusion of new technologies especially in view of the introduction of electric, shared and automated vehicles. Further research still needs to be done in all the above application areas. For example, the drivers parameters thresholds used for the driving simulator training are related to real driving situations. However, the driving behaviour at a simulator has certain differences. Thus, the proposed criteria should be redefined per simulator, in order to check the simulator validity.

Intelligent personalised algorithms may be extended to hand-over strategies between driver and vehicle for autonomous vehicles at SAE automation level 3, to be optimally following the particular driver characteristics and abilities. It is also relevant to higher levels of automation, since travellers would trust more an automated vehicle that drives according to their 'style' and are able "to comprehend". Thus, key automated vehicles parameters (such as TLC, TTC or Time Headway) could be adapted accordingly; always of course remaining within their safety margins.

References

Aksenov P, Kemperman A, Arentze T (2014) Toward personalised and dynamic cultural routing: a three-level approach. In: 12th international conference on design and decision support systems in architecture and urban planning, DDS, Elsevier, pp 257–269

Aoude GS, Desaraju VR, Stephens LH, How JP (2012) Driver behavior classification at intersections and validation on large naturalistic dataset. IEEE Trans Intell Transp Syst 13(2):724–736

Bekiaris E, Amditis A, Panou M (2003) DRIVABILITY: a new concept for modelling driving performance. Int J Cognition Technol Work 5(2):152–161. ISSN: 1435-5558. (Springer-Verlag, London Ltd.)

Buhalis D, Amaranggana A (2015) Smart tourism destinations enhancing tourism experience through personalisation of services. Inf Commun Technol Tourism

Butakov VA, Ioannou P (2015) Personalised driver/vehicle lane change models for ADAS. IEEE transaction on intelligent transportation systems 64(10):4422–4431

Butakov VA, Ioannou P (2016) Personalised driver assistance for signalized intersections using V2I communication. IEEE Trans Intell Transp Syst 17(7):1910–1919

Carr S (2010) Personalisation: a rough guide. Social Care Institute for Excellence

Edwards S et al (2008) ASK-IT Deliverable 4.6.2 'Consolidated pilot results'

Fu X, Soeffker D (2010) Modeling of individualized human driver model for automated personalised supervision. SAE technical paper, Tech Rep

Kasemsuppakorn P, Karimi AH (2009) Personalised routing for wheelchair navigation. Location based services, vol 3, pp 24–54. https://doi.org/10.1080/17489720902837936

Kauber M (2004) The emerging market of infomobility services. Elsevier 8:69–87

Kauber M, Grammling M, Stab K, Ziora A, Diederichs P, Jeziorowski M (2007) ASK-IT project D2.3.1 Accessible route guidance module

Krajzewicz D (2010) Traffic simulation with SUMO—simulation of urban mobility. In: Barcelò J (ed) Fundamentals of traffic simulation. No. 145 in international series in operations research & management science. Springer, New York, pp 269–293

Lardinois F (2013) The next frontier for google maps is personalisation. TechCrunch

Lathia N, Smith C, Froehlich J, Capra L (2013) Individuals among commuters: building personalised transport information services from fare collection systems. Pervasive Mob Comput 9(5):643–664 Elsevier

Lefèvre S, Gao Y, Vasquez D, Tseng E, Bajcsy R, Borrelli F (2014a) Lane keeping assistance with learning-based driver model and model predictive control. In: Proceeding of the 12th international symposium on advanced vehicle control, Tokyo, Japan

Lefèvre S, Sun C, Bajcsy R, Laugier C (2014b) Comparison of parametric and non-parametric approaches for vehicle speed prediction. In: Proceedings of the American control conference, Portand, Oregon, pp 3494–3499

Lefèvre S, Vasquez D, Laugier C (2014c) A survey on motion prediction and risk assessment for intelligent vehicles. Robomech J 1(1):1

Lefèvre S, Carvalho A, Gao Y, Tseng HE, Borrelli F (2015a) Driver models for personalised driving assistance. Veh Syst Dyn 53(12):1705–1720

Lefèvre S, Carvalho A, Gao Y, Tseng HE, Borrelli F (2015b) Driver models for personalised driving assistance. Veh Syst Dyn 53(12):1705–1720

Lefèvre S, Carvalho A, Borrelli F (2016) A learning-based framework for velocity control in autonomous driving. IEEE Trans Autom Sci Eng 13(1):32–42

Levine S, Koltun V (2012) Continuous inverse optimal control with locally optimal examples. In: Proceedings of the international conference on machine learning; Edinburgh, Scotland

Lin T, Tseng E, Borrelli F (2013) "Modeling driver behavior during complex maneuvers. In: Proceeding of the American control conference, Washington, DC

Linden G, Smith B, York J (2003) Amazon.com recommendations: item-to-item collaborative filtering. IEE Internet Computing, 76–80

Montini L, Rieser-Schussler N, Axhausen KW (2014) Personalisation in multi-day GPS and accelerometer data processing in: 14th Swiss transport research conference
Moraïtis P, Petraki E, Spanoudakis NI (2003) Providing advanced, personalised infomobility services using agent technology
Morris B, Doshi A, Trivedi M (2011) Lane change intent prediction for driver assistance: on-road design and evaluation. In: Proceedings of the IEEE intelligent vehicles symposium. Baden-Baden, Germany, pp 895–901
Mourtzis D, Doukas M, Psarommatis F, Giannoulis C, Michalos G (2014) A web-based platform for mass customisation and personalisation. CIRP J Manuf Sci Technol 7(2):112–128
Musicant O, Bar-Gera H (2011) Individual driver's undesirable driving events—a temporal analysis. In: Proceeding of the international conference on road safety and simulation, Indianapolis, Indiana
Nieto M, Otaegui O, Velez G, Ortega J, Cortes A (2014) On creating vision based ADAS. IET Intell Transp Syst
Nishiwaki Y, Miyajima C, Kitaoka N, Itou K, Takeda K (2007) Generation of pedal operation patterns of individual drivers in car-following for personalised cruise control. In: Proceedings of the IEEE intelligent vehicles symposium; Istanbul, Turkey, pp 823–827
Panou M (2008) Ph.D. dissertation 'Advanced personalised travellers' warning and information system. Aristotle University of Thessaloniki
Panou M, Touliou K, Bekiaris E, Gaitanidou E (2010) Pedestrian and multimodal route guidance adaptation for elderly citizens. IST-Africa
Panou MC, Bekiaris ED, Touliou A (2010) ADAS module in driving simulation for training young drivers. In: 13th international IEEE conference on intelligent transportation systems, pp 1582–1587, Funchal
Panou M, Bekiaris E, Gemou M (2012) OASIS Deliverable 4.5.2 'Pilot results consolidation'
Peters B (2009) TRAIN-ALL Deliverable 6.2 'Demonstration pilot results consolidation'
Pomerleau DA (1989) ALVINN: an autonomous land vehicle in a neural network. In: Proceedings of the advances in neural information processing systems, pp 305–313
Quercia D, Lathia N, Calabrese F, Di Lorenzo G, Crowcroft J (2010) Recommending social events from mobile phone location data. In: IEEE international conference on data mining
Salvucci D, Gray R (2004) A two-point visual control model of steering. Perception. 33(10):1233–1248
Schneiderman R (2013) Car makers see opportunities in infotainment, driver-assistance systems. IEEE Signal Process Mag 30(1):11–15
Shia V, Gao Y, Vasudevan R, Campbell K, Lin T, Borrelli F, Bajcsy R (2014) Semiautonomous vehicular control using driver modeling. IEEE Trans Intell Transp Syst 15(6):2696–2709
Spanoudakis N, Moraitis P, Petraki E, Bekiaris E, Panou M, Kalogirou K, Zochios A (2006) IM@GINE IT Deliverable 3.1 'Integrated Multi-Agent System'
Spence A, Torres C, Pachinis T, Edwards S, Hübner Y (2011) OASIS project Deliverable 2.3.1 "Elderly-friendly transport information services"
Treiber M, Hennecke A, Helbing D (2000) Congested traffic states in empirical observations and microscopic simulations. Phys Rev E 62(2):1805–1824
Tsakou G, Tsaprounis T, Agnantis K, Kalogirou K, Tantinger D, Leonidis A, Spence A, Jiménez VM, Telkamp G, Cerda AMN, Kehagias D (2009) OASIS project deliverable 1.5.1 "OASIS AmI framework and agents"
Van Der Laan JD, Heino A, De Waard D (1997) A simple procedure for the assessment of acceptance of advanced transport telematics. Transp Res Part C: Emerg Technol 5(1):1–10
Vanhulle P (2009) TRAIN ALL project Deliverable 7.4 "Exploitation business plans"
Visintainer F, Panou M, Bekiaris E, Kalogirou K, Mourouzis A, Paglé K (2007) ASK IT project Deliverable 2.6.1 "ASK-IT gateway to ADAS/IVICS"
Wang W, Xi J, Wang J (2014) Modeling and recognizing driver behavior based on driving data: a survey. Math Prob Eng 2014

Wang W, Xi J, Wang J (2015) Human-centered feed-forward control of a vehicle steering system based on a driver's steering model. In: 2015 American control conference (ACC). IEEE 2015:3361–3366

Wang W, Xi J, Liu C, Li X (2017) Human-centered feed-forward control of a vehicle steering system based on a driver's path-following characteristics. IEEE Trans Intell Transp Syst 18(6):1440–1453. https://doi.org/10.1109/tits.2016.26063

Wang W, Zhao D, Xi J, Han W (2017) A learning-based approach for lane departure warning systems with a personalised driver model. IEEE Trans Veh Technol

Xiang X, Zhou K, Zhang W-B, Qin W, Mao Q (2015) A closedloop speed advisory model with driver's behavior adaptability for ecodriving. IEEE Trans Intell Transp Syst 16(6):3313–3324

Data Is the New Oil

How Data Will Fuel the Transportation Industry—The Airline Industry as an Example

Marko Javornik, Nives Nadoh and Dustin Lange

Abstract The digital economy creates new business models. These models are based upon platforms that enable both producers and consumers to connect to the ecosystem and create mutual value. A lot of data gets generated in this process, which is core virtue of these systems. Mobility service providers, including airlines, offer mostly physical assets these days. This article intends to depict the most important concepts in terms of data driven companies and will illustrate how the airline industry will be able to profit from this transformation. While today airlines are seen the bridge between airport to airport, a digital transformation has the opportunity to link air traffic with ground transportation modes, making a travel journey seamless.

Keywords Commoditization · Data · Digitalization · Inflight–connectivity Lock–in–effect · Platform · Personalization · Seamless mobility · Targeted advertising · Two–sided market · Unutilized–travel–time · Wi-Fi

1 Introduction

History has shown that the economic world follows certain technological waves. These periods, economists call them Kondratieff cycles, last for around fifty to sixty years, whereby the exact transition from phase to phase cannot be determined accurately (Kondratieff and Dutton 1984). The *Steam Engine*, being the first cycle in the beginning of the 18th century, was superseded by the introduction of the *Railway* in the middle of that century. *Electrical Engineering* followed in the early 19th century, which was then surpassed by the invention of the *Automobile*. The fifth Kondratieff

M. Javornik · N. Nadoh (✉) · D. Lange
Comtrade Digital Services, Letališka cesta 29b, Ljubljana, Slovenia
e-mail: nives.nadoh@comtrade.com

M. Javornik
e-mail: marko.javornik@comtrade.com

D. Lange
e-mail: dustin.lange@comtrade.com

B. Müller and G. Meyer (eds.), *Towards User-Centric Transport in Europe*,
Lecture Notes in Mobility, https://doi.org/10.1007/978-3-319-99756-8_19

cycle marked the *Information Technology* by introducing the World Wide Web in 1993. It is said that the sixth cycle began after the global financial crisis in 2008 and led in a novel era of *Health Care* (Vahs and Brem 2015).

However, it can be observed that in today's world the digital transformation of products, services, companies and entire industries is an ever ongoing process. Megatrends such as Connectivity, Globalization, Individualization, New Mobility or Urbanization characterize the visions and missions of companies and governments on a global scale (Zukunftsinstitut, Megatrend Map 2015). The transformation of current business models to data-driven ones can reduce the isolation in the transportation industry and unfold opportunities for solving first- and last-mile issues. In particular, connecting air travel to ground mobility through an open marketplace, will benefit both, providers and consumers. In many ways airlines are front-runners of this transformation because they are in a very competitive market that operates in a service mode since the beginning and can therefore serve as example to other mobility services modes. Buzzwords, which epitomize the fields of computer science today and in the near future, are "big data" and "machine learning". The Internet of Things (IoT) as well as Blockchain technology are only a selection of revolutionary components that will allow a value creation elevating the customer satisfaction to a level never experienced before. The adoption of such technology will force fundamental changes to the businesses and "will accelerate the pace of the commercial world—and not just incrementally, but exponentially" (Ismail et al. 2014).

1.1 Data Is the New Oil

Imagine the following: As of today, there are about eight billion devices connected to the internet. In 2025, this number shall be outdated by 50 billion, an increase more than six-fold. A mere decade later, an amazing one trillion devices shall be connected to this system, which is a 13-digit number (Ismail et al. 2014). All of these machines will create enormous amounts of data. Taking into consideration that in 2013, a full 90% of all the world's data has been generated over the last two years, the scope of how much digital information will be produced in the future becomes obvious (Dragland 2013).

The value of data can also be outlined of present-day businesses. Five of the top ten enterprises leading Forbes "The World's Most Valuable Brands" belong to the technology industry. Apple, Google, Microsoft and Facebook (ranks 1 through 4 respectively) value altogether more than 430 billion US dollar, outperforming the 15 most valuable car manufacturers (260 billion US dollar) including Toyota, Mercedes-Benz, BMW, Honda, Volkswagen and Tesla with ease (The World's Most Valuable Brands and Forbes 2017; Rapier 2017).

Coming back to the Kondratieff cycles initially mentioned, these leading data driven tech companies generate even more than double the amount of net profits when compared to the leading incumbents in the gas and oil industry Gazprom, Exxon Mobil, Sinopec and Royal Dutch Shell. Figure 1 summarizes these findings

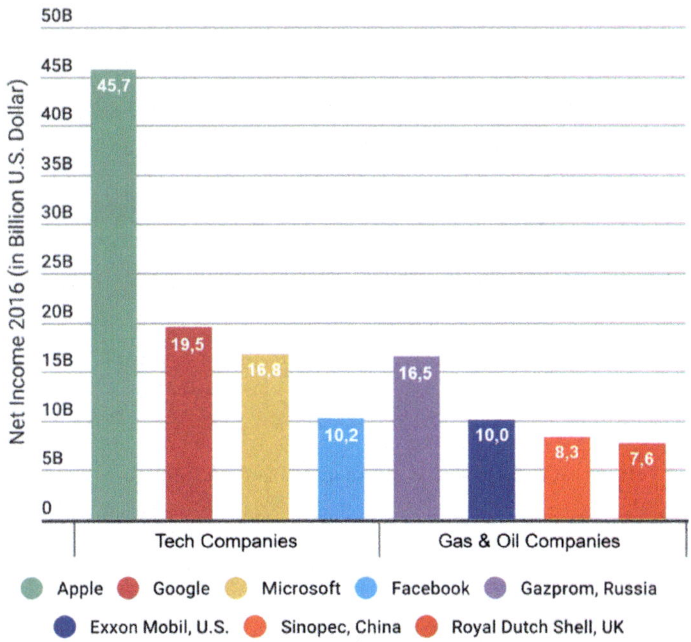

Fig. 1 Net income of major tech- and gas & oil companies in 2016, company annual reports and own research

impressively. This outcome illustrates how asset builders (gas & oil companies) are outperformed by network orchestrators (tech companies) by mastering an ecosystem without owning physical assets (Libert et al. 2014). If once gas and oil used to be the most valuable resources on the planet, surely today data is producing the most profitable businesses.

1.2 *Platform Business: Network Effects and Algorithms*

It remains to be clarified how these tech companies transform the gathered data into such tremendous amounts of value for their businesses, giving useful insights to traditional industries, as are most ground services at the moment. Frankly, the key driver for the success of wholly digital companies is the vast amount of users and data by itself. What all these companies have in common is the fact, that they were able to build a platform around their business model. The open, participative infrastructure of platforms enables fundamentally any stakeholder, be it producer, consumer or even both, to connect with the system exchanging, consuming or co-creating value by interacting with one another with the use of the platforms resources (Parker et al. 2016). This is the incredible power of a *two-sided market*. After a certain threshold

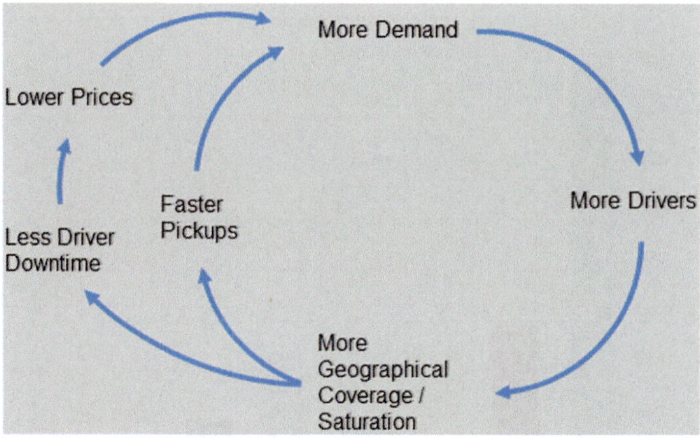

Fig. 2 Uber's virtuous cycle. Reproduced according to Platform Revolution, p. 18

of users has been surpassed, network effects favor a snowball of additional users and suppliers to enter the system. Since an ever growing community leads to a constant increase in value for the user and the system as a whole, this continuous process likewise causes a proprietary *lock-in effect* by what the dependency in the system becomes substantial and switching cost barriers get disproportionate. The example of the ride hailing service "Uber" in Fig. 2 illustrates how positive network effects created growing value in such a virtuous cycle.

Furthermore, the proliferation of user data favors the companies' algorithms, as more information allows an improvement in accuracy and predictive analytics —ultimately serving individual customers fully personalized content.

A recent study, carried out by the Department of Psychology of the University of Cambridge and the Department of Computer Science of the Stanford University, compared the accuracy of the "big five" personality traits—openness, conscientiousness, extroversion, agreeableness and neuroticism—between computer algorithms and humans (Youyou et al. 2014). For the study, the researchers used a sample size greater than 80,000 volunteers, who completed a personality questionnaire of a 100 questions. The computer predictions on the other hand were based on a generic digital footprint of the participants Facebook "Likes". The researchers' findings highlighted that Facebook's algorithms were able to predict users' personalities more accurate than human social-cognitive skills. In fact, they could link a relationship between the number of "Likes" a person's profile holds with the computer's accuracy. The results show that 10 "Likes" generated a more precise personality profile of a participant than a regular work colleague would achieve, 70 "Likes" outperformed a cohabitant or a friend, 150 "Likes" were sufficient to achieve a more accurate result than a family member would and 300 "Likes" even beat the cognitive skills of those of a spouse (Youyou et al. 2014).

Given the fact that the average user on Facebook has as much as 227 "Likes", Facebook can be treated as a close family member the user puts plenty of trust in and provides the system with a lot of predictable power (Youyou et al. 2014).

1.3 Data Ownership and Personalization

Once again, it is the sheer amounts of shared user information coupled with analytical algorithms, which enable a leading company to build a strong customer relationship by serving favorite news feeds, targeted advertisements and a global network of social connections in the case of Facebook.

It may be a self-explanatory debate where to draw a sharp boundary between legal use of data and illegal abuse of user information. For now, there does not exist a framework that regulates the use of content of data created by consumers. Companies are legitimized to make use of their users' digital footprint, as long as privacy and copyright laws of personal information are met. In fact, one does not own data just because it is about oneself (OECD 2015). It is the company's obligation to stay within legal boundaries and one can say that companies would do well not pushing too hard at consumers' expense, because otherwise adverse effects are likely to suffer consumer confidence. What is striking about the discussion how much data a company should use to monetize its business for, emphasizes a survey taken by international consulting firm Infosys. The results give evidence that consumers indeed value advertisements, which are targeted to their specific needs. As a matter of fact, the majority of U.S. consumers, namely 85%, said they were more likely to buy from a retailer again, when confronted with advertisements that were related to their interests, wants and needs. In this context, 81% indicated they would be incentivized if advertisements were based on their current location (Infosys 2013). This outcome may not come very surprisingly. Just like customer reviews can strengthen trust in products and services for other customers, personalized offers can be perceived as a form of customer appreciation, building the foundation for a lasting customer lifetime relationship.

The implications of digital information and a profound customer relationship for companies that are built on wholly digital business models, can also be applied to rather traditional industries that rely on linear revenue streams, as is the mobility industry in large parts of today's businesses. Especially the commercial airline industry, can be characterized as a prominent example to have served a value proposition that did not change exceptionally much over time, bringing rather incremental than disruptive change to airliners and its passengers (Rothkopf 2009).

2 A Storm Is Brewing Along the Horizon: Challenges of the Airline Industry

Ever since the airline industry evolved from a regulated into a deregulated market environment, monopolistic or oligopolistic structures changed and created those found under perfect competition (Rothkopf 2009). A promising market growth promised big potential for novel market entrants to enter the arena of commercial aviation, counting more than 1400 commercial airlines in the world today (Air Transport Action Group (ATAG) 2016). Generally, the worldwide travel industry is a prospering one, at least in terms of passenger numbers. International outbound travel increased by 3.9%, inbound by 4% in 2016 respectively compared to last years' numbers and the outlook for 2017 will be very similar (IATA 2016).

In 2016, the International Air Transport Association (IATA) announced 3.8 billion air travelers for the year. IATA even estimated an expected 7.2 billion passengers to travel per year by 2035, considering that globalization and rising prosperity in developing countries will give the next generation of children the opportunity to travel by air more affordably (IATA 2016; Taneja 2016). Nevertheless, airlines are facing the problem of gaining little profit with their core business of transportation. IATA forecasts total revenues of $736 billion in 2017 for the global airline industry and a net profit of $29.8 billion, which represents a profit margin of just 4.1% (IATA 2016).

Beyond that, a Low Cost Carrier (LCC) business model evolved besides the traditional Full Service Carrier (FSC) model. From cost of fare, these carriers unbundled everything but the basic seating and started offering separate elements, using an à la carte model. Ancillary service revenues have become the major profit source of present-day air carriers. Within Europe, the seat capacity of LCC captured about 36% of market share by 2015 and it is ever-expanding (OAG 2015).

This process transformed the industry internally toward an increasingly price competitive environment. Alongside particular dependencies on fuel- and labor costs, disasters of all kind causing negative effects on air travel and enormous amounts of invested capital, external competition in form of digital platform aggregators, such as Kayak or Expedia, intensify the airline contest. One could speak of a price war in which search engines took over the airlines' control, paving the race for "prices-and-profits-before-anything-else", in order to attract and retain passengers. Also, airlines attempt to keep loyal customers by means of Frequent Flyer Programs (FFP) that offer certain benefits in form of seat and cabin upgrades or entrances to airport lounges which cannot achieve this objective. Data shows that three quarters of regular travelers belong to more than one Frequent Flyer Program (FFM). Easy access to above mentioned online search and price comparison applications has left mere 34% of passengers partial to one airlines' FFP (Ewbank 2015). At the same time, customers find themselves in an increasingly unclear market environment with an ever growing number of airline offerings, likewise the choice of platform search engines becomes more complex also. What is more, nowadays air travel is conceived a hassle rather than a pleasant experience. The commoditized travel industry led to long waiting lines at check-in, flight cancellations, additional baggage fees and cramped seating.

If this is not enough, disruptive technological concepts are about to enter the theatre in the not so distant future. Autonomous cars of level 5 are expected to roll onto the streets no later than 2020 (Nikowitz 2015). The technology promises to solve major drawbacks of today's traffic congestion in big cities all over the globe and to enhance the travel commuting experience between cities. Similarly, the idea of Hyperloops, backed by the two start–ups Virgin Hyperloop One and Hyperloop Transportation Technologies (HTT), pledge to link cities in a never seen before manner no later than the beginning of the next decade (Webb 2017).

Undoubtedly, there exists not only tremendous amounts of competitive pressure within the airline industry, but also completely new concepts from adjacent sectors will set foot in the travel and tourism industry being unbiased from customer experience, as seen in the aviation industry.

2.1 *The Dream of a Seamless Customer Experience*

When looking at today's trusted sources who consumers purchase mobility products and services with, one notices that wholly digital companies such as Airbnb, TripAdvisor and Uber, or in the wider travel sphere metasearch engines such as Kayak, Trivago or Expedia are the brands customers rely on. These digital companies are based on platform business models, which allow them to scale their businesses in a way that generates very high profit margins without offering actual inventory. For consumers, it is convenient to connect to such platforms and compare fares of all airlines matching customers' search criteria. Most providers even offer additional services such as hotels, car rentals and activities. The principle of this one-stop-shopping favors market acceptance, which raises the bar of a superior customer experience and makes it ever more difficult for airlines to directly reach customers prior to their flights. Taking this into account, there is little wonder that Google Travel is estimated to be worth as much as $100 billion, whereas the brand value of the number one airline in the world, American Airlines, grew by a grand 57% in 2017 compared to the preceding year reaching sheer $9.8 billion (Wein 2017; Heigh 2017).

It is indisputable that the majority of air travelers buy tickets from online travel agencies, rather than directly on airline web sites, due to convenience reasons and a seamless shopping experience mentioned before. Unsurprisingly, the entire to-date airline experience is a journey of fragmented pieces. Once flight tickets have been purchased, at one point, passengers are asked to check in prior to their flight. At the airport, baggage needs to be checked in or dropped off also. Followed by jammed security checks, travelers have to find their way through the airport jungle to the designated departure gate. After arriving there, passengers find themselves waiting for boarding and eventually clicking seat-belts. The hassle does not end when arrived at the destination airport. Immediately after taxing, passenger hordes stumble in the direction of baggage belts. Only after receiving their belongings, the journey ends by figuring out a way to get to their final address.

Some of these stops are desired by airliners, because it guarantees direct contact with passengers and the possibility to up-sell a few services, although perceived disturbing, annoying or counterproductive by most passengers. The process of check in for example dates as far back as the beginning of the commercial aviation industry in the 1920s. Albeit paperless ticketing has become the norm nowadays, one reason airlines still interact in this way is to promote a cabin upgrade, spacious seats or Wi-Fi access (Bachman 2017).

While other industries pursue norms that enable frictionless customer interactions, air carriers prompt artificial ways to interact with customers. A doubtful strategy to generate additional revenue streams in times where consumers decide when to contact producers.

Airlines have to start rethink their approach to increase bottom line. Instead of trying to get more money from its passengers, they need to ask themselves how to add additional value for its customers. Digital companies have shown how platform models and data monetized their businesses.

Lacking behind in certain areas, the airline industry however has three potential advantages over new market entrants. Airlines already have a great customer base, they own the most important piece of a trip—the access to travel and they generate more passenger data, than they currently are able to process. (Parker et al. 2016; O'Neill 2017)

2.2 *Potential of the Unutilized–Travel–Time*

Basically, any industry or company that profits from access to information about customer needs or price fluctuations can be transformed into a platform business (Parker et al. 2016). This includes the airline industry as a service provider, which generates a lot of passenger data. Today, the focus of air carriers rests on connecting its users from the departure gate to their destination airport. Though the real adventure only starts from here. Once landed, the first action passengers tend to do is switch on their smartphones and reconnect with the digital world.

The all–time–connection surrounds every aspect of life. An exception can be found onboard airplanes. Although, worldwide passengers have a 39% probability of stepping onboard a Wi-Fi equipped airplane today, the connection quality is restricted to bandwidth shortage, a poor reliability downgrades the customer experience and high prices prevent passengers from making use of the technology (routehappy 2017). U.S. airliners have a distinct advantage when it comes to inflight entertainment and connectivity. Nearly all planes offer a chance of wireless onboard, whereas the rest of the world struggles to equip one fifth of the fleets with online connectivity. In Europe, there are 20 airlines that offer the technology to-date, the majority being FSC with ambiguous pricing strategies (compare Fig. 3).

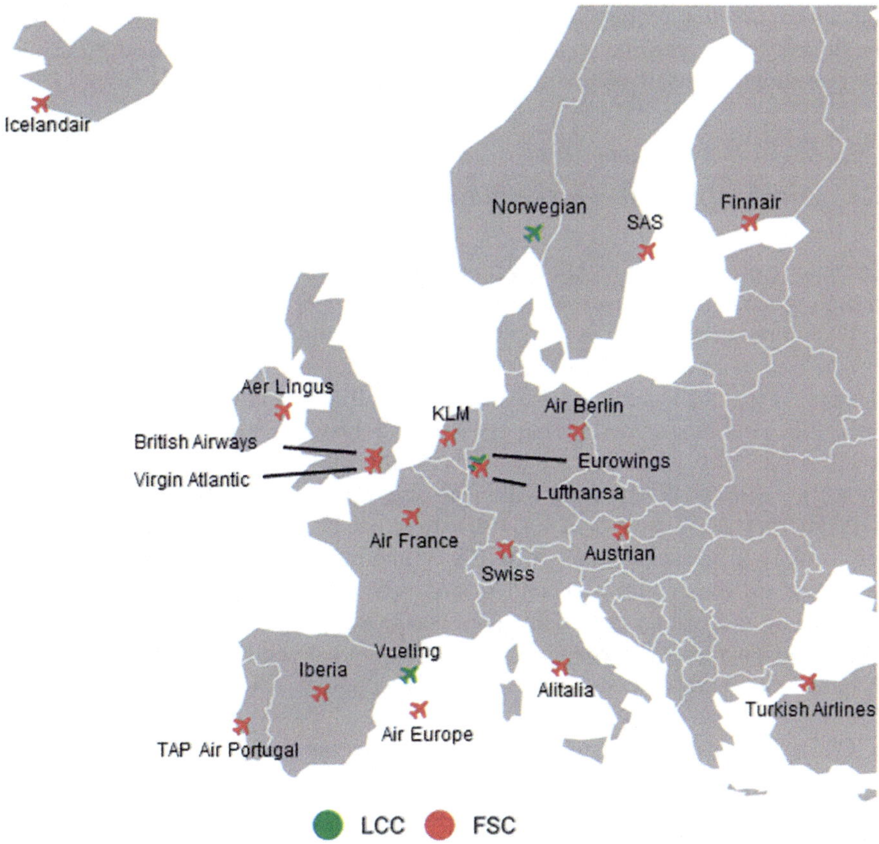

Fig. 3 European FSC and LCC offering Wi-Fi in 2017, (de Selding 2017)

No wonder that passengers rank the inflight entertainment aspect as the most negative piece throughout the entire customer journey. Not only has the cluster of unsatisfied but also overall satisfied customers voted this way according to IATA's latest passenger survey (IATA 2017).

The biggest challenge in aviation to provide a sufficient wireless connection, is the high speed of airplanes and the great benchmark to customers' ground broadband services. While a satellite connection is the predominant way to connect airplanes at the moment, the European Aviation Network (EAN) combines Inmarsat's multi–beam S–band satellite with approximately 300 Deutsche–Telekom ground towers. The air–to–ground system promises to guarantee passengers' high bandwidth requirements and to transform the passenger experience towards a fully reliable network across all 28 European Union states (Inmarsat Global Ltd 2017). This technology is going to enable European Airliners' ever growing flight traffic to bring innovative services that enable completely new revenue streams and opportunities to lift up the consumer value proposition. The majority of passengers agrees to the fact that the unutilized–travel–time aboard airplanes is outdated and access to the world of internet should be available to more people when airborne (IATA 2017).

2.3 *Airline Inflight–Marketplace*

Continuing progresses in satellite technologies will lead to an ever growing amount of high-speed data coverages. Most certainly, passengers using inflight connectivity (IFC) will rise accordingly. A greater user base will simultaneously attract more airlines to offer wireless plans onboard their fleets. With increasing supply and demand sides prices for the technology will drastically fall to a minimum. Commercial service of the EAN has started during 2017 paving the way for increased competition. It will not be long, when the race towards the market leadership of IFC will reach a point where internet connection will be an ubiquitous service offered by all air carriers. Internet access alone will not suffice to retain a loyal customer base in the long run. If most airlines cherish the idea of enabling passengers access to their favorite entertainment sources, such as Netflix or Spotify, in return for a price premium, this will only be temporarily monetizing their businesses. These services are already free with some carriers and technology improvements are about to reduce the amount of bandwidth used, reducing prices for this technology (Morris 2017; Huddleston 2017).

Clearly, releasing IFC will open multiple opportunities to establish an in-flight marketplace, which is going to power the "Amazon for Travel" phrase first mentioned by Ryanair. In–flight internet services is a billion–euro market (Koenen 2017). However, in order to retain parts of this market, Airlines need to incorporate digital strategies, which establish a unique selling proposition, serve a superior value proposition than their competitors and establish smart information architectures that maximize the benefits of passenger data gathered.

2.4 *A Successful Digital Airline Strategy*

Every novel market strategy, that contemplates to incorporate a product or service on the basis of a platform business, should ask a number of critical questions that give insights about the applicability of such plan. In any case, these include:

- Can the firm either use information or community to add value to what it sells?
- How can the value of the goods and services that are currently provided be enhanced through new data streams, interpersonal connections, and curation tools?
- Which processes that are currently managed in–house can be delegated to outside partners, whether suppliers or customers?

- How can outside partners be empowered to create products and services that will generate new forms of value for the existing customers? (Parker et al. 2016).

One of the most promising value creators connected with the prospective IFC experience both for airlines and passengers, is going to be the aspect of personalization, as seen before. *Targeted advertising* can serve this way as a powerful source of additional ancillary revenue streams for airliners, because an inflight-shopping experience may facilitate an open marketplace, any product or service provider can connect with. Passengers trust airlines with vast amounts of their data—conveniently classified by type (business or leisure), gender, age, destination and in some cases even their travel duration, payment source and similar. All this intrinsic information coupled with applicable algorithms that serve individual passenger journeys and needs, have the potential to connect travelers with companies onboard an airplane.

IFC does not end with new possibilities of matching consumers with producers, but it also opens up vast possibilities to elevate the travel experience that exceeds the service of transportation from departure gate to the destination airport by facilitating the community to add this value. A digital platform on a *peer–to–peer* system could bring travelers on the same flight together, encouraging to share rides from the airport to their designated destination. Enabling passengers to interact with each other allows a value creation process driven by its users and not predefined by airlines.

Such strategic change in mind–set has the potential to bring a tremendous amount of disruption to the airline industry. Empowering the passenger to create value to the ecosystem may improve customer awareness, drive adoption and incentivize usage, while pulling outside companies to access this network to provide further value. A successful platform strategy could link airlines with ground service providers, such as ride hailing providers, car rentals or sharing platforms that allow the development of a seamless customer journey, accompanying travelers along the entire journey.

Airlines that understand the importance of implementing open systems, in which end users get access to different service providers, will increase customer loyalty ultimately enjoying a competitive advantage with increased customer satisfaction.

A consistent travel journey relies on a connected ecosystem with the flexibility to share data among its participants and being flexible to change according to market needs.

2.5 From Air to Ground

The aspect of personalization in airplanes can also be applied to ground service providers. Traditionally, consumers bought physical tickets in cash for their preferred mode of transportation, leaving companies with no digital information about their customers. With no further information about when consumers access the mobility or where they are going, companies have little chance to predict consumers' next move or to serve them with valuable content, enriching the customer journey. By now, ticketing in ground mobility has become digital for most parts and more and more transportation providers together with cities are working on solutions to connect their services with each other, favoring multi modal mobility (https://whimapp.com/ or https://www.moovel.com/de/en). Such networks of trains, trams, buses, taxis, bicycles and pedestrians combined with sensor equipped vehicles allow a collection of vast amounts of passenger data. While cities are interested in gathering information about commuters to implement smart traffic management systems that help to predict traffic peaks and to offer individual passengers alternative modes of travel, this information also matters to third party enterprises. In this context, Open Data projects in cities become of particular relevance. A recent report of the European Data Portal depicts that transportation is among the most popular and important data domains in cities, which are made available to the public (Carrara et al. 2016). An ecosystem of data sets that may be accessed by means of application programming interfaces (API) would enable local enterprises a channel to specifically reach potential consumers on a real–time basis. With the aid of predictive analytics and anonymized route mapping, employees could be reminded to do grocery shopping on their inbound leg, preventing them from swarming out again once they arrived at home. Personalized advertising during times when traffic is at climax not only offers great potential to reach a majority of commuters, but it also increases a greater rate of successful promotion and it may counteract any additional private transport streams.

In particular interesting transformation on the ground will happen with the shared use of cars. Privately used cars represent a considerable large mode of transportation. The progress of digital technology is changing this—more efficient business models are arising that leverage cars as physical assets, for example: ride-hailing, car-sharing and ride sharing. The consequences are very significant: on one hand this means that also cars are used in a form of a service and therefore the concepts that are used in airlines and other transportation services start to apply. Secondly, with the significant growth of mobility services one can foresee that they will become a major competitive component within the city transportation system. This in turn will force other modes (such as subway, rail, busses, etc.) to become more competitive and embrace new business models and concepts that are already emerging in the airline space.

3 Summary

The most valuable brands in the digital age are companies based on platform models that leverage vast amounts of data without exception. Certainly, the importance of such data generation will further increase in the future, when more hardware and machines will be connected to the world and will interact with each other.

These two–sided markets, linking consumers with producers, are able to create positive network effects by providing an open infrastructure. This process triggers value creation, which results in a virtuous cycle of a growing user base as well as an increasing value proposition. Besides, companies' predictive analytics allow them to target potential consumers with personalized content, which may empower brands to build up long lasting customer relationships.

Airlines have offered the same or worse service ever since they have been started operating. Bringing incremental innovation, they are ripe for disruptive transformation to overcome potential threats from emerging ground service providers. As of today, the majority of plane tickets are sold by online travel agencies that offer a seamless customer journey, whereas air carriers have long pursued a strategy to fragment their passenger journey. Such fragmentation even continues due to the fact that airplanes are disconnected with the world when airborne for the most part. To reverse this trend and improve service and customer experiences, airlines will have to embrace the opportunity of evolving with the mentality of the era of digital transformation. These kind of changes often hold the possibility of creating new business models, one of them being the provision of a seamless intermodal travel experience for the customer and by doing so, resolving the first and last mile through the connection of air and ground mobility.

Big potential for European airlines offers the implementation of a wireless aviation network that spans across all EU member states putting an end to the unutilized air–travel time. A relying and strong internet connection not only will enable passengers to stream their favorite movies or listen to the latest songs, but also offers airlines opportunities to open up new revenue streams in form of open market places.

In order to build up a unique selling proposition and to regain consumer satisfaction, airlines have to transform both their business structures and the managing mind–set. A sustainable market strategy needs to focus on the creation of digital assets and empowering passengers to create value to the ecosystem. By facilitating passengers with their own data and giving them opportunities that enrich the entire travel journey through multi modal transportation possibilities, airlines will be able to lift up customer satisfaction and to enhance passenger loyalty.

References

Air Transport Action Group (ATAG) (2016) Aviation benefits beyond borders: powering global, economic growth, employment, trade links, tourism and support for sustainable development through air transport, ATAG, Geneva

Bachman J (2017) Why do airlines still make passengers check in for flights?, Bloomberg in Skift Inc. https://skift.com/2017/10/20/why-do-airlines-still-make-passengers-check-in-for-flights/

Carrara W, Engbers W, Margriet N, van Steenbergen E (2016) European data portal: open data in cities

de Selding PB (2017) In-flight connectivity: fabulous growth, but longterm profitability TBD, Space Intel Report. https://www.spaceintelreport.com/flight-connectivity-fabulous-growth-long-term-profitability-tbd/

Dragland Å (2013) Big data, for better or worse: 90% of world's data generated over last two years, SINTEF in ScienceDaily. https://www.sciencedaily.com/releases/2013/05/130522085217.htm

Ewbank JP (2015) Loyal followers: why frequent flyer programs are not loyalty programs, ascend industry, vol 2, p 23 f

Heigh A (2017) Airlines 50 2017: the most valuable airlines brands of 2017, Brand Finance, London. http://brandfinance.com/images/upload/brand_finance_airlines_50_2017_locked.pdf

Huddleston T Jr (2017) This is how netflix wants to improve In-Flight Wi-Fi on major airlines, Fortune. http://fortune.com/2017/09/26/netflix-airlines-in-flight-wifi-video-streaming/

IATA (2016) Another Strong Year for Airline Profits in 2017, Geneva. http://www.iata.org/pressroom/pr/Pages/2016-12-08-01.aspx
IATA (2016) IATA forecasts passenger demand to double over 20 years, Montreal. http://www.iata.org/pressroom/pr/Pages/2016-10-18-02.aspx
IATA (2017) Global passenger survey hightlights, IATA. http://www.iata.org/publications/store/Documents/GPS-2017-Highlights-report.pdf
Infosys (2013) Consumers worldwide will allow access to personal data for clear benefits, Infosys. https://www.infosys.com/newsroom/press-releases/Pages/digital-consumer-study.aspx
Inmarsat Global Ltd (2017) Deutsche Telekom AG, What is European aviation network? London. https://www.europeanaviationnetwork.com/european-aviation-network/
Ismail S, Malone MS, van Geest Y (2014) Exponential organizations; why new organizations are ten times better, faster, and cheaper than yours (and what to do about it). Diversion Books, New York
Koenen J (2017) The high-altitude fight for Wi-Fi in Europe, Handelsblatt Global. https://global.handelsblatt.com/companies-markets/the-high-altitude-fight-for-wi-fi-in-europe-818573
Kondratieff ND, Dutton EP (1984) Boston
Libert B, Wind YJ, Beck M (2014) What Airbnb, Uber, and Alibaba Have in Common. Harvard Bus Rev. https://hbr.org/2014/11/what-airbnb-uber-and-alibaba-have-in-common
Morris H (2017) Airline to offer free in-flight Netflix and Spotify. The Telegraph. http://www.telegraph.co.uk/travel/news/qantas-passengers-receive-free-in-flight-netflix-spotify-and-foxtel/
Nikowitz M (2015) Fully autonomous vehicles: visions of the future or still reality? epubli GmbH, Berlin
O'Neill S (2017) Lufthansa finds collecting passenger data is easier than actually using it, Skift Inc. https://skift.com/2017/09/26/lufthansa-finds-collecting-passenger-data-is-easier-than-actually-using-it/
OAG (2015) Low Cost—every little helps.... https://www.oag.com/blog/low-cost-every-little-helps
OECD (2015) Data-driven innovation: big data for growth and well-being. OECD Publishing, Paris
Parker G, van Alstyne M, Choudary SP (2016) Platform revolution; how networked markets are transforming the economy and how to make them work for you. W. W. Norton & Company, New York
Rapier G (2017) These are the 15 most valuable car brands in the world, Business Insider. http://www.businessinsider.com/15-most-valuable-car-brands-in-the-world-2017-9
Regarding this aspect see for example: https://whimapp.com/ or https://www.moovel.com/de/en
Rothkopf M (2009) Innovation in commoditized service industries—an empirical case study, Lit; Global distributor, Berlin; London
routehappy (2017) Wi-Fi report, routehappy Inc. https://www.routehappy.com/insights/wi-fi/2017
Taneja NK (2016) Airline industry: poised for disruptive innovation?. Routledge, London
The World's Most Valuable Brands, Forbes (2017). https://www.forbes.com/powerful-brands/list/
Vahs D, Brem A (2015) Innovationsmanagement, Vahs/Brem, 5. Schäffer-Poeschel Verlag, Edition
Webb S (2017) All the wonder that would be: exploring past notions of the future, Springer Berlin, Heidelberg, New York
Wein J (2017) Google travel is worth $100 Billion—even more than Priceline, Skift Inc., New York. https://skift.com/2017/09/18/google-travel-is-worth-100-billion-even-more-than-priceline/
Youyou W, Kosinki M, Stillwell D (2014) Computer-based personality judgments are more accurate than those made by humans. Proceedings of the National Academy of Sciences of the United States of America (PNAS) 112:1036–1040
Zukunftsinstitut, Megatrend Map (2015). https://www.zukunftsinstitut.de/artikel/die-megatrend-map/

Printed by Printforce, the Netherlands